CRYSTAL GROWTH
FOR BEGINNERS

Fundamentals of Nucleation, Crystal Growth and Epitaxy

CRYSTAL GROWTH FOR BEGINNERS

Fundamentals of Nucleation, Crystal Growth and Epitaxy

Ivan V Markov

Bulgarian Academy of Sciences
Sofia, Bulgaria

World Scientific
Singapore • New Jersey • London • Hong Kong

Published by

World Scientific Publishing Co. Pte. Ltd.

5 Toh Tuck Link, Singapore 596224

USA office: 27 Warren Street, Suite 401-402, Hackensack, NJ 07601

UK office: 57 Shelton Street, Covent Garden, London WC2H 9HE

Library of Congress Cataloging-in-Publication Data
Markov, Ivan V., 1941–
      Crystal growth for beginners : fundamentals of nucleation, crystal
    growth, and epitaxy / Ivan V. Markov.
         p.   cm.
    Includes bibliographical references.
    ISBN 9810215312   ISBN 9810221770 (pbk)
    1. Crystal growth.   I. Title.
    QD921.M355   1995
    548.5-dc20   94-45874
                                         CIP

British Library Cataloguing-in-Publication Data
A catalogue record for this book is available from the British Library.

First published 1995
Reprinted 1996, 1998

ISBN 978-981-02-1531-6 (hardcover)
ISBN 978-981-02-2177-5 (paperback)
ISBN 978-981-238-624-3 (ebook for institutions)
ISBN 978-981-310-467-9 (ebook for individuals)

For any available supplementary material, please visit
https://www.worldscientific.com/worldscibooks/10.1142/2145#t=suppl

PREFACE

The idea to write this book is not totally mine. It belongs partly to Dr. Svetoslav D. Toschev, a colleague and friend of mine, with whom I worked together on the kinetics of electrochemical nucleation of metals. Unfortunately, his untimely death in 1971 prevented him to carry out this project. The idea to write such a book arose out of the long experience we had at the Institute of Physical Chemistry in the theories and experiments on nucleation, crystal growth and epitaxy. In fact the first paper on crystal growth was published in Bulgaria in as early as 1927 by Ivan Stranski, who introduced the concept of half-crystal position simultaneously with W. Kossel. This turned out to be the most important concept in nucleation and crystal growth. Using this concept Stranski considered, two years later, together with K. Kuleliev the stability of the first several monolayers deposited on a foreign substrate. This work served as a basis for the later treatment by Stranski and Krastanov of the mechanism of epitaxial growth of a monovalent ionic crystal on the surface of a bivalent ionic crystal. This system was chosen because of the easier evaluation of the interatomic forces at that time. Stranski and his coworkers were the first to realise that the thickness variation of the chemical potential in ultrathin epitaxial films determines the mechanism of growth. Now the mechanism of epitaxial growth by formation of several complete monolayers followed by the growth of isolated 3D islands named after Stranski and Krastanov is well known to all researchers involved in epitaxial growth. In fact the work of Stranski and Kuleliev was the first theoretical study of epitaxial growth. In the early thirties Stranski and Kaischew introduced the concept of mean separation work in order to describe the equilibrium of small three-dimensional and two-dimensional crystals with the parent phase. This allowed them to describe the kinetics of crystal nucleation and 2D growth of perfect crystals in a quantitative way. Today these early papers are almost forgotten but they inspired further research in the field by many authors, like R. Becker and W. Döring, Max Volmer, W. K. Burton, N. Cabrera and F. C. Frank, and many others.

I got the idea to write this book when I was reading a course on Nucleation, Crystal Growth and Epitaxy at the Institute of Microelectronics and Optoelectronics in Botevgrad, Bulgaria. I was surprised to find that so many people involved *de facto* in growth and characterization of advanced materials by CVD, LPE, MOCVD, MBE, etc. have no basic knowledge in crystal growth and epitaxy. Although they were good specialists in high vacuum, surface and bulk materials characterization, device construction, etc., they did not understand the elementary processes which form the basis of the fabrication of these devices. I was further convinced to write this book after I had read a similar course of lectures at the University of Dresden a year later and after many discussions with colleagues of mine.

There are many excellent monographs, treatises and review papers on different aspects of nucleation, crystal growth and epitaxy. Most of them are listed in the reference list. I would like to bring the reader's attention to the monographs of Max Volmer, *Kinetik der Phasenbildung*, and Y. Frenkel, *Kinetic Theory of Liquids*. The only books which cover an appreciable part of the problems treated in this book are the monograph of A. A. Chernov in Vol. 3 of *Modern Crystallography* and *The Theory of Transformations in Metals and Alloys*, 2nd edition, by J. W. Christian, which I read with great pleasure. All these books are, however, aimed more or less at researchers with some preliminary knowledge of the matter. It turned out that there was not a textbook which could give the basic knowledge on nucleation, crystal growth and epitaxy from a unified point of view and on a level accessible to graduate students or even undergraduate students who have just begun to do research. The reader will need some knowledge in elementary crystallography and chemical thermodynamics. The mathematical description should not give trouble at all even to undergraduate students. In fact in all cases in which more complicated mathematical treatment is required, problems with lower dimensionality are considered instead. Thus the mathematical treatment is considerably simplified and the physical meaning is easier to grasp. A typical case is the consideration of Herring's formula. In some cases the use of some specific mathematical methods like the Euler equation in Chap. 1 and special functions like the Bessel functions and the elliptic integrals could not be avoided. However, the reader should not accept them with "grief." Any good textbook will be able to help the reader who is unfamiliar with the mathematical methods. All of the above is what determined the title of the book.

The book is naturally divided into four chapters: Equilibrium, Nucleation, Crystal Growth and Epitaxial Growth. In the first chapter all the information necessary to understand the material in the remaining chapters is given. Thus the mean separation work which determines the equilibrium of a 2D island with the parent phase is defined in Chap. 1 and is used in Chap. 3 to derive an expression for the rate of propagation of curved steps. The second chapter deals with all the problems connected with nucleation. The only exception is the theory of one-dimensional nucleation which is included in Chap. 3 because it is intimately connected with the propagation of single height steps. The concept which unifies the whole presentation is that of the separation work from a half-crystal (kink) position. One could think of it as the chemical potential of the particular crystal of a monolayer of the deposit (taken with negative sign). By using this concept it is shown that the only difference between crystal growth and epitaxial growth is of a thermodynamic nature. The chemical potential in ultrathin epitaxial films differs from that of the bulk crystal. The kinetics of growth of both single crystals and epitaxial films are one and the same.

I did not discuss the text with any of my colleagues. That is why I take the sole responsibility for any misinterpretations or errors and I will be very grateful to anybody who detects them and brings them to my notice. On the other hand, I am extremely grateful to V. Bostanov, A. Milchev, P. F. James, H. Böttner, T. Sakamoto and S. Balibar who gave me their kind permission to reproduce figures from their papers and supplied me with the corresponding photographs. I am also greatly indebted to Professor D. D. Vvedensky from the Imperial College, London, for his kindness to grant me his permission to use the beautiful picture of a Monte Carlo simulation of Si(001) growth for the cover of the book. The book has been written at the Institute of Physical Chemistry of the Bulgarian Academy of Sciences with the only exception of Sec. 3.2.4, which deals with the growth of Si(001) vicinal surfaces by step flow. I decided to include this section in order to illustrate the propagation of single height steps by one-dimensional nucleation. This section was written during my stay at the Department of Materials Science and Engineering of the National Tsing Hua University in Hsinchu, Taiwan, Republic of China, where I was invited as a visiting professor. The final preparation of the manuscript was also carried out at the National Tsing Hua University. I would like to express my sincere gratitude to Professor L. J. Chen and the staff of the department for their kind hospitality and assistance. In this respect I would also like to acknowledge the financial support of the Science Council of the Republic of China.

Ivan V. Markov
Sofia, Bulgaria
Hsinchu, Taiwan

CONTENTS

CHAPTER 1

CRYSTAL–AMBIENT PHASE EQUILIBRIUM

1.1. Equilibrium of Infinitely Large Phases

The equilibrium between two infinitely large phases α and β is determined by the equality of their chemical potentials μ_α and μ_β. The latter represent the derivatives of the Gibbs free energies with respect to the number of particles in the system at constant pressure P and temperature T, $\mu = (\partial G/\partial n)_{P,T}$, or, in other words, the work which has to be done in order to change the number of particles in the phase by unity. In the simplest case of a single component system we have

$$\mu_\alpha(P,T) = \mu_\beta(P,T) \ . \tag{1.1}$$

The above equation means that the pressures and the temperatures in both phases are equal. The requirement $P_\alpha = P_\beta = P$ is equivalent to the condition that the boundary dividing both phases is flat or, in other words, the phases are infinitely large. This question will be clarified in the next section where the equilibrium of phases with finite sizes will be considered.

Let us assume now that the pressure and the temperature are infinitesimally changed in such a way that the two phases remain in equilibrium, i.e.

$$\mu_\alpha + d\mu_\alpha = \mu_\beta + d\mu_\beta \ . \tag{1.2}$$

It follows from (1.1) and (1.2) that

$$d\mu_\alpha(P,T) = d\mu_\beta(P,T) \ . \tag{1.3}$$

1

Recalling the properties of the Gibbs free energy ($dG = -S dT + V dP$) we can rewrite (1.3) in the form

$$-s_\alpha dT + v_\alpha dP = -s_\alpha dT + v_\beta dP \; , \tag{1.4}$$

where s_α and s_β are the molar entropies, and v_α and v_β are the molar volumes of the two phases in equilibrium with each other.

Rearranging (1.4) gives the well-known equation of *Clapeyron*:

$$\frac{dP}{dT} = \frac{\Delta s}{\Delta v} = \frac{\Delta h}{T \Delta v} \; , \tag{1.5}$$

where $\Delta s = s_\alpha - s_\beta$, $\Delta v = v_\alpha - v_\beta$, and $\Delta h = h_\alpha - h_\beta$ is the enthalpy of the corresponding phase transition.

Let us consider first the case when the phase β is one of the condensed phases, say, the liquid phase, and the phase α is the vapor phase. Then the enthalpy change Δh will be the enthalpy of evaporation $\Delta h_{ev} = h_v - h_l$, and v_l and v_v will be the molar volumes of the liquid and the vapor phases, respectively. The enthalpy of evaporation is always positive and the molar volume of the vapor v_v is usually much greater than that of the crystal v_l. In other words, the slope dP/dT will be positive. We can neglect the molar volume of the liquid with respect to that of the vapor and assume that the vapor behaves as an ideal gas, i.e. $P = RT/v_v$. Then Eq. (1.5) attains the form

$$\frac{d \ln P}{dT} = \frac{\Delta h_{ev}}{RT^2} \; , \tag{1.6}$$

which is well known as the equation of *Clapeyron–Clausius*. Replacing Δh_{ev} with the enthalpy of sublimation Δh_{sub} we obtain the equation which describes the crystal–vapor equilibrium.

Assuming Δh_{ev} (or Δh_{sub}) does not depend on the temperature, Eq. (1.6) can be easily integrated to

$$\frac{P}{P_0} = \exp \left[-\frac{\Delta h_{ev}}{R} \left(\frac{1}{T} - \frac{1}{T_0} \right) \right] \; , \tag{1.7}$$

where P_0 is the equilibrium pressure at some temperature T_0.

In the case of the crystal–melt equilibrium the enthalpy Δh is equal to the enthalpy of melting Δh_m which is always positive and the equilibrium temperature is the melting point T_m.

As a result of the above considerations we can construct the phase diagram of our single component system in coordinates P and T (Fig. 1.1). The enthalpy of sublimation of crystals Δh_{sub} is greater than the enthalpy

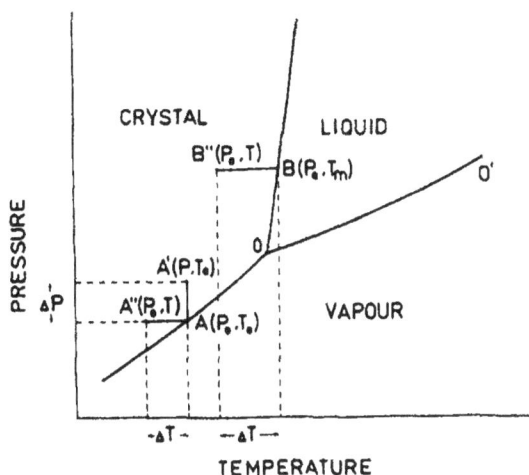

Fig. 1.1. Phase diagram of a single component system in P–T coordinates. O and O' denote the triple point and critical point, respectively. The vapor phase becomes supersaturated or undercooled with respect to the crystalline phase if one moves along the line AA' or AA". The liquid phase becomes undercooled with respect to the crystalline phase if one moves along the line BB". ΔP and ΔT are the supersaturation and undercooling.

of evaporation Δh_{ev} of liquid and hence the slope of the curve in the phase diagram giving the crystal–vapor equilibrium is greater than the slope of the curve of the liquid–vapor equilibrium. On the other hand, the molar volume v_l of the liquid phase is usually greater than that of the crystal phase v_c (with some very rare but important exceptions, for example, in the cases of water and bismuth), but the difference is small so that the slope dP/dT is great, in fact much greater than that of the other two cases, and is also positive with the exception of the cases mentioned above. Thus the P–T space is divided into three parts. The crystal phase is thermodynamically favored at high pressures and low temperatures. The liquid phase is stable at high temperatures and high pressures and the vapor phase is stable at high temperatures and low pressures. Two phases are in equilibrium along the lines and the three phases are simultaneously in equilibrium at the so-called triple point O. The liquid–vapor line terminates at the so-called critical point O', beyond which the liquid phase does not exist any more because the surface energy of the liquid becomes equal to zero and the phase boundary between both phases disappears.

1.2. Supersaturation

When moving along the dividing lines the corresponding phases are in equilibrium, i.e. Eq. (1.1) is strictly fulfilled. If the pressure or the temperature is changed in such a way that we deviate from the lines of the phase equilibrium, one or the other phase becomes stable. This means that its chemical potential becomes smaller than the chemical potentials of the phases in the other regions. Any change of the temperature and/or pressure which results in a change of the region of stability leads in turn to transition from one phase to another. Thus a decrease of temperature or an increase of pressure leads to crystallization or liquefaction of the vapor; the decrease of temperature leads to solidification of liquid. Figure 1.2 shows the variation of the chemical potentials of the crystal and vapor phases with pressure at a constant temperature. The chemical potential of the vapor increases with pressure following a logarithmic law which corresponds to a shift along the line AA' in Fig. 1.1. At the same time the chemical potential of the crystal phase is a linear function of pressure, its slope being given by the molecular volume v_c. Both curves intersect at the equilibrium pressure P_0. At pressures smaller than P_0 the chemical potential of the crystal is greater than that of the vapor and the crystal should sublimate. In the opposite case $P > P_0$, the vapor should crystallize. The difference of the chemical potentials, which is a function of the pressure, represents the thermodynamic driving force for crystallization to occur. It is called *supersaturation* and is defined as the difference of the chemical potentials $\Delta\mu$ of the infinitely large mother and new phases at the particular values of pressure and temperature. In other words, we have (Fig. 1.2)

$$\Delta\mu = \mu_v(P) - \mu_c(P) \ . \tag{1.8}$$

Bearing in mind Eq. (1.1), or $\mu_v(P_0) = \mu_c(P_0)$, we can rewrite Eq. (1.8) in the form

$$\Delta\mu = [\mu_v(P) - \mu_v(P_0)] - [\mu_c(P) - \mu_c(P_0)] \ .$$

For small deviations from equilibrium the above equation turns into

$$\Delta\mu \cong \int_{P_0}^{P} \frac{\partial\mu_v}{\partial P} dP - \int_{P_0}^{P} \frac{\partial\mu_c}{\partial P} dP = \int_{P_0}^{P} (v_v - v_c) dP \cong \int_{P_0}^{P} v_v dP \ .$$

Treating the vapor as an ideal gas ($v_v = kT/P$) we obtain upon integration

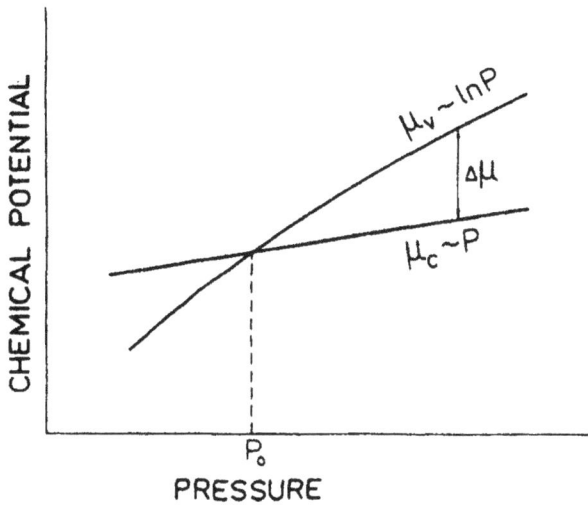

Fig. 1.2. Dependence of the chemical potentials of the vapor, μ_v, and the crystal, μ_c, on pressure when one moves along the line AA' in Fig. 1.1. P_0 denotes the equilibrium pressure.

$$\Delta\mu = kT \ln \frac{P}{P_0} \; , \qquad (1.9)$$

where P_0 is the equilibrium vapor pressure of the infinitely large crystal phase at the given temperature.

Without going into details we can write an expression for the supersaturation in the case of crystallization from solutions, when the solutions are considered as ideal, in the form

$$\Delta\mu = kT \ln \frac{C}{C_0} \; , \qquad (1.10)$$

where C and C_0 are, respectively, the real and equilibrium concentrations of the solute. In fact a more rigorous treatment requires the consideration of multicomponent systems. For more details see Chernov [1984].

Figure 1.3 shows the variation of the chemical potentials of the crystal and liquid phases with temperature at a constant pressure (the line BB'' in Fig. 1.1). The supersaturation which in this case is frequently called *undercooling* is again defined as the difference of the chemical potentials of the infinitely large liquid and crystal phases, μ_l and μ_c, respectively, at a given temperature:

$$\Delta\mu = \mu_l(T) - \mu_c(T) \; . \qquad (1.11)$$

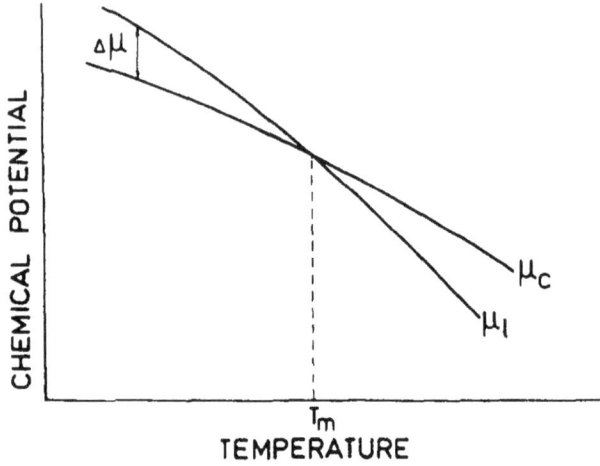

Fig. 1.3. Dependence of the chemical potentials of the liquid phase, μ_l, and the crystal phase, μ_c, on temperature when one moves along the line BB″ in Fig. 1.1. The melting point is denoted by T_m.

Following the same procedure as above we obtain

$$\Delta\mu = [\mu_l(T) - \mu_l(T_m)] - [\mu_c(T) - \mu_c(T_m)]$$

$$\cong \int_{T_m}^{T} \frac{\partial \mu_l}{\partial T} dT - \int_{T_m}^{T} \frac{\partial \mu_c}{\partial T} dT = \int_{T}^{T_m} \Delta s_m dT \ .$$

Assuming the entropy of melting $\Delta s_m = s_l - s_c$ is independent of the temperature one obtains after integration

$$\Delta\mu = \Delta s_m(T_m - T) = \frac{\Delta h_m}{T_m} \Delta T \ . \tag{1.12}$$

Obviously, Eq. (1.12) is also applicable to the case of crystallization of undercooled vapor after the enthalpy of melting is replaced by the enthalpy of sublimation (the line AA″ in Fig. 1.1).

Finally, in the particular case of electrocrystallization of metals the supersaturation is given by

$$\Delta\mu = ze\eta \ , \tag{1.13}$$

where z denotes the valence of the neutralizing ions, $e = 1.60219 \times 10^{-19}$C, is the elementary electric charge and $\eta = E - E_0$ is the so-called *overvoltage* or *overpotential* given by the difference of the equilibrium potential E_0 of the deposited metal in the solution and the electrical potential E applied from outside [Kaischew 1946/1947].

It was Max Volmer [1939] who introduced the term "Überschreitung" or "step across" for both the supersaturation and the undercooling to denote the transition through the line of coexistence of the two phases. Thus, the difference of the chemical potentials of the infinitely large new and mother phases appears as a measure of deviation from the phase equilibrium and as the thermodynamic driving force for the phase transition to occur.

1.3. Equilibrium of Finite Phases

In the previous section we considered the equilibrium of two phases in a single component system assuming that the phases are sufficiently (or infinitely) large or, in other words, the phase boundary between them is flat. This is not obviously the case at the beginning of the process of phase transition which is of interest to us. Thus in the cases of transition from vapor to crystal, from vapor to liquid or from liquid to vapor phases, the process of formation of the new phase always goes through the formation of small crystallites, droplets or bubbles. In this section we will clarify two questions: (i) the mechanical equilibrium of small particles with their ambient phase or, in other words, the interrelation of the pressures in the two phases when the phase boundary is not flat, and (ii) the thermodynamic equilibrium of small particles or their equilibrium vapor pressure as a function of their size. In fact we will derive and interpret the equations of Laplace and Thomson–Gibbs.

1.3.1. *Equation of Laplace*

We consider a vessel with a constant volume V containing vapor with pressure P_v and a liquid droplet with a radius r and an inner pressure P_l, both at one and the same constant temperature T. The condition for equilibrium is given by the minimum of the Helmholz free energy of the system $F(V, T)$:

$$dF = -P_v dV_v - P_l dV_l + \sigma dS = 0 , \qquad (1.14)$$

where V_v and V_l are the volumes of the vapor phase and the droplet, σ is the surface tension of the liquid and S is the droplet surface area. The value of the surface tension σ of the infinitely large liquid phase is usually taken. As $V = V_v + V_l = $ const and $dV_v = -dV_l$, Eq. (1.14) can be rewritten after rearrangement in the form

$$P_l - P_v = \sigma \frac{dS}{dV_l} .$$

Bearing in mind that $S = 4\pi r^2$ and $V_l = 4\pi r^3/3$,

$$\frac{dS}{dV_l} = \frac{d(4\pi r^2)}{d(4\pi r^3/3)} = \frac{2}{r} ,$$

and the above equation turns into

$$P_l - P_v = \frac{2\sigma}{r} , \tag{1.15}$$

which is known as the *Laplace* equation. The latter states that the pressure in a small droplet is always higher than the pressure of the surrounding vapor. The difference $2\sigma/r$ is called the *Laplace* or *capillary pressure* and is equal to zero when the phase boundary is flat ($r \to \infty$). Then $P_l = P_v = P_\infty$, as stated in Sec. 1.1. Here the notation P_∞ for the equilibrium pressure is used instead of P_0 to emphasize the fact that the dividing surface is flat, i.e. it has an infinite radius of curvature.

The physical meaning of Eq. (1.15) becomes clearer if we derive it from balance of forces. The overall force exerted on the droplet from the outside is a sum of the force $4\pi r^2 P_v$ exerted from the vapor phase and the force $8\pi r\sigma$ due to the surface tension. It is equal to the force due to the internal pressure $4\pi r^2 P_l$, i.e.

$$4\pi r^2 P_v + 8\pi r\sigma = 4\pi r^2 P_l ,$$

and Eq. (1.15) results. Thus the Laplace pressure is clearly due to the surface tension of the small droplet.

In the case of an arbitrary surface with principal radii of curvature r_1 and r_2, Eq. (1.15) reads

$$P_l - P_v = \sigma \left(\frac{1}{r_1} + \frac{1}{r_2} \right) . \tag{1.16}$$

1.3.2. *Equation of Thomson–Gibbs*

In order to solve the problem of thermodynamic equilibrium we consider the same system as before at constant pressure P and temperature T. In this case the variation of the Gibbs free energy $G(P, T, n_v, n_l, S)$ of the system reads ($dP = 0, dT = 0$)

$$\Delta G = \mu_v dn_v + \mu_l dn_l + \sigma dS = 0 \, , \tag{1.17}$$

where n_v and n_l are the numbers of moles in the vapor and liquid phases, respectively.

When writing the expression for the Gibbs free energy we take for the chemical potential of the atoms in the small droplet the value which is valid for the bulk phase and compensate the difference between the bulk liquid and the small droplet by the surface energy σS. Besides, we again ascribe to the surface tension its bulk value. As the system is closed, $n_v + n_l = $ const and $dn_v + dn_l = 0$. Solving (1.17) together with $dn_v = -dn_l$ gives

$$\mu_v - \mu_l = \sigma \frac{dS}{dn_l} \, .$$

With $n_l = 4\pi r^3 / 3 v_l$ it turns into the famous equation of *Thomson-Gibbs*

$$\mu_v - \mu_l = \frac{2\sigma v_l}{r} \, . \tag{1.18}$$

Comparing Eqs. (1.15) and (1.18) it becomes immediately clear that the product $(P_l - P_v)V_l = n\Delta\mu$ is just equal to the work which is gained when a liquid droplet is formed from the unstable vapor phase or, in other words, when $n = V_l/v_l$ atoms or molecules are transferred from the vapor phase with higher chemical potential to the liquid phase with lower chemical potential [Gibbs 1928].

As μ_l is the chemical potential of the bulk liquid phase the difference $\mu_v - \mu_l$ is just equal to the supersaturation $\Delta\mu$ (see Eq. (1.9)) and

$$P_r = P_\infty \exp\left(\frac{2\sigma v_l}{r k T}\right) \, . \tag{1.19}$$

It follows that the equilibrium vapor pressure of a small liquid droplet with radius r is higher than that of the infinitely large liquid with a flat surface. The physical reason is easy to understand if we imagine that an atom on a curved convex surface is more weakly bound to the remaining atoms than in the case of an atom on a flat surface.

Obviously, we have the opposite case when considering a vapor bubble with radius r in an overheated liquid. Following a similar procedure we find

$$P_r = P_\infty \exp\left(-\frac{2\sigma v_l}{r k T}\right) \, . \tag{1.20}$$

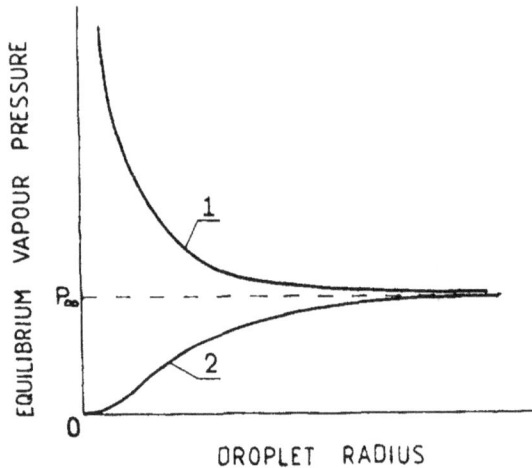

Fig. 1.4. Dependence of the equilibrium vapor pressures of a small liquid droplet (curve 1, Eq. (1.19)) and in a vapor bubble in an overheated liquid (curve 2, Eq. (1.20)). P_∞ denotes the equilibrium vapor pressure of the infinitely large liquid phase.

The dependence of the equilibrium vapor pressure of a liquid droplet and of a gas bubble on their size is shown in Fig. 1.4.

In order to derive an analogous equation for a small crystallite we have to know its equilibrium shape. The latter is defined as the shape at which the crystal has a minimal surface energy at a given constant volume. For a liquid droplet the equilibrium shape is obviously a sphere. We shall now accept that Eq. (1.19) is valid for small crystallites with the only exception being that the radius of the droplet is replaced by the radius of the sphere inscribed in the crystallite.

1.4. Equilibrium Shape of Crystals

When considering the equilibrium of a small crystal with its ambient phase (vapor, solution or melt) there exists, obviously, a shape which is the most favorable from a thermodynamic point of view in the sense that the work of formation of such a crystal is the minimal one at the given crystal volume. The work of formation of a small crystal consists of two parts: a volume part $(P_c - P_v)V_c = n(\mu_v - \mu_c)$ which is gained when transferring n atoms or molecules from the ambient (vapor) phase with higher chemical potential μ_v to the crystal phase with lower bulk chemical potential μ_c when the crystal phase is the stable one ($\mu_c < \mu_v$), and a surface part σS which is spent to create a new phase-dividing surface. The volume part depends obviously only on the volume of the crystal or on the number of the atoms transferred. At a constant volume the surface part depends only on the crystal shape. Then the condition for minimum of the Gibbs free energy change connected with the crystal formation at a constant volume which determines the equilibrium shape is reduced to the condition for minimum of the surface energy. The equilibrium shape of a liquid droplet is evidently a sphere. The case of a crystal is more complicated as the latter is confined by crystal faces with different crystallographic orientations and, respectively, different specific surface energies. This means that the surface energy depends on the crystallographic orientation and in that sense it is anisotropic.

In general there are three ways to create a new surface. In the first one two homogeneous phases are cut (or cleaved) into two parts each and then the different halves are put into contact. New surface can be formed also by the transfer of atoms from one phase to another forming convex (during growth) or concave (during evaporation or dissolution) form. Finally new surface can be created by stretching out an old one (by stretching out the bulk crystal). When two liquid phases are involved the above three methods do not differ. The work spent to create reversibly and isothermally a unit area of a new surface is called *specific surface free energy*. When a new surface is formed by the two first methods chemical bonds are broken. Thus the work for creation of a new surface or, in other words, the specific surface free energy, is equal as a first approximation to the sum of the energies of the broken bonds per unit area. When applying the third method the number of broken bonds remains unchanged but the surface area per dangling bond is changed, which in turn leads to a change of the surface energy.

It follows from the above that the more closely packed the given crystal face is the smaller is the density of the unsaturated bonds and thus its specific surface free energy. Let us consider for example the specific surface free energies of the faces of a crystal with a simple cubic lattice. The latter does not exist in nature, with the only exception being one of the crystal modifications of the metal polonium, but is widely used in theoretical considerations. It is well known under the name *Kossel crystal*. When determining the specific surface energy we will take into account the bonds between first, second and third nearest neighbor atoms. The procedure of the determination of σ of the face (hkl) involves the construction of a column in the shape of a prism with a form of the base which follows, for convenience, the symmetry of the crystal face, i.e. square for (100), hexagonal for (111) faces of cubic crystals, etc. [Honnigmann 1958]. Then the energy necessary for the detachment of this column from the crystal face Ψ_{hkl} is divided by the doubled area Σ_{hkl} of the contact because two surfaces are involved. Thus $\sigma_{hkl} = \Psi_{hkl}/2\Sigma_{hkl}$. Following Fig. 1.5(a), a column of atoms with a square base is detached from the (100) face of a Kossel crystal and two surfaces are created. The value of the surface energy is then

$$\sigma_{100} = \frac{\psi_1 + 4\psi_2 + 4\psi_3}{2b^2} = \frac{1}{b^2}\left(\frac{1}{2}\psi_1 + 2\psi_2 + 2\psi_3\right) ,$$

where ψ_1, ψ_2 and ψ_3 are the works required to break the bonds between the first, second and third neighbors, respectively, and b^2 is the area per atom, b being the interatomic distance.

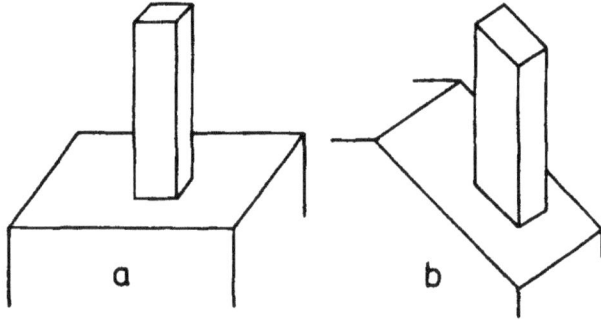

Fig. 1.5. For the determination of the specific surface energies σ_{hkl} of the faces (a) (100) and (b) (110) of a crystal with simple cubic lattice.

For the rombohedral face (110) (see Fig. 1.5(b)) the column has a rectangular base and

$$\sigma_{110} = \frac{2\psi_1 + 6\psi_2 + 4\psi_3}{2b^2\sqrt{2}} = \frac{1}{b^2}\left(\frac{1}{\sqrt{2}}\psi_1 + \frac{3}{\sqrt{2}}\psi_2 + \sqrt{2}\psi_3\right) .$$

The shortest first neighbor bonds have the greatest contribution to the energy. The energies of the second neighbor bonds are probably not greater than 10% of the energies of the first neighbor bonds for metallic and covalent bonds. The contribution of the third neighbor bonds could be neglected. It follows from above that $\sigma_{100} < \sigma_{110}$. Performing the same calculations for the faces (111) and (211) we will find that $\sigma_{100} < \sigma_{110} < \sigma_{111} < \sigma_{211}$, etc.

1.4.1. *Theorem of Gibbs–Curie–Wulff*

Following Gibbs [1878] and Curie [1885] we can derive an expression for the equilibrium shape of a single crystal proceeding from the general condition of the minimum of the Helmholz free energy of the system at $T = $ const and $V = $ const:

$$dF = 0, \qquad dV = 0 . \tag{1.21}$$

We assume first that the crystal is a polyhedron confined by a limited number of different crystal faces with areas Σ_n to which a series of *discrete* values of the specific surface energies σ_n correspond. Then the equilibrium condition (1.21) reads

$$dF = -P_v dV_v - P_c dV_c + \sum_n \sigma_n d\Sigma_n = 0 , \tag{1.22}$$

where P_c is the inner pressure of the crystal phase, P_v is the pressure of the vapor phase, and V_v and V_c are the volumes of the vapor phase and the crystal phase, respectively.

Bearing in mind that $V = V_v + V_c = $ const or $dV_v = -dV_c$ the above equation is reduced to

$$-(P_c - P_v)dV_c + \sum_n \sigma_n d\Sigma_n = 0 . \tag{1.23}$$

The volume of the crystal can be considered as a sum of the volumes of pyramids constructed on the crystal faces with a common apex in an arbitrary point within the crystal. Then

$$V_c = \frac{1}{3} \sum_n h_n \Sigma_n$$

and

$$dV_c = \frac{1}{3} \sum_n (\Sigma_n dh_n + h_n d\Sigma_n) ,$$

where h_n are the heights of the pyramids.

On the other hand, every change of the volume with accuracy to infinitesimals of second order is equal to a shift of the surfaces Σ_n by a distance dh_n so that

$$dV_c = \sum_n \Sigma_n dh_n .$$

Combining the last two equations gives

$$dV_c = \frac{1}{2} \sum_n h_n d\Sigma_n \ . \tag{1.24}$$

Substituting (1.24) into (1.23) gives

$$\sum_n \left[\sigma_n - \frac{1}{2}(P_c - P_v)h_n \right] d\Sigma_n = 0 \ .$$

As the changes $d\Sigma_n$ are independent of each other every term in the brackets is equal to zero and

$$P_c - P_v = 2\frac{\sigma_n}{h_n} \ . \tag{1.25}$$

The difference $P_c - P_v$ does not depend on the crystallograpic orientation and for the equilibrium shape one obtains

$$\frac{\sigma_n}{h_n} = \text{ const} \tag{1.26}$$

or

$$\sigma_1 : \sigma_2 : \sigma_3 \cdots = h_1 : h_2 : h_3 \cdots \ . \tag{1.26'}$$

The relationship (1.26) expresses the geometrical interpretation given later by Wulff (1901) known as the *Wulff rule* or *Gibbs–Curie–Wulff theorem*. It states that *in equilibrium the distances of the crystal faces from a point within the crystal (called a Wulff's point) are proportional to the corresponding specific surface energies of these faces*. According to this rule we can construct the equilibrium shape by the following procedure: We draw vectors normal to all possible crystallographic faces from an arbitrary point. Then distances proportional to the corresponding values of the specific surface energies σ_n are marked on the vectors and planes normal to the vectors are constructed through the marks. The resulting closed polyhedron is the equilibrium form. Crystal faces with the lowest surface energies belong to it. Crystal faces which only touch the apexes of this polyhedron or are situated even further do not belong to the equilibrium form.

The proportionality constant in (1.26) is determined by the difference of the pressures in both phases. As shown in the previous section, $(P_c - P_v)v_c = \Delta\mu$, where $v_c = V_c/n_c$ is the molar volume of the crystal phase. The condition $P_c - P_v = \text{const}$ is thus equivalent to the statement that the difference of the chemical potentials or, in other words, the supersaturation

$\Delta\mu = \mu_v - \mu_c$, has one and the same value all over the crystal surface. Then

$$\Delta\mu = \frac{2\sigma_n v_c}{h_n} \qquad (1.27)$$

and hence the supersaturation determines the scale or the size of the crystal. As seen Eq. (1.27) has the familiar form of the Thomson–Gibbs equation. As the Wulff point has been arbitrarily chosen we can take it at the center of the crystal.

In the same way we can derive the Gibbs–Curie–Wulff theorem for a crystal formed on a foreign substrate (Fig. 1.7) [Kaischew 1950, 1951, 1960]. In this case the crystal lies with one of its faces with specific surface energy σ_m on the substrate, the latter having a specific surface energy σ_s. An interfacial boundary is formed between the crystal and the substrate. In order to find its specific energy we will perform the following imaginary experiment.

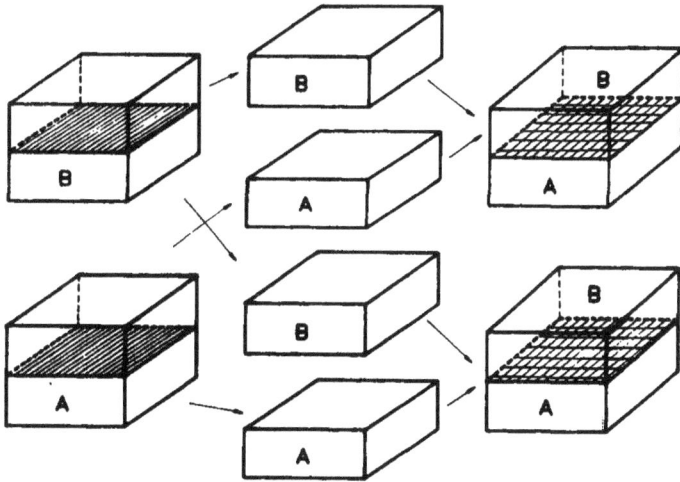

Fig. 1.6. For the determination of the specific energy of the interface between two isomorphic crystals A and B (after Kern *et al.* [1979]).

We consider two crystals A (substrate) and B (deposit) of equal dimensions (Fig. 1.6) [Kern *et al.* 1979]. We cleave them reversibly and isothermally and produce two surfaces of A, each with area Σ_A, and two surfaces of B, each with area $\Sigma_B = \Sigma_A$. In doing so, we expend energies

U_{AA} and U_{BB}. We then put the two halves of A in contact with the two halves of B and produce two interfacial boundaries AB, each with area $\Sigma_{AB} = \Sigma_B = \Sigma_A$. The work gained is $-2U_{AB}$. The excess energy of the boundary AB required to balance the energy change accompanying the above process is $2U_i$. Thus we have

$$2U_i = U_{AA} + U_{BB} - 2U_{AB} .$$

Clearly when the two crystals are indistinguishable from each other, $U_{AA} = U_{BB} = U_{AB}$ and the excess energy $U_i = 0$. Using the definition of the specific surface energy ($\sigma_{hkl} = U_{hkl}/2\Sigma_{hkl}$) one obtains the well-known relation of *Dupré* [1869]:

$$\sigma_i = \sigma_A + \sigma_B - \beta , \tag{1.28}$$

where the specific interfacial energy $\sigma_i = U_i/\Sigma_{AB}$ is defined as the excess energy of the boundary per unit area and the specific adhesion energy $\beta = U_{AB}/\Sigma_{AB}$ is defined as the energy per unit area to disjoin two different crystals. Note that β accounts for the binding between the two crystals and does not depend on the lattice misfit. The latter will be taken into account in Chap. 4 when considering the epitaxial growth of thin films.

When a crystal is formed on a foreign substrate a surface energy $\sigma_s \Sigma_m$ is lost and surface energy $\sigma_i \Sigma_m$ is expended, S_m being the area of contact. Then instead of (1.22) one has to write

$$dF = -P_v dV_v - P_c dV_c + \sum_{n \neq m} \sigma_n d\Sigma_n + (\sigma_i - \sigma_s) d\Sigma_m = 0 . \tag{1.29}$$

Following the same procedure as above we find

$$\frac{\sigma_n}{h_n} = \frac{\sigma_m - \beta}{h_m} = \text{const} \tag{1.30}$$

or

$$\sigma_1 : \sigma_2 : \sigma_3 \cdots \sigma_m - \beta = h_1 : h_2 : h_3 \cdots h_m , \tag{1.30'}$$

where h_m is the distance from the Wulff point to the contact plane (Fig. 1.7).

It follows that the distance h_m from the Wulff point to the contact plane is proportional to the difference of the corresponding specific surface energy and specific adhesion energy. Obviously, when the substrate catalytic potency is equal to zero, $\beta = 0$, the distance h_m will have its "homogeneous" value in the absence of substrate. In this case we speak of *complete*

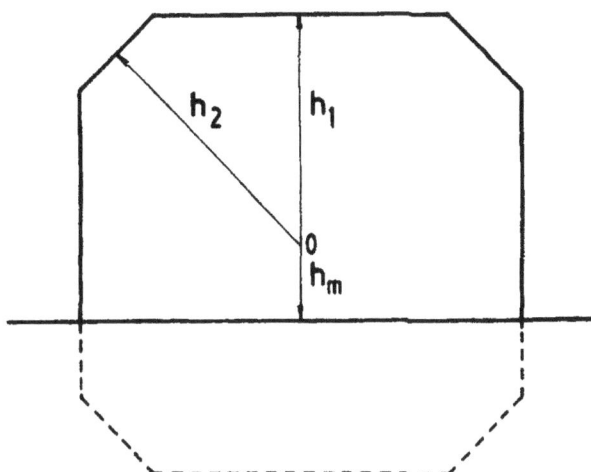

Fig. 1.7. Equilibrium shape of a crystal on a foreign substrate. Wulff's point is denoted by 0. The distance from Wulff's point to the contact plane is denoted by h_m. The distances h_1 and h_2 to the free surfaces remain the same as in a free polyhedron. The equilibrium shape in the absence of a foreign substrate is shown by the additional dashed line.

nonwetting. At the other extreme, $\beta = \sigma_A + \sigma_B = 2\sigma$ ($\sigma_A = \sigma_B = \sigma$), we have the case of *complete wetting* and the crystal will be reduced to a two-dimensional monolayer island. In all intermediate cases, $0 < \beta < 2\sigma$, we have *incomplete wetting* and the height of the crystal will be smaller than its lateral size.

1.4.2. Polar diagram of the surface energy

We have considered by now a crystal confined by discrete faces with small Miller indices. Let us now imagine a crystal face which is slightly deviated (by a small angle θ) from one of the small index faces, say, the cubic one (100) of a Kossel crystal as shown in Fig. 1.8(a). Such a face is called *vicinal*. It is clear for geometrical reasons that it consists of terraces and steps. For simplicity we accept that the steps are monatomic and equidistant. The specific surface energy of such a face is the sum of the surface energy of the terraces σ_0 and the energy of the steps or the edge energy $\varkappa \cong b\sigma_0$, which can be evaluated in the same way as the surface energy just by counting the number of broken bonds per unit length. If one neglects the interaction between the steps as a first approximation for the specific surface energy of such a vicinal face one obtains [Landau 1969]

$$\sigma(\theta) = \frac{\varkappa}{b}\sin(\theta) + \sigma_0\cos(\theta) \ . \tag{1.31}$$

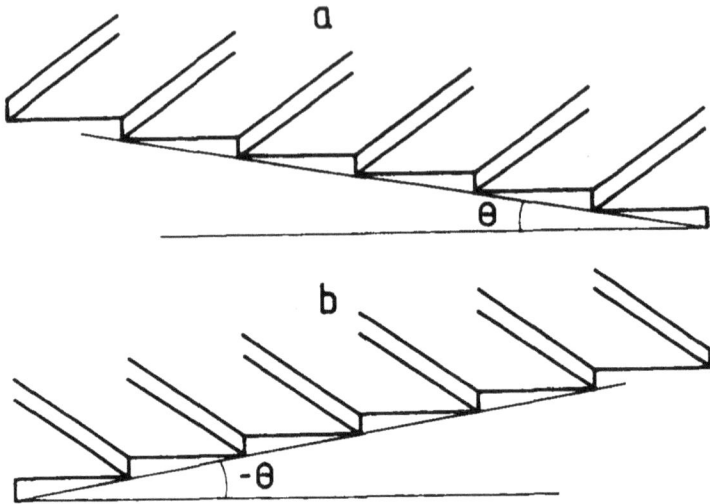

Fig. 1.8. Vicinal surfaces tilted by an angle (a) θ and (b) $-\theta$ from a small index (singular) face.

We consider now a vicinal face which is symmetric to the first one, i.e. tilted by an angle of $-\theta$ (Fig. 1.8(b)). Its surface energy is

$$\sigma(-\theta) = -\frac{\varkappa}{b}\sin(\theta) + \sigma_0\cos(\theta) . \qquad (1.32)$$

Graphic representations of the $\sigma(\theta)$ functions (1.31) and (1.32) are given in Fig. 1.9(a). As seen they are continuous everywhere with the exception of the point $\theta = 0$ where $(d\sigma/d\theta)_{\theta \geq 0} = \varkappa/b$ and $(d\sigma/d\theta)_{\theta \leq 0} = -\varkappa/b$. In other words, the $\sigma(\theta)$ dependence has a *singular* point at $\theta = 0$ with its derivative making a jump of $2\varkappa/b$. The same singular points exist at $\theta = \pm\pi/2, \pm\pi$, etc. In Fig. 1.9(b) the same functions are plotted in polar coordinates. A contour consisting of circular segments and possessing singular points at $\theta = 0$, $\pi/2$, and $3\pi/2$ results. In the three-dimensional case (Fig. 1.9(c)) a body is obtained which consists of 8 spherical segments and has 6 sharp singular points. This plot is called a *polar diagram* of the surface energy.

When constructing the above polar diagram only the bonds between the first neighbor atoms have been taken into account. The bonds between the second neighbors in the Kossel crystal are directed at angle $\pi/4$ with respect to the first neighbor bonds. We can perform the same considerations as above [Chernov 1984] accounting for the second neighbor bonds only. Thus a polar diagram which is inscribed in the first one (the second neighbor

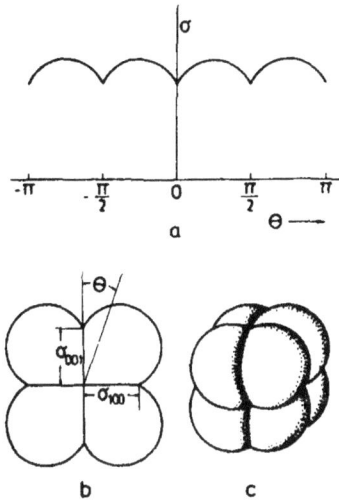

Fig. 1.9. Polar diagram of the specific surface energy — dependence of the surface energy on angle θ in (a) orthogonal coordinates, (b) polar coordinates (two-dimensional representation), and (c) spherical coordinates (three-dimensional representation). First neighbor interactions only are taken into account (after Chernov [1984]).

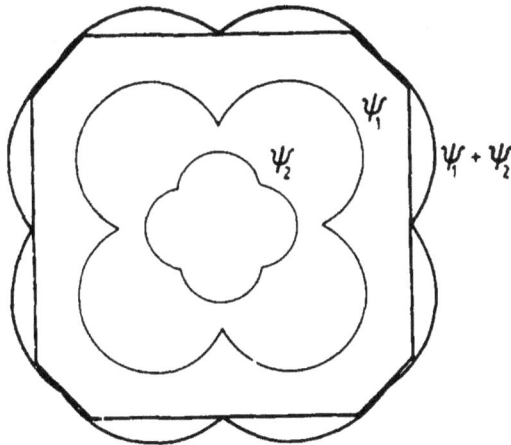

Fig. 1.10. Polar diagram of the specific surface energy (the outermost contour denoted by $\Psi_1 + \Psi_2$, two-dimensional representation) taking into account the first (Ψ_1) and second (Ψ_2) nearest neighbor interactions. The contours denoted by Ψ_1 and Ψ_2 give the polar diagrams as calculated by taking into account separately the first and second nearest neighbors. The closed contour consisting of straight lines drawn through the singular points of the contour $\Psi_1 + \Psi_2$ gives the equilibrium shape of the crystal.

bonds are much weaker than the first neighbor ones) and is rotated at an
angle of $\pi/4$ with respect to the latter is obtained (Fig. 1.10). The sum of
the two curves gives the polar diagram when accounting simultaneously for
both the first and the second neighbors. As seen, new, shallower minima
appear which correspond to faces which are analogous to the (110) faces in
the three-dimensional Kossel crystal. The contribution of third neighbor
bonds is insignificant and will not affect considerably the shape of the polar
diagram. In any case accounting for the more distant neighbors leads to
more complicated polar diagram.

1.4.3. *Herring's formula*

Let us now derive the condition for the equilibrium shape of the crystal
accounting for the anisotropy of the surface energy or, in other words,
the $\sigma(\theta)$ dependence. This is a question of utmost importance as it is
unambiguously connected with the problem of the equilibrium structure
(the roughness) of the crystal surfaces. The three-dimensional problem
is somewhat complicated from a mathematical point of view and for this
reason we will consider the simpler case of a "two-dimensional" crystal
which represents a cross section of a three-dimensional one. On the other
hand, the 2D case is very important for understanding two-dimensional
nucleation and, in turn, layer growth of smooth crystal faces.

When treating the problem we will follow exactly the approach of
Burton, Cabrera and Frank [1951]. The crystal volume V_c will be replaced
by the crystal surface area S_c and the specific surface energy $\sigma(\theta)$ by the
specific edge energy $\varkappa(\theta)$. Then, instead of (1.21), we write

$$\Phi = \min, \qquad S_c = \text{const} , \qquad (1.33)$$

where

$$\Phi = \int_L \varkappa(\theta) dl \qquad (1.34)$$

is the edge energy of the "two-dimensional crystal," the integration being
carried out over the whole periphery L of the latter, and

$$S_c = \int_S ds \qquad (1.35)$$

is the surface area of the crystal where ds is the surface area of a curvilinear
sector.

Let r and φ be the polar coordinates of a point M on the crystal boundary L (Fig. 1.11) and let x and y be the corresponding orthogonal coordinates. We construct a tangent T to the crystal boundary L at the point M and a perpendicular ON with length n from the origin O to the tangent. The latter makes an angle θ with the abscissa. The line element of the crystal boundary in a parametric form $x = x(t)$ and $y = y(t)$ reads

$$dl = \left(x'^2 + y'^2\right)^{\frac{1}{2}} dt ,$$

where $x' = dx/dt$ and $y' = dy/dt$.

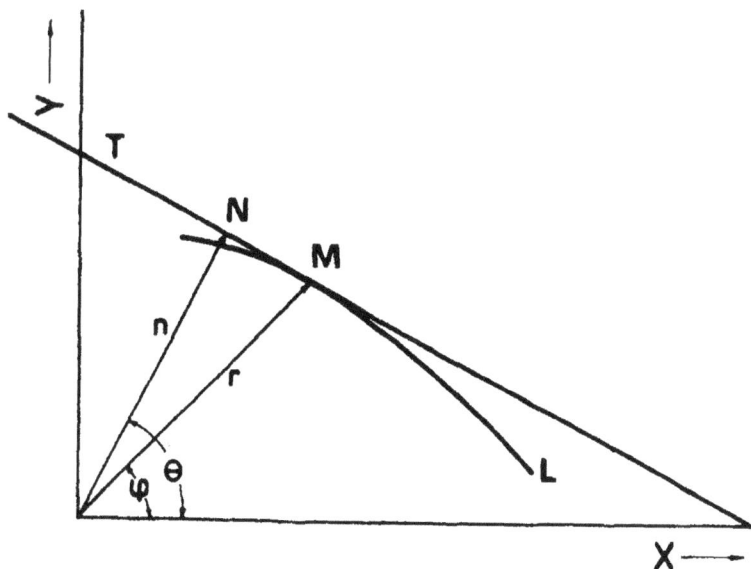

Fig. 1.11. The derivation of Herring's formula in the two-dimensional case: L and T denote the crystal boundary and the tangent to it at a point M, r and φ are the polar coordinates of a point M of the crystal surface, and n and θ are the polar coordinates of the point N belonging to the polar diagram $\varkappa(\theta)$ (after Burton, Cabrera and Frank [1951]).

In the same way the area of the curvilinear sector ds is

$$ds = \frac{1}{2}\left(xy' - yx'\right) dt .$$

If we choose the angle θ as the parameter the integrals (1.34) and (1.35) can be rewritten in the form

Crystal–Ambient Phase Equilibrium

$$\Phi = \int_0^{2\pi} \varkappa(\theta) \left(x'^2 + y'^2 \right)^{\frac{1}{2}} d\theta \ , \tag{1.36}$$

$$S_c = \frac{1}{2} \int_0^{2\pi} (xy' - yx')d\theta \ . \tag{1.37}$$

From Fig. 1.11 we find that the perpendicular n from the origin to the tangent T is given by

$$n = x \cos\theta + y \sin\theta \ .$$

Then we can find the point N for any point M on the crystal boundary. Conversely, making use of the transformations

$$x = n \cos\theta - n' \sin\theta \ , \tag{1.38'}$$

$$y = n \sin\theta + n' \cos\theta \ , \tag{1.38''}$$

we can construct the crystal contour once we know the polar diagram $n(\theta)(n' = dn/d\theta)$.

Then Eqs. (1.36) and (1.37) can be written in terms of $n(\theta)$:

$$\Phi = \int_0^{2\pi} \varkappa(\theta)(n + n'')d\theta = \min \ , \tag{1.39}$$

$$S_c = \frac{1}{2} \int_0^{2\pi} n(n + n'')d\theta = \text{const} \ , \tag{1.40}$$

where $n'' = d^2n/d\theta^2$.

Following the method of Lagrange we multiply Eq. (1.40) by an indefinite scalar λ and sum up Eqs. (1.39) and (1.40) to obtain

$$\int_0^{2\pi} \left[\varkappa(\theta)(n + n'') - \frac{1}{2}\lambda n(n + n'') \right] d\theta = \min \ . \tag{1.41}$$

The condition for minimum is still preserved when we multiplied it by a constant and added a constant. The solution of (1.41) will give us a function $n = n(\theta)$ which satisfies both conditions $\Phi = \min$ and $S_c = \text{const}$. To solve the problem we will use the method of Euler for finding an extremum [Arfken 1973]. It states that if we have a functional of the kind

$$\int_a^b F(x,y,y',y'')dx = \min ,$$

where $y = y(x)$, $y' = dy/dx$ and $y'' = d^2y/dx^2$, the equation which satisfies it has the form

$$\frac{\partial F}{\partial y} - \frac{\partial}{\partial x}\left(\frac{\partial F}{\partial y'}\right) + \frac{\partial^2}{\partial x^2}\left(\frac{\partial F}{\partial y''}\right) = 0 .$$

This is the well-known equation of Euler where the functions y, y' and y'' are taken upon differentiation as independent variables without accounting for their dependence on x.

Applying the above method, from (1.41) one obtains

$$\frac{\partial^2}{\partial \theta^2}[\varkappa(\theta) - \lambda n(\theta)] + [\varkappa(\theta) - \lambda n(\theta)] = 0 . \qquad (1.42)$$

Thus the equilibrium shape of the crystal is governed by a nonlinear equation of second order which satisfies the conditions $\Phi = \min$ and $S_c = $ const. Its solution $n(\theta)$ will give us the rule for constructing the equilibrium shape of our two-dimensional crystal (the crystal countour) on the base of the polar diagram of the edge energy and Eqs. (1.38).

Actually we have to solve the much simpler linear differential equation of second order. With the substitution $u = \varkappa(\theta) - \lambda n(\theta)$, Eq. (1.42) turns into $u'' + u = 0$. Its solution reads [Kamke 1959]

$$u = C\sin(x - \phi)$$

or

$$n(\theta) = \frac{1}{\lambda}\varkappa(\theta) - C\sin(\theta - \phi) ,$$

where C and ϕ are constants. The second term on the right-hand side is a periodic function with period 2π. However, different crystals have different symmetry and thus different period. For example, cubic crystals have a period of $\pi/2$, hexagonal crystals have a period $\pi/3$, etc. In order to get rid of this restriction we put $C = 0$ and obtain

$$n(\theta) = \frac{1}{\lambda}\varkappa(\theta) . \qquad (1.43)$$

By analogy with the previous case the proportionality constant which multiplies the crystal volume is $\lambda = \Delta\mu/s_c$, where s_c is the area of an atom in the two-dimensional crystal and

$$\Delta\mu = \frac{\varkappa(\theta)s_{\mathrm{c}}}{n(\theta)} \ . \tag{1.44}$$

Equation (1.44) which appears as a generalized Thomson–Gibbs equation is the Gibbs–Curie–Wulff theorem for two-dimensional crystals in which $n(\theta)$ is the radius vector of the polar diagram $\varkappa(\theta)$.

Carrying out the differentiation in Eq. (1.42) gives

$$n + n'' = \frac{1}{\lambda}[\varkappa(\theta) + \varkappa''(\theta)] \ ,$$

where $\varkappa''(\theta) = d^2\varkappa(\theta)/d\theta^2$.

Realizing that

$$n + n'' = \frac{(x'^2 + y'^2)^{3/2}}{x'y'' - y'x''} = R$$

is in fact the principal radius of curvature of the polar diagram, R, then with $\lambda = \Delta\mu/s_{\mathrm{c}}$ one obtains

$$\Delta\mu = \frac{s_{\mathrm{c}}}{R}\left[\varkappa(\theta) + \frac{d^2\varkappa(\theta)}{d\theta^2}\right] \ .$$

By analogy, for a three-dimensional crystal with principal radii of curvature R_1 and R_2 and polar angles θ_1 and θ_2 for the equilibrium shape, one obtains an expression with s_{c} and $\varkappa(\theta)$ respectively replaced by v_{c} and $\sigma(\theta)$ (for more rigorous derivation see Chernov [1984]) which is known as the formula of *Herring* [Herring 1951, 1953]:

$$\Delta\mu = \frac{v_{\mathrm{c}}}{R_1}\left(\sigma + \frac{d^2\sigma}{d\theta_1^2}\right) + \frac{v_{\mathrm{c}}}{R_2}\left(\sigma + \frac{d^2\sigma}{d\theta_2^2}\right) \ . \tag{1.45}$$

In the same manner the generalized Gibbs–Curie–Wulff theorem for three-dimensional crystals reads

$$\Delta\mu = \frac{2\sigma(\theta)v_{\mathrm{c}}}{n(\theta)} \ . \tag{1.46}$$

Equations (1.44) and (1.46) give us the practical rule for the construction of the equilibrium shape. First, we ascribe a particular value to the scale parameter $\Delta\mu/2v_{\mathrm{c}}$ which determines the size of the crystal. Then we draw the radius vector $n(\theta)$ from the central point at an arbitrarily selected crystallographic direction θ and find the cross-sectional point with the polar diagram (Fig. 1.12). We then construct through it a plane normal to $n(\theta)$ and repeat this procedure for the whole contour of the diagram. A family

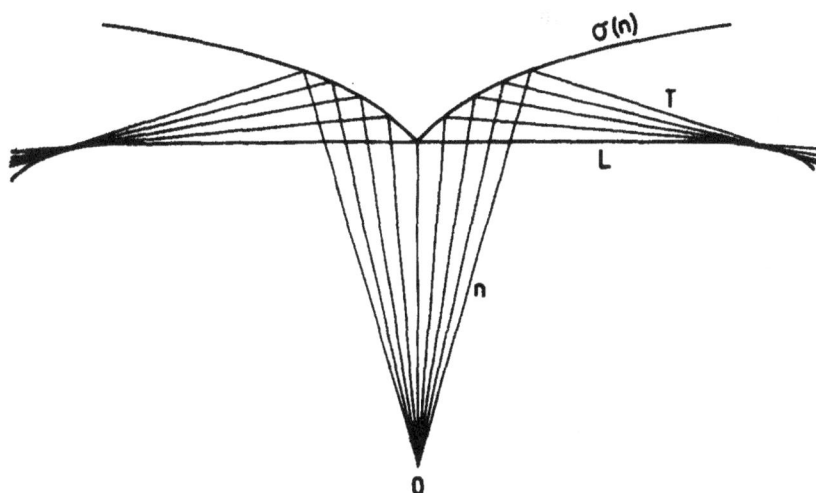

Fig. 1.12. Construction of the equilibrium shape based on the polar diagram of the surface energy following Eq. (1.46) (after Chernov [1984]).

of planes results and its inner envelope is in fact the equilibrium form of the crystal.

Let us go back to Herring's formula (1.45) and consider more closely the quantity $\sigma_n^* = \sigma + d^2\sigma/d\theta_n^2$ $(n = 1, 2)$ which is usually called the *surface stiffness*. At the singular points the first derivative of σ with respect to θ experiences a jump (equal to $2\varkappa/b$ in the above oversimplified case). Hence, the second derivatives and in turn the surface stiffnesses σ_n^* have infinitely large values. The left-hand side of Eq. (1.45) has a finite value, and obviously the condition for the right-hand side to have a finite value is for the principal radii R_1 and R_2 to be infinitely large. Hence at the singular points the curvature of the corresponding crystal faces will be equal to zero or, in other words, the crystal faces will be flat. This is the reason that the flat faces are often called *singular faces*. Immediately aside of the singular points the second derivatives of σ and in turn the surface stiffnesses σ_n^* acquire finite positive values, and hence also do the radii R_1 and R_2. Therefore, the crystal surface will be rounded. It will consist of terraces and steps or, in other words, it will be atomically rough. Finally, there are regions where $\sigma_n^* = 0$ and so are the radii R_1 and R_2 (the ratios σ_n^*/R_n again having finite values). These are obviously the edges and apexes of the crystal. Negative values of the surface stiffnesses have no physical meaning as this means negative radii R_1 and R_2 and hence concave regions, which cannot exist on the equilibrium shape.

As discussed above the supersaturation must have one and the same value over every point of the surface of a crystal with an equilibrium shape. It follows that the supersaturation around a crystal without an equilibrium shape will vary from one point to another. Facets with areas smaller than that required by the equilibrium condition will have larger chemical potentials and hence the supersaturation over them will be lower than the current one in the system, and vice versa. If such a crystal is immersed in a supersaturated ambient phase and given enough time to equilibrate, smaller facets will dissolve to become larger facets and larger facets will grow to become smaller facets up to the moment the supersaturation attains one and the same value all over the crystal surface and the equilibrium shape is reached. As will be shown below the above conclusion is valid for crystallites sufficiently small ($kTR_n/\sigma_n^* v_c \ll 1$) so that the supersaturation difference which is the driving force for the equilibrium to be reached is sufficiently large.

Finally, Eq. (1.45) can be expressed in the form

$$\Delta\mu = v_c \left(\frac{\sigma_1^*}{R_1} + \frac{\sigma_2^*}{R_2} \right) \tag{1.47}$$

or

$$P_c - P_v = \frac{\sigma_1^*}{R_1} + \frac{\sigma_2^*}{R_2} \ . \tag{1.48}$$

We conclude that Herring's formula is a generalization for finite crystallites of Laplace's equation (1.16) which relates the liquid surface tension to the pressure difference on both sides of the curved liquid surface. The Laplace pressure is determined in this case by the surface stiffnesses which govern the crystal curvature, rather than by the specific surface energies. Thus Eq. (1.48) explains the term surface stiffness. It is a measure of the resistivity of the crystal faces against bending (roughening) when a pressure (force per unit area) is applied on them. Flat facets require infinitely high pressures in order to be "bent."

1.4.4. *Stability of crystal surfaces*

In fact we have just concluded that the structure of a given crystal surface is determined by the corresponding value of the surface stiffness. When the latter is infinite the corresponding crystal face is flat and atomically smooth. When the surface stiffness has some finite positive value, the crystal surfaces

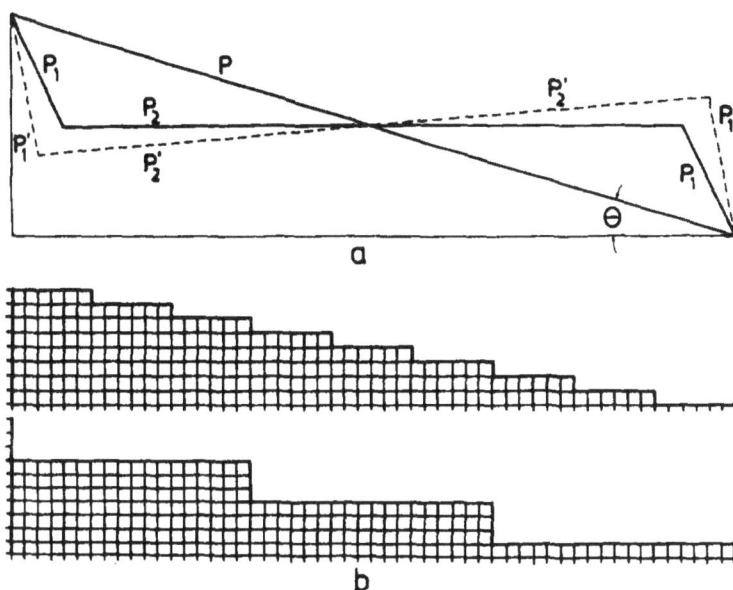

Fig. 1.13. A scheme illustrating the stability of a vicinal crystal surface against faceting: (a) Schematic view in terms of the slope $p = \tan\theta$ of the original vicinal face which breaks down into facets with slopes $p_1 = \tan\theta_1$ and $p_2 = \tan\theta_2$. The dashed lines with slopes p_1' and p_2' give possible deviations from the faceted surface with slopes p_1 and p_2, respectively (see text). (b) Side view of a vicinal surface of a Kossel crystal whose slope is accommodated either by single height steps or by facets (after Cabrera and Coleman [1963]).

are rounded and, in the near vicinity of the singular faces, should consist of terraces divided by steps. We accepted that the steps are of monomolecular height. The width of the terraces or the density of the steps depends on the value of the polar angle. However, we can have different structures of a vicinal face at one and the same value of the polar angle. Thus if the steps are of a double height the terraces should be twice as wide; when the steps have a triple height the terraces will have a triple width, etc., at one and the same value of the polar angle (Fig. 1.13(b)). So one cannot determine unambiguously the real structure of the corresponding vicinal surface on geometrical reasons only. Moreover, under real conditions the surface energy and in turn the surface stiffness can change their values. This is usually the case when some impurity atoms are adsorbed on the crystal surface. The impurity atoms saturate the unsaturated dangling bonds on the crystal surface and decrease the surface energy. The larger the concentration of the impurity atoms is the smaller will be the specific energy of a particular crystal surface. Another reason for changing of the surface energy is surface reconstruction [Mönch 1979]. As a result the structure of the crystal surface should change. So our next task is to find the real structure of the crystal surfaces or, in other words, the condition of the stability of a given crystal surface. The problem of the stability of the crystal surfaces was first considered by Chernov [1961] and later discussed by Cabrera and Coleman [1963]. In this chapter we follow the presentation of the latter.

We consider an infinite vicinal crystal surface inclining at an angle θ to the nearest singular face (Fig. 1.13) and consisting of terraces and monomolecular steps. It is easy to realize such a face bearing in mind that single crystal wafers which are cut and polished under the crystallographic orientation of one of the singular faces are always inclined to the latter at some very small angle. In general such a face can be represented by $z = z(x, y)$, where $z = 0$ determines the singular face. Then the orientation at a point (x, y) will be determined by two independent components $p = -dz/dx$ and $q = -dz/dy$. We consider for simplicity the case in which the steps are parallel to the y axis (which is thus normal to the surface of the sheet), i.e. $q = 0$, and the vicinal face is described by $z = z_0 - px$, where $p = \tan\theta$. Let the face area be denoted by Σ_0. Then the area of the reference singular face is $\Sigma = \Sigma_0 \cos\theta$.

The surface energy of the face Σ_0 is $\Phi_0 = \sigma(\theta)\Sigma_0$. In terms of the component p it can be written as

$$\Phi_0 = \sigma(\theta)\Sigma_0 = \sigma(p)(1 + p^2)^{1/2}\Sigma = \xi(p)\Sigma , \qquad (1.49)$$

where $(1 + p^2)^{1/2} = 1/\cos\theta$ and $\xi(p) = \sigma(p)(1 + p^2)^{1/2}$.

Let us now assume that two new faces with slightly differing orientations

$$p_1 = p + \delta p_1 \text{ and } p_2 = p + \delta p_2 \qquad (1.50)$$

are formed. The total projected areas of these faces are Σ_1 and Σ_2, respectively. If the face Σ_0 is the stable one its surface energy Φ_0 should be smaller than the surface energy of the newly formed profile $\Phi = \xi(p_1)\Sigma_1 + \xi(p_2)\Sigma_2$ at constant volume. In other words, the condition $\Delta\Phi = \Phi - \Phi_0 > 0$ should be fulfilled. The condition of constant volume is reduced in our case to that of constant area of the cross section shown in Fig. 1.13(a). It can be easily obtained from $\Delta S = S_1 - S_2 = 0$, where S_1 and S_2 are the areas under the profiles, with slopes p and p_1 and p_2, respectively. One obtains

$$p_1\Sigma_1 + p_2\Sigma_2 = p\Sigma \ . \tag{1.51}$$

Substituting (1.50) into (1.51) and making use of the relation

$$\Sigma_1 + \Sigma_2 = \Sigma \ , \tag{1.52}$$

the condition for constant volume (1.51) turns into

$$\Sigma_1\delta p_1 + \Sigma_2\delta p_2 = 0 \ . \tag{1.53}$$

The functions $\xi(p_i)$ $(i = 1, 2)$ can be expanded as a Taylor series up to parabolic terms:

$$\xi(p_i) = \xi(p) + \xi'(p)\delta p_i + \frac{1}{2}\xi''(p)(\delta p_i)^2 \ , \tag{1.54}$$

where $\xi'(p) = d\xi/dp$ and $\xi''(p) = d^2\xi/dp^2$.

Substituting (1.54) into the expression for $\Delta\Phi$ and making use of (1.52) and (1.53) give

$$\Delta\Phi = \frac{1}{2}\xi''(p)\left[\Sigma_1(\delta p_1)^2 + \Sigma_2(\delta p_2)^2\right] \ . \tag{1.55}$$

The term in the square brackets is always positive and the condition $\Delta\Phi > 0$ is satisfied when

$$\xi''(p) > 0 \ . \tag{1.56}$$

Bearing in mind that $d\sigma/dp = (d\sigma/d\theta)(d\theta/dp)$, $\theta = \tan^{-1}(p)$, $d\theta/dp = -1/(1+p^2)$ and $\xi''(p) = \sigma^*/(1+p^2)^{3/2}$, the condition (1.56) reduces to

$$\sigma^* = \sigma + \frac{d^2\sigma}{d\theta^2} > 0 \ . \tag{1.56'}$$

It follows that when the surface stiffness of the original vicinal face which consists of terraces divided by monomolecular steps is positive it will be stable. Otherwise, it will break down into terraces divided by macrosteps or, in the limiting case, into separate crystal faces preserving the overall slope of the original face with respect to the singular face.

A schematic plot of $\xi(p)$ is given in Fig. 1.14. We have in principle three possibilities. In the first one (Fig. 1.14(a)) the second derivative of $\xi(p)$ is everywhere positive between the singular minima at $p = 0$ and $p = p_0$. This means that all the possible surfaces between $p = 0$ and $p = p_0$ will be stable and will not break into facets as long as there are no species adsorbed on them. In the second case (Fig. 1.14(b)) the second derivative of $\xi(p)$ is everywhere negative. This means that only the singular surfaces with

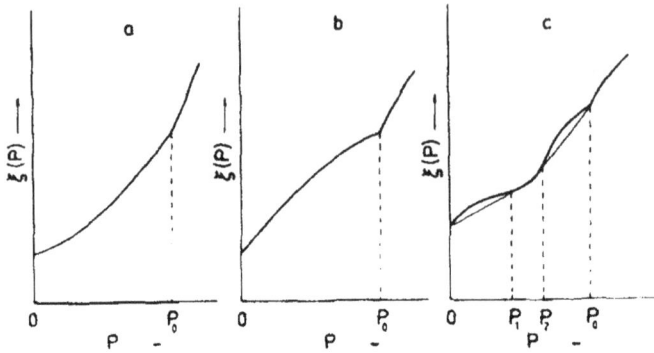

Fig. 1.14. Schematic plot of the parameter $\xi(p) = \sigma(p)(1 + p^2)^{1/2}$ vs the slope p : (a)$\xi''(p) > 0$, (b) $\xi''(p) < 0$, (c) $\xi''(p) > 0$ between p_1 and p_2 and $\xi''(p) < 0$ between 0 and p_1 and between p_2 and p_0 (after Cabrera and Coleman [1963]).

orientations p and p_0 will be stable. If a surface with orientation p such that $0 < p < p_0$ is formed it will break down into facets with orientations $p = 0$ and $p = p_0$. In the general case (Fig. 1.14(c)) in between 0 and p_0 there are regions with $\xi''(p) > 0$ and $\xi''(p) < 0$. It follows that only the surfaces at the singular minima whose orientations are between p_1 and p_2, where the condition $\xi''(p) > 0$ is fulfilled, can exist on the crystal surface. All others are not stable and should break down into facets.

The physical meaning of this result is simple. If the new crystal surfaces with orientations p_1 and p_2 have smaller specific surface energies than the original face with orientation p and a decrease of the surface energy overcompensates the increase of the surface area the crystal surface will break down into facets. In the reverse case it will be stable.

The stability of the facets can be examined in the same way as for the original surface. In this case we allow two new surfaces with areas Σ_1' and Σ_2' and orientations $p_1' = p_1 + \delta p_1$ and $p_2' = p_2 + \delta p_2$ to form (Fig. 1.13(a)). From the condition of a positive change of the surface energy $\Delta\Phi' = \xi(p_1')\Sigma_1' + \xi(p_2')\Sigma_2' - [\xi(p_1)\Sigma_1 + \xi(p_2)\Sigma_2]$ at constant volume ($\Sigma_1' + \Sigma_2' = \Sigma = \Sigma_1 + \Sigma_2$ and $p_1\Sigma_1 + p_2\Sigma_2 = p\Sigma = p_1'\Sigma_1' + p_2'\Sigma_2'$) we find that the facets will be stable when the corresponding surface stiffnesses σ_1^* and σ_2^* of the facets are positive. Thus an unstable smooth surface with negative surface stiffness should break down into facets with positive surface stiffnesses.

Faceting of crystal surfaces was observed long ago. Early works are reviewed by Moore [Moore 1963]. In principle, faceting could be caused by any reason which diminishes the specific surface energy. In particular, this is the adsorption of surface active species. Thus the adsorption of Ag on stepped Ge (111) surfaces leads to the formation of (544) and (111) facets at Ag coverage of about 0.2 monolayers at $T \cong 400°C$ [Suliga and Henzler 1983]. At an increased coverage of Ag greater than 0.7 monolayers (211) facets are formed. The above results were refined in a later paper [Henzler, Busch and Friese 1990]. A Ge crystal has been cut and polished with 3° misorientation ($p = 0.0524$) with respect to the (111) surface. Vicinal surfaces with monatomic steps and 18 atomic spacings wide terraces were obtained which can be described by the Miller indices (19,17,17). Silver was then deposited at room temperature and the crystal was annealed at 400°C for 10 min. After cooling down to room temperature again the following changes were observed. At a coverage of the Ag atoms of 1/4 monolayer (with respect to the surface coverage of (111) Ge) the initial (19,17,17) surface broke down to facets (13,11,11) with greater slopes and narrower terraces and flat wider portions with (111) orientation. The fractions of the overall surface of both facets were estimated to be 0.75 and 0.25, respectively. At a Ag coverage of 1/2 monolayer half of the surface was found to have (10,8,8) orientation and the other half has (111) orientation. After desorption of the silver and annealing of the Ge crystal the initial orientation (19,17,17) was recovered.

Clean Si (100) surfaces misoriented by 5° in the [011] direction contain steps of average spacing 29 Å and height 2.7 Å (double steps). After deposition of As beyond a critical surface coverage of 0.38 monolayers, which is independent of the temperature, the terraces become about 100 Å wide and the steps are 9 Å (6 monolayers) high [Ohno and Williams 1989a]. The process is reversible. After desorption of the arsenic the surface recovers its initial structure. The same is observed for Si (111) surfaces misoriented by 6° in the [1$\bar{1}$0], [2$\bar{1}\bar{1}$] and [$\bar{2}$11] directions. At coverages of As higher than a critical one of 0.16 monolayers the single steps turn into double steps [Ohno and Williams 1989b].

Besides, it has been found that Si (112) surfaces are stable when heated up to 800°C regardless of carbon contamination, but break down to (111), (113), (525) and (255) surfaces when heated between 950–1150°C in the presence of carbon on the crystal surface. The extent of faceting increases if carbon is introduced before heating beyond 950°C. Annealing at 1250°C removes the carbon and restores the initial surface structure [Yang and Williams 1989].

It is worth noting that the As-induced faceting of Si (100) and (111) surfaces takes place at As$_2$ pressures comparable to those employed when

GaAs is grown on Si single crystal wafers. It is thus unsafe to assume that the surface of the substrate preserves its original structure under the conditions of the experiment. The same is valid when the necessary precautions to lower the carbon contamination in the vacuum chamber are not taken into account. Works on faceting of metal surfaces are summarized by Somorjai and van Hove [1989].

1.5. Atomistic Views on Crystal Growth

1.5.1. *Equilibrium of infinitely large crystal with the ambient phase—The concept of half-crystal position*

The above considerations were purely macroscopic in the sense that thermodynamic macroscopic quantities have been used for the description of the equilibrium between different phases. The elementary processes of attachment and detachment of individual building units (atoms, ions or molecules) to and from the crystal surfaces have not been taken into account. This is one of the reasons the earlier ideas of Gibbs have not been fully comprehended until 1927 when, simultaneously, Kossel [1927] and Stranski [1927, 1928] introduced the concept of work of separation of a building unit from the so-called half-crystal position. In this section we will consider the problem on a microscopic atomic level.

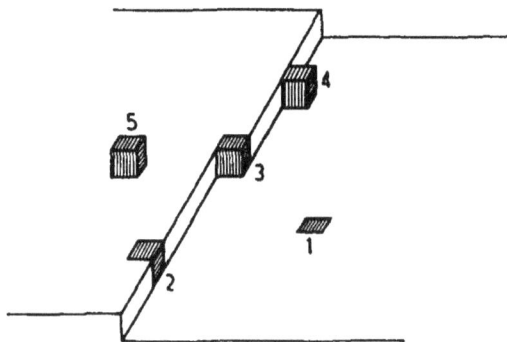

Fig. 1.15. The most important sites an atom can occupy on a crystal surface: 1 — atom embedded into the outermost crystal plane, 2 — atom embedded into the step edge, 3 — atom in a half-crystal (kink) position, 4 — atom adsorbed at the step, 5 — atom adsorbed on the crystal face.

Let us consider, for example, the cubic face (100) of a Kossel crystal containing one monatomic step (Fig. 1.15). The step can be defined as the boundary between some region of the surface and an adjacent region whose height differs by one interplanar spacing. Atoms can occupy different sites on this surface — incorporated into the face (site 1), the step (site 2), into corner position 3, or adsorbed at the step (site 4) or on the crystal surface (site 5). Depending on their positions the atoms are differently bound to the crystal surface. Thus an atom adsorbed on the crystal surface is bound by one bond to the crystal and has five unsaturated dangling bonds. On the contrary, an atom incorporated into the face has five of its bonds saturated and one unsaturated. Moreover, the detachment of these atoms leads to a change in the number of the unsaturated dangling bonds or, in other words, to the specific surface energy. The only exception is the atom in position 3 which has an equal number of saturated and unsaturated bonds. Then no change of the surface energy will take place when the latter is detached from this peculiar position. As seen an atom in this position is bound to a half-atomic row, half-crystal plane and half-crystal block. This is the reason this position is called *a half-crystal* or *kink* position. By repetitive attachment or detachment of atoms to and from this position the whole crystal (if it is large enough to exclude the size effects) can be built up or disintegrated into single atoms.

The work $\varphi_{1/2}$ necessary to detach an atom from a half-crystal position depends on the symmetry of the crystal lattice but is always equal to the work required to break half of the bonds of an atom situated in the bulk of the crystal (Table 1.1). Thus for a Kossel crystal

$$\varphi_{1/2} = 3\psi_1 + 6\psi_2 + 4\psi_3 .$$

Table 1.1. The number of the first, second and third neighbors of an atom in a half-crystal position.

Crystal lattice	Number of neighbors		
	First	Second	Third
Simple cubic	3	6	4
Face-centered cubic	6	3	12
Body-centered cubic	4	3	6
Hexagonal closed packed	6	3	1
Diamond	2	6	6

If we denote by Z_1, Z_2 and Z_3 the coordination numbers of the first, second and third coordination spheres in the corresponding crystal lattice, then

$$\varphi_{1/2} = \frac{1}{2}\left(Z_1\psi_1 + Z_2\psi_2 + Z_3\psi_3\right) . \qquad (1.57)$$

When a sufficiently large crystal is in equilibrium with the ambient phase the half-crystal position is statistically occupied and unoccupied with equal frequency. This means that the probability of attachment of atoms from the ambient phase to the kink position is equal to the probability of their detachment. It follows that the equilibrium of the infinitely large crystal with the ambient phase is determined by the half-crystal position and $\varphi_{1/2}$ may be taken as approximately equal to the enthalpy of evaporation Δh_{ev}. In other words, it is namely the work of separation from half-crystal (kink) position which determines the equilibrium vapor pressure of infinitely large crystal and in turn its chemical potential. Thus for crystals with monatomic vapors [Stern 1919; Kaischew 1936]

$$\mu_c^{\infty} = \mu_0 + kT\ln P_{\infty} = -\varphi_{1/2} + kT\ln[(2\pi m)^{3/2}(kT)^{5/2}/h^3] \qquad (1.58)$$

holds, where μ_c^{∞} is the chemical potential of the infinitely large bulk crystal, m is the atomic mass and h is Planck's constant.

As seen from the above equation, at $T = 0$, the chemical potential is equal to the separation work from the half-crystal position taken with opposite sign. It is namely this property of the half-crystal position which makes it unique in the theory of crystal growth. For the history of the discovery of the half-crystal position the reader is referred to the historical review of Kaischew [1981].

The second very important property of the half-crystal position becomes evident if we write the expression for the work of separation from it in the form

$$\varphi_{1/2} = \varphi_{lat} + \varphi_{nor} ,$$

where φ_{lat} denotes the lateral bonding with the half-crystal plane and half-atomic row and φ_{nor} denotes the normal bonding with the underlying half-crystal block.

This division has two advantages. First, it reflects the properties of the particular crystal face. Let us consider, for example, the most closely packed faces (111) and (100) of the fcc lattice. We will restrict ourselves to first neighbor interactions. In order to detach an atom from a half-crystal position on the (111) face we have to break three lateral bonds and three normal bonds, whereas on the (100) face we have to break two lateral bonds and four normal bonds. In both cases we have to break six bonds, but we could conclude that the (100) face has a greater adsorption potential than the (111) face.

Another very important consequence of this division is connected with the epitaxial growth of thin films. In fact if we replace the underlying crystal block by another block of different material the lateral bonding will remain approximately the same if we assume additivity of the bond energies. However, the normal bonding or the bonding across the interface will change [Stranski and Kuleliev 1929]. Then for a Kossel crystal the separation work from the half-crystal position will be ($\psi_1 \equiv \psi$)

$$\psi'_{1/2} = 2\psi + \psi' = 3\psi - (\psi - \psi')$$

or

$$\varphi'_{1/2} = \varphi_{1/2} - (\psi - \psi') \, , \tag{1.59}$$

where ψ' is the energy to break a bond between unlike atoms.

It is immediately obvious that when $\psi < \psi'$, $\varphi'_{1/2} > \varphi_{1/2}$ and the equilibrium vapor pressure of the first monolayer on the foreign substrate is smaller than the equilibrium vapor pressure of the bulk crystal, i.e. $P'_\infty(1) < P_\infty$. Then at least one monolayer can be deposited at any vapor pressure higher than $P'_\infty(1)$. This means that deposition will take place even when $P'_\infty(1) < P < P_\infty$, i.e. at undersaturation with respect to the bulk crystal. In the opposite case ($\psi > \psi'$), $\varphi'_{1/2} < \varphi_{1/2}$ and $P'_\infty(1) > P_\infty$. This means that the deposition requires the existence of a supersaturation in the system. The atoms of the second monolayer feel more weakly the energetic influence of the substrate and the latter will have negligible effect on the atoms of the third monolayer. It follows that in this particular case the chemical potential is not constant but depends on the number of the monolayers or, in other words, on the film thickness. In this case we speak of epitaxy which will be considered in more detail in Chap. 4. It will be shown that the thickness dependence of the chemical potential can be easily derived by assuming that the lateral bonding remains the same and accounting only for the difference in bonding across the interface. As will be discussed in Chap. 4 the latter leads to different modes of growth of the thin epitaxial films (for a review see Markov and Stoyanov [1987]).

1.5.2. Equilibrium finite crystal–ambient phase — The concept of mean separation work

As was mentioned above, we can build up or dissolve a crystal by repetitive attachment or detachment of building units only when the crystal is large enough so that the role of the edges can be ignored. If this is not the case the half-crystal position is no longer a repetitive step and it does not determine the equilibrium of the crystal with its vapor phase. To solve this problem Stranski and Kaischew [1934a,b,c,d] considered the dynamic equilibrium of a small crystal with its vapor and concluded that for a small particle to be in equilibrium with its own ambient phase the probability of building up a whole new crystal plane should be equal to the probability of its dissolution. So as a measure of the equilibrium of a finite crystal with its surrounding, they introduced the so-called "mean separation work" which is defined as the energy per atom of disintegration of a whole crystal plane into single atoms. This quantity must have one and the same value for all the crystal planes belonging to the equilibrium form.

Consider, for example, a Kossel crystal with edge length $l_3 = n_3 a$, where n_3 is the number of atoms in the edge of the 3D crystal and a is the atomic spacing. The energy per atom for disintegration of a whole lattice plane into single atoms will be (following Figs. 1.16(a)–(c))

$$\bar{\varphi}_3 = [3\psi(n_3 - 1)^2 + 4\psi(n_3 - 1) + \psi]/n_3^2 = 3\psi - 2\psi/n_3 .$$

On the other hand, $3\psi = \varphi_{1/2}$ (Table 1.1) so that

$$\bar{\varphi}_3 = \varphi_{1/2} - 2\psi/n_3 . \tag{1.60}$$

It follows that the mean work of separation goes asymptotically to the work of separation from kink position as the crystal size is increased. Then a crystal can be considered large enough if $n_3 > 70$ or $l_3 > 2 \times 10^{-6}$ cm ($a = 3 \times 10^{-8}$ cm).

As $\bar{\varphi}_3$ determines the equilibrium with the vapor phase we can write in analogy with Eq. (1.58)

$$\mu_c = \mu_v = \mu_0 + kT \ln P_1 = -\bar{\varphi}_3 + kT \ln[(2\pi m)^{3/2}(kT)^{5/2}/h^3] . \tag{1.61}$$

Then

$$\Delta\mu = \mu_v - \mu_c^\infty = kT \ln(P_1/P_\infty) = \varphi_{1/2} - \bar{\varphi}_3 = 2\psi/n_3 . \tag{1.62}$$

Obviously, this is the same Gibbs–Thomson equation (1.19). We can define the specific surface energy of the Kossel crystal confining ourselves to first neighbor interactions as the energy to create two surfaces of area a^2 each:

$$\sigma = \psi/2a^2 . \tag{1.63}$$

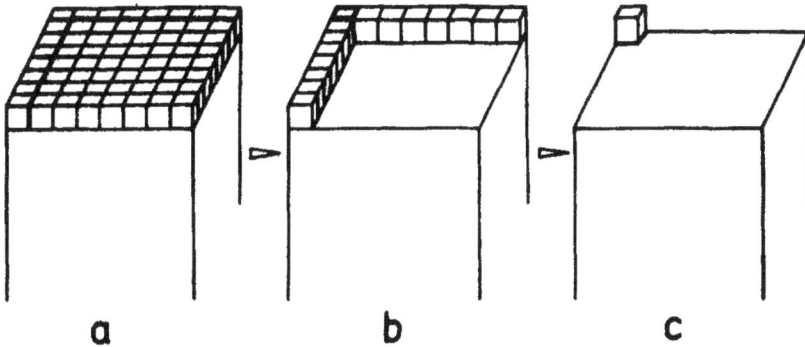

Fig. 1.16. For the evaluation of the mean separation work $\bar{\varphi}_3$ which determines the equilibrium of a finite 3D crystal with the vapor phase according to Stranski and Kaischew (1934b). First, (a) $(n-1)^2$ atoms are detached in such a way that two edge rows of atoms are left. The detachment of each atom requires the breaking of three bonds, then (b) the two remaining rows of atoms, each consisting of $n-1$ atoms, are detached with the exception of the corner atom. The detachment of the atoms requires the breaking of two bonds per atom, and finally (c) the last atom at the corner is detached which requires the breaking of only one bond (after Stranski and Kaischew [1934b]).

Substituting for ψ from Eq. (1.63) into (1.62) one obtains

$$\Delta\mu = 4\sigma v_c / l_3 , \tag{1.64}$$

which is exactly the Thomson–Gibbs equation as given by (1.19) ($l_3 = 2r$ and $v_c = a^3$ for a Kossel crystal).

1.5.3. *Equilibrium 2D crystal–ambient phase*

Stranski and Kaischew considered further the case when a 2D crystal formed on one of the faces of a 3D crystal is in equilibrium with the ambient phase. In analogy with the 3D case they suggested that the probability of building up a whole new atomic row with a length $l_2 = n_2 a$ (Fig. 1.17) should be equal to the probability of its disintegration into single atoms. The equilibrium 2D crystal vapor phase is now determined by the corresponding *mean work of separation* $\bar{\varphi}_2$, which in this particular case is equal to the energy per atom for disintegration of a whole edge row of atoms. Assuming for simplicity a square-shaped crystal with n_2 atoms in the edge the mean work of separation reads

$$\bar{\varphi}_2 = 3\psi - \psi/n_2 = \varphi_{1/2} - \psi/n_2 . \tag{1.65}$$

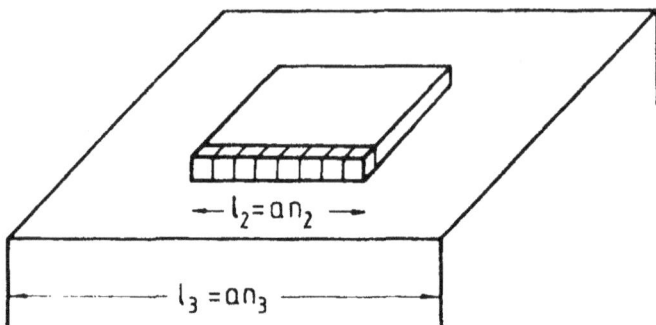

Fig. 1.17. For the evaluation of the mean separation work $\bar{\varphi}_2$ which determines the equilibrium with the ambient phase of a finite 2D crystal with edge length l_2 on the surface of a 3D crystal with edge length l_3. n_2 and n_3 denote the numbers of atoms on the edges of the 2D and 3D crystals, respectively (after Stranski and Kaischew [1934b]).

Then the supersaturation required for the formation of a 2D crystal on the surface of a 3D crystal is

$$\Delta\mu = RT\ln(P/P_\infty) = \psi/n_2 \ . \tag{1.66}$$

Using the definition of the specific edge energy

$$\varkappa = \psi/2a \ , \tag{1.67}$$

one obtains the familiar equation of Thomson–Gibbs for the two-dimensional case:

$$\Delta\mu = 2\varkappa a^2/l_2 \ . \tag{1.66'}$$

Comparing Eqs. (1.64) and (1.66) ($\varkappa = \sigma a$) leads to the conclusion that in equilibrium the edge length of the 2D crystal should be shorter than that of the 3D one by a factor of 2 at one and the same supersaturation, i.e. $l_2 = l_3/2$.

1.5.4. *Equilibrium shape of crystals—Atomistic approach*

The introduction of mean works of separation enabled Stranski and Kaischew [1935] to give a new atomistic approach to the determination of the equilibrium shape of the crystals. The basic idea is that atoms whose energy of binding with the crystal is smaller than the mean work of separation cannot belong to the equilibrium shape because the corresponding vapor pressure will be higher than the equilibrium vapor pressure in the system. Then in order to derive the equilibrium shape one starts from a crystal with an arbitrary simple form in which all atoms whose separation works are smaller than $\varphi_{1/2}$ are successively removed from the crystal surface. Precisely at that moment all the crystal faces belonging to the equilibrium shape appear. Then the areas of the faces are varied (whole crystal planes are removed or added) up to the moment when the mean separation works $\bar{\varphi}_3$ of all the crystal planes have one and the same value. During the last operation all faces which do not belong to the equilibrium form disappear (see Honnigmann 1958).

Taking into account more distant neighbor atoms in the calculation of the mean separation works, facets with higher specific surface energy appear on the equilibrium shape just like as shown in Fig. 1.10. Thus when accounting only for the first neighbors in a Kossel crystal the equilibrium shape consists only of the cubic faces (100). Taking into account the second neighbors leads to the appearance of the (110) and (111) faces in addition to the (100) faces. Then by comparing the theoretical predictions with experimental observations one can make conclusions concerning the influence of the radius of action of the interatomic forces on the equilibrium shape.

The atomistic approach of Stranski and Kaischew can be illustrated by finding the equilibrium shape of a 3D crystal lying on a foreign substrate. We consider for simplicity a cubic crystal with a square base with lateral edge $l = na$ and height $h = n'a$, where n and n' are the numbers of atoms on the horizontal and vertical edges, respectively (Fig. 1.18). Following the above procedure the mean separation work calculated from the side crystal face is

$$\bar{\varphi}'_3 = 3\psi - \frac{\psi - \psi'}{n'} - \frac{\psi}{n} \ .$$

The mean separation work calculated from the upper base is given by Eq. (1.60). The condition for the equilibrium shape is that the chemical potentials of the different faces or, in other words, their mean separation works, have to be equal ($\bar{\varphi}'_3 = \bar{\varphi}_3$). The latter leads to the relation [Kaischew 1950]

$$\frac{h}{l} = \frac{n'}{n} = 1 - \frac{\psi'}{\psi} \ . \tag{1.68}$$

Substituting ψ and ψ' with the specific surface energy and the specific adhesion energy, respectively, gives

$$\frac{h}{l} = 1 - \frac{\beta}{2\sigma} = \frac{\sigma + \sigma_i - \sigma_s}{2\sigma} \ , \tag{1.69}$$

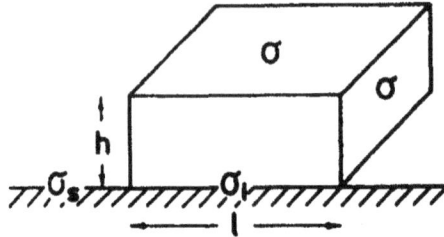

Fig. 1.18. A cubic crystal of lateral extent l and height h on a foreign substrate. The specific surface energies of the substrate, σ_s , the upper and lateral faces of the deposit crystal, σ, and the substrate–deposit interface, σ_i, determine the equilibrium shape ratio h/l.

where σ_i is the specific interfacial energy expressed by the relation of Dupré (1.28). The same result can be obtained if one starts from the classical thermodynamic conditions $\Phi = l^2(\sigma + \sigma_i - \sigma_s) + 4lh\sigma = \min$ and $V_c = l^2 h = $ const [Bauer 1958].

1.5.5. *Equilibrium vapor pressure of a 2D crystal on a foreign substrate*

It is also of interest to treat the question of equilibrium vapor pressure of a two-dimensional crystal formed on the surface of a crystal of different material. Assuming for simplicity a square shape the mean separation work estimated from the crystal edge reads

$$\bar{\psi}'_2 = 2\psi + \psi' - \frac{\psi}{n_2} = \varphi'_{1/2} - \frac{\psi}{n_2} \; ,$$

where the term $\varphi'_{1/2} = 2\psi + \psi'$ is in fact the work of separation of an atom from the half-crystal position of the semi-infinite adlayer on the foreign substrate [Stranski and Kuleliev 1929].

Bearing in mind that $\varphi_{1/2} = 3\psi$ for the supersaturation and hence for the equilibrium vapor pressure one finds

$$\Delta\mu = kT \ln \frac{P}{P_\infty} = \psi - \psi' + \frac{\psi}{n_2} \; .$$

The difference $\varphi_{1/2} - \varphi'_{1/2} = \psi - \psi'$ of the binding energies can be either positive ($\psi > \psi'$) or negative ($\psi < \psi'$) as discussed above. This means that the equilibrium vapor pressure of the 2D crystal can be either higher or lower than the equilibrium vapor pressure of the bulk crystal and the deposition can be carried out at supersaturation or undersaturation, respectively.

1.6. Equilibrium Structure of Crystal Surfaces

1.6.1. Classification of crystal surfaces

The process of growth of crystals takes place at the crystal–ambient phase interface where the latter can be vapor, melt or solution. Obviously, the equilibrium structure of this interface or, in other words, its roughness, determines the crystal shape on one hand, and the mechanism of growth and in turn its rate of growth on the other.

Let us consider, for example, an atomically smooth crystal face belonging to a perfect defectless crystal. The formation of a new lattice plane requires the existence of monatomic steps which offer half-crystal positions. As a source such steps can serve randomly appearing two-dimensional formations of the new lattice layer with closed contours. Initially they are unstable and have the tendency to dissolve into the mother phase. When such formations which serve as "two-dimensional nuclei" of the new layer exceed some critical size their further growth is thermodynamically favored and they cover completely the crystal face. After that the steps vanish and the initial state is restored. Then the formation of new lattice plane requires the formation of new 2D nuclei and the process is repeated. Hence the growth of a defectless atomically smooth crystal face is a periodic process involving successive 2D nucleation and lateral growth (Fig. 3.23) which is usually observed in crystal growth by Molecular Beam Epitaxy (MBE) [Harris, Joyce and Dobson 1981a, 1981b; Neave, Joyce, Dobson and Norton 1983]. The formation of 2D nuclei is connected, however, with definite energetic difficulties and requires overcoming a critical supersaturation. Then the rate of growth of a defectless crystal surface will be a nonlinear (in fact exponential) function of the supersaturation.

However, experimental data showed that crystals can grow at supersaturations as low as 0.01%, in marked discrepancy with the nucleation theory of crystal growth. The problem was solved in 1949 when at a discussion meeting of the Faraday Society in Bristol, Frank [1949a,b] proposed the *spiral mechanism* of crystal growth. He suggested that continuous growth of crystals at low supersaturation can be attributed to the presence of crystal defects, particularly screw dislocations (see Fig. 3.9). The latter offer nonvanishing monatomic steps with kink positions along them, thus making the 2D nucleation unnecessary. In 1951 Burton, Cabrera and Frank published their famous paper "The growth of crystals and the equilibrium structure of their surfaces" [Burton *et al.* 1951]. Considering the growth of a crystal face in the presence of screw dislocations they found that the distance between consecutive coils of the spiral is directly proportional to the linear size of the critical 2D nucleus which is determined by the existing supersaturation. Then the slopes of the growth pyramids which are formed at the emergency points of the screw dislocations are directly proportional to the supersaturation. One can conclude that, in general, the growth rate will again be a nonlinear function of the supersaturation.

Finally if the crystal face is atomically rough it offers a great number of kink positions. Building particles arriving from the mother phase can be incorporated to the crystal lattice practically at any place, which makes the 2D nucleation as well as the presence of screw dislocations unnecessary. No thermodynamic hindrance exists any more; the process is fast and the growth rate is simply proportional to the flux of atoms from the ambient phase and thus should be a linear function of the supersaturation.

Thus crystal growth means incorporation of building units which arrive from the ambient phase to the half-crystal positions. Then the rate of growth of a given crystal face in a direction normal to its surface is proportional to the density of growth sites (half-crystals or kink positions) which the face offers to the building units from the ambient phase. This density depends on the crystallographic orientation of the face on the one hand, and temperature on the other.

Burton and Cabrera (1949) classified the crystal surfaces with respect to their capability of growth into *close-packed* and *non-close-packed* or *stepped* surfaces. The question has been further elucidated by Hartman [1973] and others [Honigmann 1958; Cabrera and Coleman 1963].

The crystal surfaces are thus divided into three groups: F (*flat*), S (*stepped*) and K (*kinked*) surfaces depending on whether they are parallel to at least two most dense rows of atoms, one most dense row of atoms, or are not parallel to any of the most dense rows of atoms at all, respectively [Hartman 1973]. F faces are, for example, the (100) face of Kossel crystals and fcc crystals which are parallel to two most dense rows of atoms, the (111) face of fcc and the (0001) face of hcp crystals which are parallel to three most dense rows of atoms, etc. (see Fig. 1.19). Typical examples of S and K faces are the (110) and (111) faces of sodium chloride (or Kossel) crystals. It is clear that when a crystal face is parallel to more than one most dense row of atoms, the number of saturated shortest, and hence strongest, chemical bonds parallel to the crystal surface is greatest. The number of the unsaturated bonds is minimal and so is the specific free energy of the face (see the arrows in Fig. 1.19). When the crystal face is parallel to one most dense row of atoms only, it intersects the other one and all the chemical bonds parallel to the latter become unsaturated. Thus the number of the unsaturated bonds reaches its highest value when the crystal face intersects all the densest rows of atoms and hence such a face offers more growth sites (kink positions) than the S and F faces.

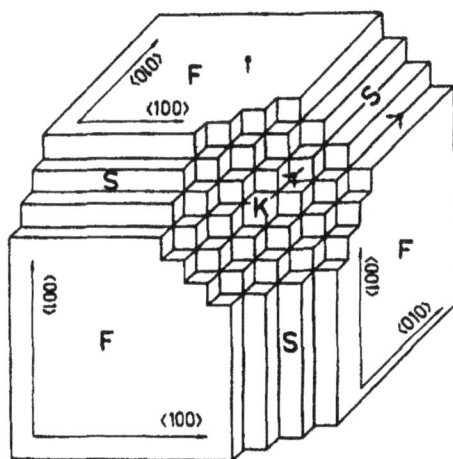

Fig. 1.19. A schematic representation of a Kossel crystal illustrating F (flat) and S (stepped) surfaces depending on whether they are parallel to two and one most dense rows of atoms, respectively, and a K (kinked) surface which is not parallel to a most dense row of atoms at all. The long arrows give the directions of the most dense rows of atoms (the directions of the first neighbor bonds). The short arrows represent the unsaturated first neighbor bonds of an atom belonging to the corresponding crystal face (after Hartman [1973]).

As the K faces offer kink sites with much greater density than the S and F faces they will grow faster than the latter. The S faces also offer kink sites along the steps but their density is smaller than on the K faces. Finally, the F faces of perfect crystals do not offer kink sites at all. Then at small enough supersaturations to prevent 2D nucleation the rate of growth in a direction normal to the particular faces will be highest for the K faces, smaller for the S faces and zero for the F faces. It follows that the K faces should disappear first, followed by the S faces and finally the crystal will be enclosed during growth by the F faces only and will cease to grow at all at small enough supersaturation.

As has been shown in the previous section the crystal faces can be divided also into singular and vicinal faces. Singular minima correspond to the singular faces and the latter can be any of the low index faces irrespective of whether they are F, S or K faces. Finally, the vicinal faces which are slightly tilted with respect to one of the main (singular) crystal planes offer to the arriving building units a train of parallel steps divided by smooth terraces. This classification is of practical interest. When crystals are cut for the preparation of substrates for epitaxial growth with particular crystallographic orientations the angle of cutting with respect to the orientation selected is never equal to zero. Thus crystal wafers prepared by cutting always offer vicinal surfaces for crystal growth. Besides, during spiral growth (Fig. 3.9) the side faces of the growth pyramids represent in fact vicinal surfaces. The same is true for the case when the crystal face grows through formation and growth of 2D nuclei. Pyramids of growth are formed by successive formation of 2D nuclei one on top of the other and their side faces represent again vicinal surfaces (see Fig. 3.2). This is the reason why we consider first the equilibrium structure of single height steps.

1.6.2. *Equilibrium structure of a step*

We consider for simplicity a single step of monatomic height and infinite length on the surface of a simple cubic crystal. At $T = 0$ the step will be perfectly straight. As the temperature is increased kinks separated by smooth parts will begin to appear. The kinks can be conventionally divided into *positive* and *negative* kinks depending on whether a new row of atoms begins or ends (Fig. 1.20). If the step follows on average the direction of a most dense row of atoms the number of positive and negative kinks will be equal. The total number of kinks increases when the step deviates from this direction. The kinks can be monatomic as well as polyatomic. For simplicity we consider first monatomic kinks only. We rule out also the so-called "overhangs" (Fig. 1.20).

The energy required to form a kink is $\omega = \psi/2$. Indeed, as seen in Fig. 1.21, in order to produce a hole and an adsorbed atom in the initially straight step we break 3 lateral bonds and create one bond, thus expending a net amount of energy of 2ψ. The transfer of the next atom (to separate the kinks by a smooth part) is not connected with the change of energy. Thus we expend 2ψ energy to form four kinks (two positive and two negative) or $\omega = \psi/2$ per kink. This holds for crystal-vapor interface. If the ambient phase is a melt we can calculate the energy to form a kink on the base of

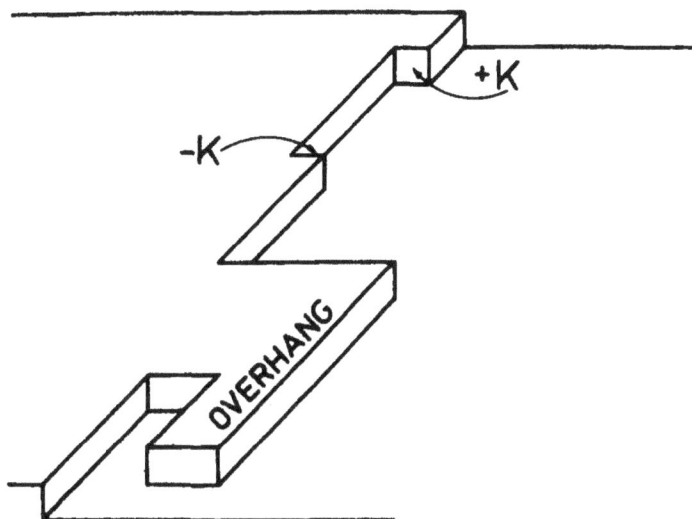

Fig. 1.20. Positive K^+ and negative K^- kinks along a step at $T \neq 0$. The overhang is also shown.

the so-called lattice model of the melt. It is assumed that the atoms in the liquid form the same lattice as in the crystal but the energies to disrupt the bonds between two neighboring atoms in the crystal, ψ_c, in the melt, ψ_m, and belonging to both the crystal and the melt, ψ_{cm}, are different. We perform the same considerations as before but now we add the energy required to transfer two liquid atoms from the melt into the created holes in the step. Then the energy ω to form a kink per crystal–melt bond ψ_{cm} is

$$\omega = \frac{1}{2}(\psi_c + \psi_m) - \psi_{cm} \ . \tag{1.70}$$

In the case of crystal–vapor interface $\psi_m = \psi_{cm} = 0$ and $\omega = \psi_c/2 \equiv \psi/2$.

It is worth noting that the work to create a kink is in fact the work to create one more dangling bond or, in other words, to elongate the step by one interatomic spacing. Thus, the creation of kinks leads to a change of the specific edge energy of the step or of the surface energy of the crystal face.

We denote by n_+ and n_- the densities of single positive and negative kinks, respectively, and by n_0 the density of the smooth parts where there are no jumps at all. Their sum

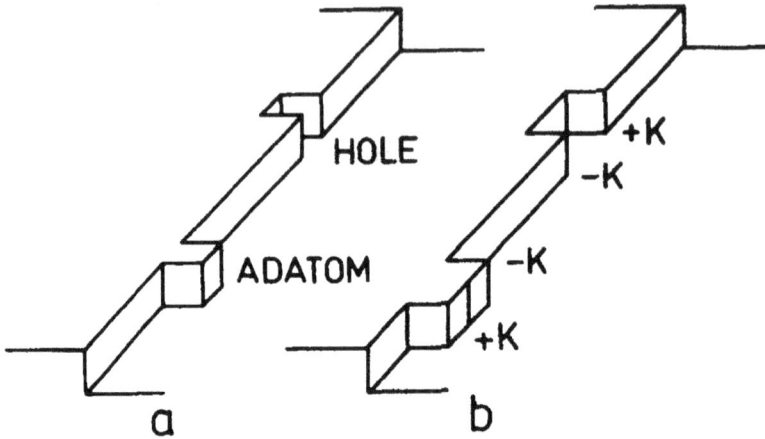

Fig. 1.21. For the determination of the activation energy of formation of a kink: (a) First we transfer an atom embedded into the step (site 2 in Fig. 1.15) to an adsorption position at the step (site 4 in Fig. 1.15). A hole and an adatom are formed. (b) The hole is widened by removing a second atom and transferring it next to the first one. As a result four kinks, two positive and two negative, are formed (after Burton and Cabrera [1949]).

$$n_+ + n_- + n_0 = n = 1/a \tag{1.71}$$

is just equal to the number of atoms per unit length of the step where a is the first neighbor distance.

If the energy to form a kink is $\omega = \psi/2$ we can write [Burton, Cabrera and Frank 1951]

$$n_+/n_0 = n_-/n_0 = \eta$$

or

$$n_+ n_-/n_0^2 = \eta^2, \qquad \eta = \exp(-\omega/kT) = \exp(-\psi/2kT) . \tag{1.72}$$

If the average orientation of the step deviates from the direction of the most dense row of atoms by a small angle φ then

$$n_+ - n_- = \varphi/a . \tag{1.73}$$

By solving the system (1.71)–(1.73) at $\varphi = 0$ for the mean distance between the kinks, δ_0, one obtains

$$\delta_0 = \frac{1}{n_+ + n_-} = a\left(1 + \frac{1}{2\eta}\right) \approx a\left(1 + \frac{1}{2}\exp(\omega/kT)\right) . \tag{1.74}$$

Let us evaluate this quantity. The enthalpy of evaporation of, say, silver is $\Delta H_e = 60720$ cal/mole; $\psi = \Delta H_e/6$ or $\omega = \Delta H_e/12$. At $T = 1000$K, $\omega/kT = 2.53$ and $\delta_0 = 7a$. In other words at the crystal–vapor interface of silver we have kinks of any sign, on average, at every 7 atomic distances, i.e. the step will be rough. In the case of Si crystal–melt interface at the temperature of melting, $T_m = 1685$ K, $\Delta H_m/kT_m = 3.3$, $\omega/kT_m = 1.66$ and $\delta_0 = 3.5a$. An interesting example is the vicinal surface of Si(001). As will be shown in Sec. 3.5 steps called S_A and S_B with $w = 0.15$ eV and 0.01 eV, respectively, alternate. Using Eq. (1.74) we find that very rough and very smooth steps coexist on the Si(001) vicinal surface.

All this means that whereas at $T = 0$ the step is perfectly straight without kinks of any sign at any temperature higher than zero, it will contain kinks or, in other words, it will be rough. This is due to the decrease of the Gibbs free energy of the step with increasing temperature due to the increase of entropy. For the Gibbs free energy of the step we can write the usual expression (we neglect the PV term)

$$G_{st} = U_{st} - TS_{st} , \tag{1.75}$$

where

$$U_{st} = (n + n_+ + n_-)\psi/2 \tag{1.76}$$

is the potential energy of the unsaturated bonds at the step (n is the number of unsaturated bonds in a direction normal to that of the step and n_+ and n_- are the numbers of unsaturated bonds parallel to the step, every unsaturated bond having energy $\omega = \psi/2$).

The entropy is determined by the number of possible ways of distribution of the kinks and smooth parts so that (if we neglect the kink–kink interaction)

$$S_{st} = k \ln \left(\frac{n!}{n_+!n_-!n_0!} \right) . \tag{1.77}$$

Solving again the system (1.71)–(1.73) for the simpler case of a step parallel to the direction of the most dense atomic row ($\varphi = 0$) we obtain

$$\frac{n_+}{n} = \frac{n_-}{n} = \frac{\eta}{1 + 2\eta}, \qquad \frac{n_0}{n} = \frac{1}{1 + 2\eta} . \tag{1.78}$$

Substituting (1.78) into (1.75), (1.76) and (1.77) and using the Stirling formula $\ln N! = N \ln N - N$ gives

$$G_{st} = -nkT \ln[\eta(1 + 2\eta)] . \tag{1.79}$$

This expression was obtained under the assumption that the kinks are monatomic. If we relax this restriction for G_{st} one obtains (see Eq. (3.157')) [Burton, Cabrera and Frank 1951]

$$G_{st} = -nkT \ln \left(\eta \frac{1 + \eta}{1 - \eta} \right) . \qquad (1.80)$$

It is easy to see that $G_{st} > 0$ when η is greater than some critical value $\eta_r = \sqrt{2} - 1$. Beyond this value $G_{st} < 0$. This means that a critical temperature determined by the condition $G_{st} = 0$ exists:

$$\frac{kT_r}{\psi} = 0.57 \quad \text{or} \quad \frac{\Delta H_e}{kT_r} = 0.88Z \qquad (1.81)$$

below which the steps are rough but still exist. Bearing in mind that a step divides a half-crystal plane (with vacancies in it) from a dilute adlayer it is clear that at temperatures higher than T_r both regions are mutually dissolved, the steps no longer exist, and the surface of the face becomes atomically rough. This phenomenon is similar to the mutual dissolution of a liquid and a vapor phase at the critical point at which the phase boundary between them disappears.

1.6.3. *Equilibrium structure of F faces*

We have seen in the previous section that the disappearance of a monatomic step leads to the roughening of the stepped surface. So the next logical question is: Can F faces roughen if steps are not initially present? The answer is *yes* but the problem is much more complicated.

The roughness of a crystal face can be defined as

$$R = \frac{U - U_0}{U_0} , \qquad (1.82)$$

where U_0 is the internal (potential) energy of the reference flat face at $T = 0$, which is proportional to the number of the unsaturated bonds normal to the face ($U_0 = \psi/2$ per atom), and U is the internal energy of the face at $T > 0$. The latter is proportional to the number of both normal and lateral bonds. Thus the roughness is given by the ratio of the unsaturated lateral and normal bonds. If we exclude the overhangs shown in Fig. 1.24 from our considerations (this is the well-known solid-on-solid or SOS model) the number of the normal bonds does not change upon roughening and $U_0 =$ const.

At $T = 0$ the face is atomically smooth and all the surface atoms are on one and the same level. Unsaturated lateral bonds do not exist and $R = 0$. At $T > 0$ some atoms can leave the uppermost atomic plane, leaving vacancies in it, and adsorb on it, thus giving rise to unsaturated lateral bonds and hence to some degree of roughness. When there are atoms on different levels jumps analogous to the kinks in the step appear. In the case of steps, however, the problem is much simpler because the total number of kinks and smooth parts is equal to the number of bonds per unit length of the step and the appearance of a kink or a smooth part at any particular point does not depend on the situation in the neighbor points. In the 2D case of a crystal surface the situation is completely different. If we consider, for example, the (100) face of a Kossel crystal we will see that the number of the lateral bonds is twice the number of atoms per unit area. Hence the number of jumps between atoms on different levels is greater than the number of atoms. This means that the existence of a jump at a given site depends on the existence of jumps in neighboring sites although the existence of atoms on different levels is independent. As we are interested in the number and distribution of jumps or unsaturated lateral bonds we have to deal with this *cooperative* phenomenon which means that the appearance of jumps between atoms on different levels is interdependent. This is quite a difficult problem and an exact solution exists only for the simplest case of a surface with a square atomic mesh and atoms on two levels. More general solutions can be found if some approximations are used.

Burton, Cabrera and Frank [1951] first realized that the two-level problem is analogous to the 2D Ising model in the theory of ferromagnetism. The latter deals with a square mesh of spins which can be directed either up or down (Fig. 1.22). The energy of interaction of two neighboring spins can be taken to be either $+1$ or -1 depending on whether they are parallel or antiparallel, respectively. Such a system shows critical behavior in the sense that beyond some critical temperature (the well-known temperature of Curie) all the spins are randomly oriented (Fig. 1.22(a)). At lower temperatures all the spins are equally directed thus giving rise to ferromagnetic state (Fig. 1.22(b)).

Consider now a crystal face with atoms on two levels. If an atom has a neighbor on the same level the energy of interaction between them will be $-\psi$. Otherwise the bond will be unsaturated and the energy of interaction will be zero. This is equivalent to parallel and antiparallel spins, respectively. Hence, if all the atoms are situated on one and the same level and the face is atomically smooth, all the lateral bonds are

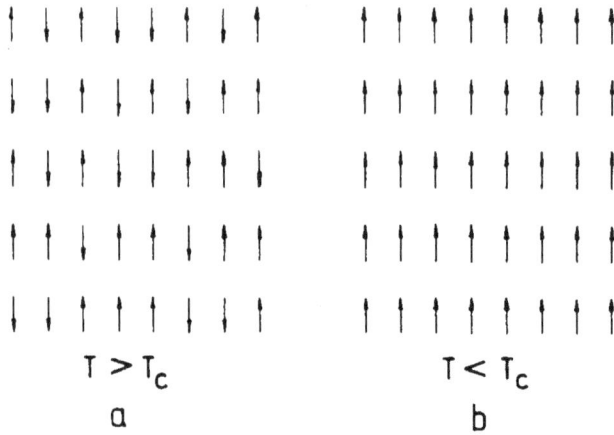

Fig. 1.22. A square mesh of spins as an illustration of the two-dimensional Ising model at temperatures (a) higher and (b) lower than the critical temperature T_c.

saturated and this state is equivalent to the ferromagnetic one. Obviously, a critical temperature analogous to that of Curie will exist above which approximately half of the atoms will be on the upper level and the other half on the lower level. Using the exact solution of Onsager [1944] for this "simple" case, Burton, Cabrera and Frank deduced that the critical temperature is given by

$$\exp(-\psi/2kT_r) = \sqrt{2} - 1 \quad \text{or} \quad kT_r/\psi = 0.57 \qquad (1.83)$$

for the (001) face of a Kossel crystal.

Obviously, the roughening of a crystal face taking place on more than two levels and without square symmetry cannot be treated in this way. Some approximation should be used. So we consider first the two-level model of Jackson [1958] and then treat in the same way the multilevel model of Temkin [1964, 1968] (see also Bennema and Gilmer [1973]).

1.6.3.1. *Model of Jackson*

We consider a flat, atomically smooth face with N adsorption sites per unit area at the equilibrium temperature T_e. Let N_A atoms be adsorbed on this face so that the surface coverage is $\theta = N_A/N$. Every atom has Z_1 lateral bonds. For instance, $Z_1 = 4$ for the (100) face of Kossel and fcc crystals, $Z_1 = 6$ for the (111) face of fcc crystals, etc.

The relative Gibbs free energy of the crystal face is (the PV term is again neglected)

$$\Delta G_f = \Delta U_f - T\Delta S_f , \qquad (1.84)$$

where ΔU_f is again the internal energy due to the unsaturated lateral bonds and ΔS_f is the configurational entropy of distribution of N_A atoms over N adsorption sites ($S_f(T = 0) = 0$). In order to calculate ΔU_f we use the so-called approximation of Bragg–Williams [1934], also known as "mean-field" approximation. In this particular case the latter consists of the following. An adatom on the surface can have $1, 2, 3, \ldots, Z_1$ first neighbors. If we assume that the atoms are randomly distributed and clustering is ruled out, we can accept that approximately every adatom will have on average $Z_1\theta$ first neighbors and respectively $Z_1(1 - \theta)$ unsaturated bonds. Then

$$\Delta U_f = N_A Z_1(1 - \theta)\frac{\psi}{2} = N Z_1\theta(1 - \theta)\frac{\psi}{2} . \qquad (1.85)$$

The entropy can be calculated in the usual way:

$$\Delta S_f = k \ln \left(\frac{N!}{N_A!(N - N_A)!}\right) = -kN\theta \ln \theta - kN(1 - \theta)\ln(1 - \theta) , \qquad (1.86)$$

where we again made use of Stirling's formula.

Then for the Gibbs free energy one obtains

$$\Delta G_f/NkT_e = \alpha\theta(1 - \theta) + \theta \ln \theta + (1 - \theta)\ln(1 - \theta) . \qquad (1.87)$$

A graphical representation of $\Delta G_f/NkT_e$ is given in Fig. 1.23 for different values of the parameter α:

$$\alpha = \frac{Z_1\psi}{2kT_e} = \frac{Z\psi}{2kT_e}\frac{Z_1}{Z} = \frac{\Delta H_e}{kT_e Z}\frac{Z_1}{Z} = \frac{\Delta S_e}{k}\frac{Z_1}{Z} , \qquad (1.88)$$

where Z is the coordination number of an atom in the bulk crystal.

As seen all curves are symmetric and have maximum or minimum at $\theta = 1/2$. The second derivative

$$\frac{d^2}{d\theta^2}\frac{\Delta G}{NkT_e} = -2\alpha + \frac{1}{\theta} + \frac{1}{1 - \theta}$$

is negative at $\theta = 1/2$ when $\alpha > 2$ and is positive when $\alpha < 2$. This means that at $\alpha < 2$ the surface free energy has a minimum at $\theta = 1/2$. On the contrary at $\alpha > 2$ the Gibbs free energy has a maximum at $\theta = 1/2$ and two equally deep minima at values of θ very close to 0 and 1. The two minima correspond to two equivalent configurations — the first one ($\theta \cong 0$)

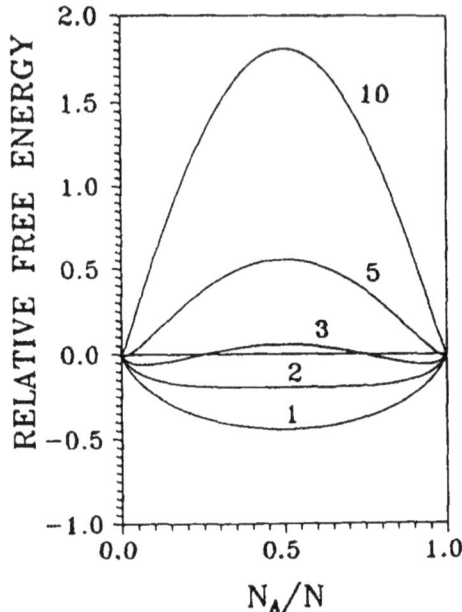

Fig. 1.23. Dependence of the relative Gibbs free surface energy $\Delta G_f/NkT_e$ on the surface coverage $\theta = N_A/N$ at different values of the parameter α denoted by the figure at each curve (after Jackson [1958]).

consisting of small density of adatoms on a flat crystal face and the second one ($\theta \cong 1$) representing a flat crystal face with a few vacancies in it. If we accept that the maximum surface roughness is defined by $\theta = 1/2$ the crystal face will be rough at $\alpha < 2$ and smooth at $\alpha > 2$.

We can now answer the question of what the structure of the crystal–ambient phase interface will be at the temperature of transition melting or sublimation. For many metals the relative entropy of melting $\Delta S_m/R$ has a typical value around 1.2 and their surfaces will be rough at T_m or near to it. The different crystal faces will have nearly one and the same density of kinks and the crystals will grow rounded from their melts. On the other hand, the relative entropy of evaporation $\Delta S_e/R$ has a typical value over 10 and $\alpha > 2$. The F faces will be smooth and the crystals will grow well polygonized from the vapor phase. The above refers to metals. In the case of some organic crystals the entropy of melting is large and they grow polygonized from their melts.

The model of Jackson was further generalized by Chen, Ming and Rosenberger [1986] to account for the nonlinear behavior of energy to break a bond between first neighbors due to many body interactions with more distant neighbors. The analysis leads to lower values of the α factor for a series of metals (Cu, Pb, Zn), thus increasing the tendency to roughening at temperatures lower than predicted by the Jackson model.

1.6.3.2. *Model of Temkin*

The main shortcomings of the Jackson model aside from the use of the mean field approximation is that the roughness of the face is restricted to two levels only and the result is applicable only at the equilibrium temperature (or very close to it). This is the reason why Temkin [1964, 1968] developed further this approach allowing the crystal face to roughen at an arbitrary depth and arbitrary temperature, i.e. during the processes of growth and dissolution ($\Delta\mu \neq 0$).

The Bragg-Williams approximation again permits one to solve the more general case of multilevel roughening [Temkin 1968; Bennema and Gilmer 1973]. At $T = 0$ the crystal face is completely smooth as shown in Fig. 1.24(a). At some higher temperature the face is rough and the roughness is not confined to two levels as in Jackson's model but can go from $-\infty$ to $+\infty$. In fact we consider the SOS model ruling out the overhangs (Fig. 1.24(b)). Every crystal layer (with number n) consists of N_{ns} solid atoms belonging to the crystal and N_{nf} atoms belonging to the fluid phase. Then

$$N_{ns} + N_{nf} = N = \text{const.} \tag{1.89}$$

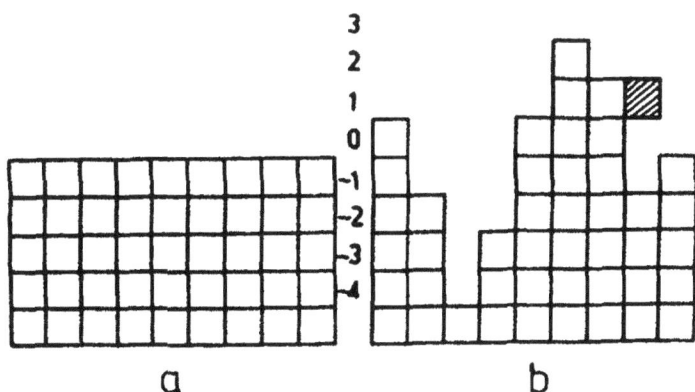

Fig. 1.24. Schematic representation of the SOS (Solid-On-Solid) model of an F face: (a) smooth face at $T = 0$, (b) rough face at $T > 0$. Overhangs as shown by the shadowed block are forbidden. The figures denote the number of the corresponding crystal layer.

We define the corresponding surface coverages as before:

$$\theta_n = \frac{N_{ns}}{N} \quad \text{and} \quad 1 - \theta_n = \frac{N_{nf}}{N} . \tag{1.90}$$

Then the fraction of fluid atoms is $1 - \theta_n$. As n varies from $-\infty$ to $+\infty$ the problem of finding all the θ_n is subject to the boundary conditions

$$\theta_{-\infty} = 1 \quad \text{and} \quad \theta_{+\infty} = 0 . \tag{1.91}$$

As we exclude the overhangs from consideration and hence $\theta_n \leq \theta_{n-1}$, (1.91) means that we go from completely solid phase to completely fluid phase. In order to find a solution for θ_n we will follow the approach of Mutaftschiev [Mutaftschiev 1965; Bennema and Gilmer 1973].

The Gibbs free energy of a rough surface with respect to the smooth one is given by [Mutaftschiev 1965]

$$\Delta G_f = \Delta G_v + \Delta U_f + T \Delta S_f . \tag{1.92}$$

The first term accounts for the interchange of atoms between the crystal and fluid phases. This interchange is connected with the difference between the chemical potential of the atoms in the crystal, μ_c, and that in the fluid, μ_f: $\Delta\mu = \mu_c - \mu_f$, which is just the supersaturation. Then for ΔG_v we can write (see Eq. (1.90))

$$\Delta G_v = (\mu_c - \mu_f) \sum_{-\infty}^{0} N_{nf} + (\mu_f - \mu_c) \sum_{1}^{\infty} N_{ns}$$

$$= N\Delta\mu \left(\sum_{-\infty}^{0} (1 - \theta_n) - \sum_{1}^{\infty} \theta_n \right) ,$$

the terms in the brackets giving the net amount of atoms leaving the crystal or joining the crystal from the ambient phase. If the roughening takes place without interchange of atoms between the two phases then both sums cancel each other and $\Delta G_v = 0$. This is strictly valid only at the equilibrium temperature $T = T_e$ or, in other words, at $\Delta\mu = 0$.

The second term in (1.92) is completely analogous to the internal energy in Jackson's model and gives simply the number of the unsaturated lateral bonds. Then

$$\Delta U_f = Z_1(\psi/2)N \sum_{-\infty}^{\infty} \theta_n(1 - \theta_n) .$$

It is immediately seen that in the case of a two-level model the sum is reduced to one term only which is precisely equal to the first term in Jackson's expression (1.85).

The configurational entropy ΔS_f can be calculated by the same way as in Jackson's model:

$$\Delta S_f = k \ln \prod_{-\infty}^{\infty} \frac{(N\theta_n)!}{(N\theta_{n+1})!(N\theta_n - N\theta_{n+1})!} \ .$$

Using the boundary conditions (1.91) and Stirling's formula an expression completely analogous to that of Jackson is obtained:

$$T\Delta S_f = -kTN \sum_{-\infty}^{\infty} (\theta_n - \theta_{n+1}) \ln(\theta_n - \theta_{n+1}) \ .$$

Then for the relative Gibbs free energy $\Delta G_f / NkT$ one obtains

$$\frac{\Delta G_f}{NkT} = \beta \left(\sum_{-\infty}^{0} (1 - \theta_n) - \sum_{1}^{\infty} \theta_n \right) + \alpha \sum_{-\infty}^{\infty} \theta_n (1 - \theta_n)$$

$$+ \sum_{-\infty}^{\infty} (\theta_n - \theta_{n+1}) \ln(\theta_n - \theta_{n+1}) \ , \tag{1.93}$$

where

$$\beta = \frac{\Delta\mu}{kT} \tag{1.94}$$

and α is again given by (1.88) in which T_e is replaced by T.

This expression is not simply a generalization of Jackson's for the case of many levels although it is immediately seen that Jackson's expression is automatically obtained if we put $\beta = 0$, $\theta_{-1} = 1$, $\theta_0 = \theta$ and $\theta_{+1} = 0$. The roughness of the crystal face is considered in a *nonconservative* system as we do not keep the number of the solid atoms constant but allow interchange between phases. Thus although the Temkin model is a thermodynamic model it considers surface roughening in the process of growth or dissolution of a crystal, whereas Jackson considers a system in equilibrium ($T = T_e$ and hence $\beta = 0$).

The stability of the crystal–fluid interface is determined by the condition for minimum Gibbs free energy:

$$\frac{\partial}{\partial \theta_n} \left(\frac{\Delta G}{NkT} \right) = 0 \ ,$$

which leads to the following master equation for θ_n:

$$\frac{\theta_n - \theta_{n+1}}{\theta_{n-1} - \theta_n} = \exp(2\alpha\theta_n - \alpha + \beta) . \tag{1.95}$$

Using the substitution $z_n = 2\alpha\theta_n - \alpha + \beta$, Eq. (1.95) turns into

$$\frac{z_n - z_{n+1}}{z_{n-1} - z_n} = \exp(z_n) . \tag{1.96}$$

Expressing $z_{n-1} - z_n$ through $z_n - z_{n+1}$ and substituting the latter into the equation for z_n, (1.96) can be written in the form

$$z_{n+1} = z_n - (z_0 - z_1)\exp\left(\sum_{m=1}^{n} z_m\right) , \tag{1.97'}$$

$$z_{-(n+1)} = z_{-n} + (z_0 - z_1)\exp\left(-\sum_{m=0}^{n} z_{-m}\right) . \tag{1.97''}$$

The boundary conditions $\theta_{-\infty} = 1$ and $\theta_\infty = 0$ turn into

$$z_{-\infty} = \beta - \alpha, \qquad z_\infty = \beta + \alpha . \tag{1.98}$$

Let us consider first the simpler case of $\beta = 0$, i.e. the multilevel generalization of Jackson's model. Two symmetric solutions of Eqs. (1.97) subject to the boundary conditions (1.98) are possible [Temkin 1968]. The first solution is

$$z_0 = -z_1, \quad z_{-1} = -z_2, \quad z_{-2} = -z_3, \text{ etc. }, \tag{1.99}$$

which corresponds to

$$\theta_0 = 1 - \theta_1, \quad \theta_{-1} = 1 - \theta_2, \quad \theta_{-2} = 1 - \theta_3, \text{ etc.}$$

The second solution is

$$z_0 = 0, \quad z_{-1} = -z_1, \quad z_{-2} = -z_2, \text{ etc. }, \tag{1.100}$$

which corresponds to

$$\theta_0 = 1/2, \quad \theta_{-1} = 1 - \theta_1, \quad \theta_{-2} = 1 - \theta_2, \text{ etc.}$$

Comparing both solutions it is immediately seen that the second one (1.100) corresponds to a higher value of the Gibbs free energy as $\theta_0 = 1/2$ and some degree of roughness always exists, whereas the first solution (1.99) permits θ_0 to be close to unity and θ_1 close to zero. Hence solution (1.99) provides a minimum in the surface energy whereas solution (1.100) corresponds to an inflection point. If we go back to Jackson's model for which θ_{-1}, θ_{-2}, $\theta_{-3} = 1$ and θ_1, θ_2, $\theta_3 = 0$, we will see that the second solution $\theta_0 \equiv \theta \cong 0.5$ corresponds to the maximum of the relative Gibbs free energy at values of the parameter α greater than 2. The first solution $\theta_0 = 1 - \theta_1$ corresponds to either a flat face with negligible density of adatoms ($\theta \cong 0$) or a flat face with negligible density of vacancies ($\theta \cong 1$).

Figure 1.25 shows the dependence of θ_n on n for different values of the parameter α. As seen the interface becomes more and more smeared with decreasing α. On the contrary, when $\alpha = 3.31$ the interface consists of one layer with some vacancies in it and a few adatoms on top of it. In other words, at $\alpha > 3.3$ the interface is flat and atomically smooth.

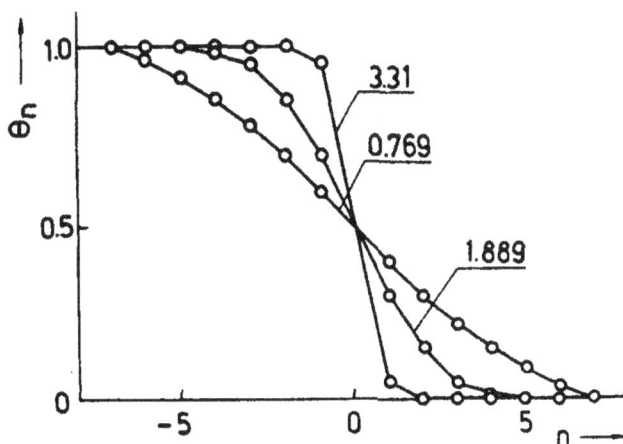

Fig. 1.25. Dependence of the surface coverage on the number n of the layer at different values of the parameter α denoted by the figure at each curve at equilibrium ($\beta = 0$) (after Temkin [1964, 1968]).

We consider further the more general case of $\beta \neq 0$ and construct a plot of $\ln \beta$ vs α (Fig. 1.26). As seen the whole field is divided into two parts, A and B. In part A Eq. (1.95) (or (1.97)) has two solutions: one ground state solution (1.99) and one saddle point solution (1.100). On the dividing line both solutions coincide. In region B the master equation (1.95)

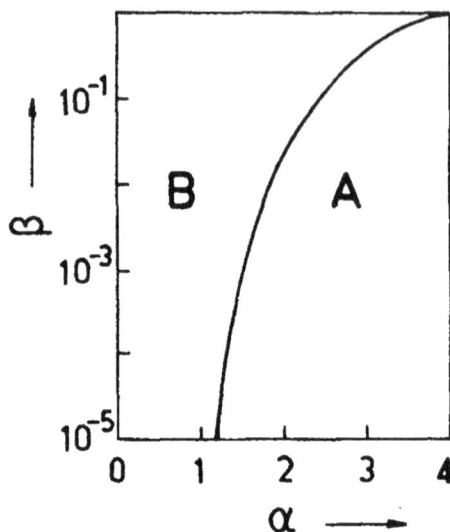

Fig. 1.26. Phase diagram in coordinates $\ln \beta$ and α showing the state of the crystal surface as follows from the model of Temkin. In the region denoted by A the system of equations (1.95) has two solutions, one of which is the ground state solution corresponding to a smooth surface and the other is a saddle point. Both solutions coincide on the dividing line, and in the region denoted by B no solution exists. The latter can be interpreted as disappearance of the crystallographic orientation of the face or roughening of the face (after Temkin [1964, 1968]).

has no solution at all which is interpreted as a loss of the crystallographic orientation of the crystal face. In other words, it becomes so rough that it cannot be distinguished as a crystal face with definite crystallographic orientation any more. The latter means that under conditions of region B the atoms arriving from the ambient phase can be incorporated at any site of the crystal surface without the necessity to overcome a thermodynamic energy barrier. At values of α and β such that we are in region A the ground state is a more or less flat surface and its growth requires the formation of 2D nuclei or the presence of screw dislocations.

1.6.3.3. *Criterion of Fisher and Weeks*

Although the models of Jackson and Temkin predict roughening of the crystal surfaces above some critical temperature they do not correctly account for the thermal fluctuations in the system because of the mean field approximation used.

As shown above the model of Temkin predicts that under conditions of region B in Fig. 1.26 the crystal surface cannot be distinguished any more as having a definite crystallographic orientation. In other words, the crystal surface becomes delocalized with respect to the crystal lattice. This statement contains in itself the main difference between a liquid and a crystal surface. The flat crystal surface has a definite crystallographic orientation and is said to be *localized* or *immobile*. On the contrary, the liquid surface is delocalized in the sense that it has not a definite crystallographic orientation. At low enough temperatures the crystal surfaces are more or less smooth and so are the steps on them. At higher temperatures thermal fluctuations become important and the steps become more and more rough, the step Gibbs free energy G_{st} tending to zero. It was shown [Swendsen 1978; Weeks 1980; Fisher and Weeks 1983; Jayaprakash, Saam and Teitel 1983] that G_{st} vanishes with temperature following the law

$$G_{st} \sim \exp\left(-\frac{C}{(T_r - T)^{1/2}}\right) , \qquad (1.101)$$

i.e. in a very smooth manner.

The step free energy G_{st} is closely connected with the specific surface energy and thus with the surface stiffness. It has been shown in the previous section that the surface stiffness is infinite for a flat crystal surface whereas it has a finite value for a rounded "rough" surface. Theoretical treatment of the temperature behavior of surface stiffness [Chui and Weeks 1978; Fisher and Weeks 1983] resulted in an expression for the roughening temperature which naturally connects the latter with the surface stiffness:

$$kT_r = \frac{2}{\pi}\sigma^*(T_r)d_{hkl}^2 , \qquad (1.102)$$

where $\sigma^*(T_r)$ is the surface stiffness at the transition temperature and d_{hkl} is the interplanar distance parallel to the interface. The more closely packed a given crystal face is, the larger the interplanar distance and the higher the roughening temperature will be. Thus for a fcc lattice, $d_{111} = a_0/\sqrt{3}$, $d_{100} = a_0/2$ and $(d_{111}/d_{100})^2 = 4/3 = 1.33$. Hence, in equilibrium at some finite temperature the most closely packed surfaces will be flat, whereas others will be rounded. Slight deviation from equilibrium will lead to growth of the rounded regions and their subsequent disappearance. The crystal will be confined by low index planes only. If again equilibrated the rounded regions should reappear.

The theoretical criterion (1.102) has been experimentally verified in the case of ^{4}He crystals [Wolf *et al.* 1985; Keshishev *et al.* 1981; Babkin *et al.* 1984; Avron *et al.* 1980; Gallet *et al.* 1986, 1987; Nozières and Gallet 1987], which is the ideal choice for such kind of investigations. The main advantages are the following. The liquid helium-4 is superfluid below 1.76 K, i.e. its viscosity is nearly zero. Moreover its thermal conductivity is practically infinite. Besides, helium can be easily purified, the impurity concentration being as low as 1×10^{-9}at. %. The latter is of utmost importance as the adsorption of impurities changes drastically the surface energy. Crystal helium has a very high thermal conductivity also. Thus, in contrast to other crystals, the heat and mass transport are very fast. Then, the equilibrium shape is reached in a very short time.

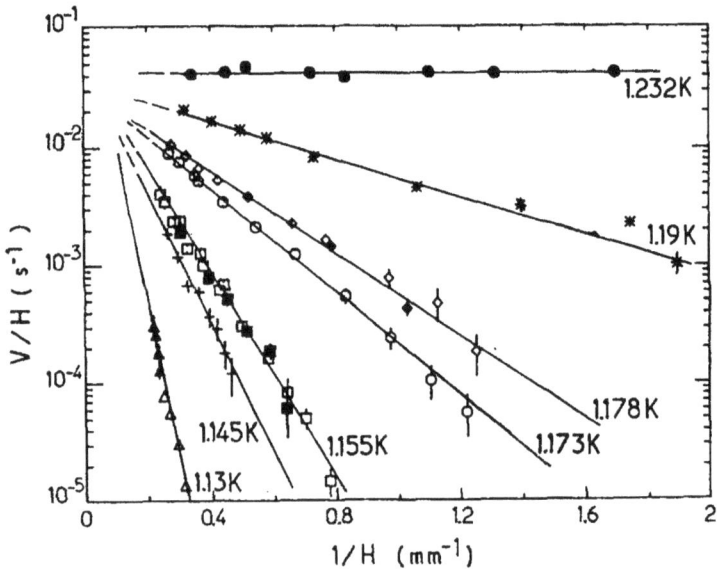

Fig. 1.27. Semi-logarithmic plot of the ratio of the rate of growth V in (0001) direction and the level difference H (the latter is proportional to the supersaturation $\Delta\mu$) as a function of the reciprocal supersaturation $1/H \cong 1/\Delta\mu$ for various temperatures. The straight lines show that $V \cong \Delta\mu \exp(-K_2/\Delta\mu)$ and hence the growth takes place by two-dimensional nucleation (see Chap. 2). The slopes of the straight lines give in fact the squares of the specific edge energies of the steps surrounding the 2D nuclei. As seen the slope becomes equal to zero at $T = 1.232$ K which shows that the specific energy of the steps also becomes equal to zero. The latter means that the crystal surface is no longer smooth but is atomically rough. (P. E. Wolf, F. Gallet, S. Balibar, E. Rolley and P. Nozières, *J. Phys.* **46** (1987). By permission of Les Editions de Physique and courtesy of S. Balibar.)

As mentioned above the structure of the crystal surface affects the mechanism of growth. Wolf *et al.* [1985] investigated the latter and found that at temperatures beyond 1.232 K the dependence of the growth rate R on $\Delta\mu$ is linear, which can be considered as a direct indication of atomically rough surface. On the contrary, at temperatures below 1.232 K a nonlinear dependence is established which proved to be exponential as required by the 2D nucleation mechanism of growth (Fig. 1.27). The slope of the logarithmic plot gives directly the energetic barrier for 2D nucleus formation, and at $T = 1.232$ K it becomes equal to zero. The latter means that the free energy of the steps becomes equal to zero, the surface becomes atomically rough, and 2D nucleation is no more necessary for the crystals to grow. What is more interesting is that the dependence of the Gibbs free energy of the steps (estimated from growth experiments) on temperature decays exponentially, going smoothly to the roughening temperature T_r (Fig. 1.28). Thus an excellent quantitative agreement between theory and experiment is achieved.

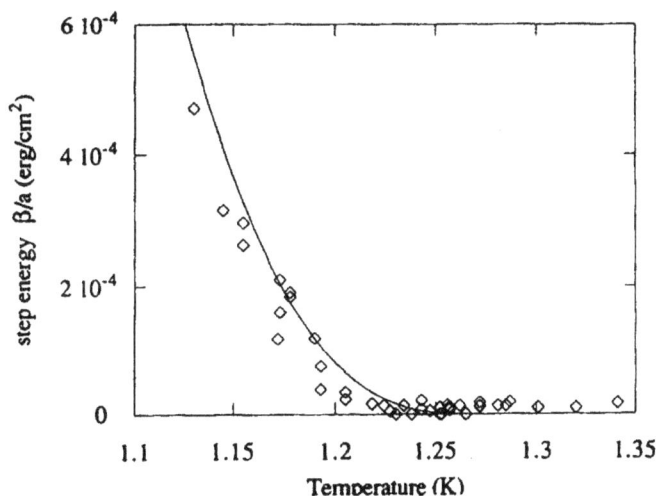

Fig. 1.28. Variation of the step free energy with temperature as deduced from plots shown in Fig. 1.27. As seen the step free energy vanishes within the experimental accuracy near $T = T_r = 1.28$ K. (F. Gallet, S. Balibar and E. Rolley, *J. Phys.* **48**, 369 (1987). By permission of Les Editions de Physique and courtesy of S. Balibar.)

Let us consider Eq. (1.102) in more detail. The interplanar distance d_{hkl} can be identified as the step height. Then the product $\sigma^*(T_r)d_{hkl}$ can be treated as the energy of the step and the product $\sigma^*(T_r)d_{hkl}^2$ as the energy to form a kink or the energy ω to form a dangling bond [Chernov 1989]. Assuming $\sigma^* \cong \sigma$ as a first approximation one can now use Eq. (1.102) to find the critical temperature T_r and predict the mechanism of growth. Thus for Ge (111) the surface energy of the crystal–melt interface as found from nucleation experiments is equal to 251 erg.cm^{-2} [Skripov 1977] and that of the crystal–vapor interface is equal to 1100 erg.cm^{-2} (Swallin [1962]; see also Kern *et al.* [1979]). Then with $d_{hkl} = 3.26 \times 10^{-8}$ cm for the critical temperature for roughening, T_r, the values 1150 K and 5000 K for the crystal–melt and crystal–vapor interfaces, respectively, are obtained. Bearing in mind that the melting point of Ge is $T_m = 1210$ K, it follows that the most closely packed surface (111) and hence all the other surfaces of Ge should be rough when in contact with the melt below the melting point and the Ge crystals should grow rounded from the melt. In the other case, the roughening temperature is well above the melting point and the Ge crystals should grow well polygonized from a vapor phase.

1.6.4. *Kinetic roughness*

All criteria of surface structure derived so far are of thermodynamic nature. The roughness of the crystal face is due to entropy effects which decrease the Gibbs free energy of the crystal surface. However, there can also be the so-called *kinetic* roughness which can take place at temperatures below the thermodynamic critical temperature. When the supersaturation is high enough the rate of formation of 2D nuclei becomes very large so that new nuclei can be formed before the complete coverage of the crystal face by the preceding one. Several layers grow simultaneously and we observe what is called a *multilayer* growth. Then if the density of 2D nuclei is very large it may happen that the mean distance between their edges (which in turn are rough) becomes comparable with the interatomic distance [Chernov 1973]. Arriving atoms can thus be incorporated practically at any site. Obviously, kinetic roughness can be observed when the specific edge free energy and the work of formation of 2D nuclei are very small. This question will be considered in more detail in Chap. 3.

CHAPTER 2

NUCLEATION

2.1. Thermodynamics

Gibbs was the first to realize that the formation of a new phase requires as a necessary prerequisite the appearance of small clusters of building units (atoms or molecules) in the volume of the supersaturated ambient phase (vapors, melt or solution). He considered these nuclei as small liquid droplets, vapor bubbles or small crystallites, or, in other words, small complexes of atoms or molecules which have the same properties as the corresponding bulk phases with the only exception being their small linear sizes. Although oversimplified this picture has been a significant step towards the understanding of the transitions between different states of aggregation, because when phases with small sizes are involved the surface-to-volume ratio turns out to be large compared with that of macroscopic entities. Then the fraction of the Gibbs free energy of systems containing small particles which is due to the surface energy becomes considerable. Moreover, this approach allows a description of phases with finite sizes in terms of such macroscopic thermodynamic quantities as specific surface and edge energies, pressure, etc. That is why the theory of formation of new phases as developed by Gibbs [1928], Volmer [1926, 1939], Farkas [1927], Stranski and Kaischew [1934], Becker and Döering [1935], Frenkel [1955], and others is known as the *capillary or classical* theory of nucleation. As will be shown in this chapter the classical theory is valid at small or moderate supersaturations, in contrast to the atomistic theory which is applicable at extremely high supersaturations where the nuclei consist of very small number of building units of the order of unity.

In any thermodynamic system, even stable ones, local deviations from the normal state or *fluctuations* should have place which are less probable in the sense that they increase the thermodynamic potential of the system. If one considers a homogeneous molecular system (liquid or vapor) there are always small fluctuations of the density in the sense of small molecular aggregates which are well compatible with the given state of aggregation. Such density fluctuations can be called, after Frenkel [1955], "homophase" fluctuations. On the other hand, there might be the so-called "heterophase" fluctuations which could lead to visible transition to another state of aggregation. Their concentration should increase considerably near the phase equilibrium determined by the equality of the chemical potentials $\mu_\alpha = \mu_\beta$. If the initial bulk phase α is the stable one ($\mu_\alpha < \mu_\beta$) these density fluctuations are "lifeless" in the sense that they grow to negligible sizes and decay without revealing a tendency to unlimited growth. If, however, the initial phase α is unstable ($\mu_\alpha > \mu_\beta$) the tendency to growth prevails after exceeding a certain critical size. It is just these density fluctuations or clusters that are called the critical nuclei of the new phase. In order to form such clusters some free energy should be expended. In other words, the system should overcome an activation barrier whose height is given by the work of formation of the critical nuclei.

When considering the change of the thermodynamic potential connected with the formation of nuclei of the new phase one assumes that the shape of the nuclei is just the equilibrium shape as determined in the previous chapter. Arbitrary shapes could be accounted for as well but it is the equilibrium shape that ensures minimal work for nucleus formation and thus determines the most probable path. Moreover, when one considers the transition from one condensed phase to another, say crystalline from crystalline or amorphous ones, formation of the nuclei will be accompanied by the appearance of elastic stresses due to the different molar volumes of the two phases. The contribution of these stresses could be significant [Hilliard 1966; Christian 1981] and often greater than the contribution due to the nucleus shape. In the following presentation the contribution of these strains will not be accounted for. For more details the reader is referred to the monograph of Christian [1981]. In the case of epitaxial growth of thin films, however, the substrate and deposit crystals have as a rule different lattice parameters and lateral stresses in both crystals appear as a consequence. Contributions due to these elastic strains cannot be avoided if we want to understand the phenomenon and these should be added to the change of the thermodynamic potential, in addition to the volume and surface terms discussed in the previous chapter.

In the present chapter the nucleation of single component systems will be considered. Nucleation in binary systems was first treated by Reiss [1950] following the approach of Frenkel [1955]. The problem was considered further by many authors [Wilemski 1975a, 1975b; Temkin and Shevelev 1981, 1984; Shi and Seinfeld 1990; Zeng and Oxtoby 1991] and the interested reader is referred to the original papers. Cahn and Hilliard [1958, 1959] considered nucleation in a two-component incompressible fluid and found that at small supersaturations the classical theory results. However, on approaching the spinodal the work for nucleus formation tends to zero as the energy of the interface between the nucleus and the ambient phase vanishes at the spinodal. The radius of the critical nucleus tends to infinity but the density of the nucleus tends to that of the ambient phase. The theory of Cahn and Hilliard was further developed by Hoyt [1990] for the case of multicomponent systems. For a review see Uhlmann and Chalmers [1966].

The problems of nucleation, both thermodynamic and kinetic, are considered in numerous monographs and review papers [Volmer 1939; Defay *et al.* 1966; Kaischew 1980; Turnbull 1956; Frenkel 1955; Dunning 1955; Hirth and Pound 1963; Nielsen 1964; Hollomon and Turnbull 1953; Toschev 1973; Stoyanov and Kashchiev 1981; Stoyanov 1979; Zettlemoyer 1969; Nucleation Phenomena 1966, 1977; Skripov 1977; James 1982; Christian 1981; Oxtoby 1992] and the reader interested in different aspects of particular phase transitions is referred to them.

2.1.1. *Homogeneous formation of nuclei*

We consider first the simplest case of formation of liquid nuclei in the bulk of a vapor phase. The simplicity is obviously due to the isotropic surface tension σ of the liquid which leads to spherical equilibrium shape of the small liquid entities. We consider a volume containing n_v moles of a vapor with chemical potential μ_v which is a function of the temperature T and pressure P. The thermodynamic potential of the initial state of the system at $T = $ const and $P = $ const is thus given by

$$G_1 = n_v \mu_v .$$

A liquid droplet with bulk chemical potential μ_l is formed from n_l moles of the vapor phase and the thermodynamic potential of the system vapor–liquid droplet reads

$$G_2 = (n_v - n_l)\mu_v + n_l\mu_l + 4\pi r^2\sigma .$$

In this equation σ is the surface energy of the flat surface. The dependence of the surface energy on the droplet size has been discussed by Kirkwood and Buff [1949] and by Tolman [1949]. They found that the surface energy should in general decrease with decreasing droplet size. Assuming Lennard–Jones interatomic forces Benson and Shuttleworth [1951] have found a decrease of 15% for the surface energy of a close-packed cluster consisting of 13 atoms as compared with the energy of a flat surface. In the analysis that follows we will neglect the curvature dependence of the surface energy.

The change of the Gibbs free energy upon the formation of the droplet is then

$$\Delta G = G_2 - G_1 = -n_l(\mu_v - \mu_l) + 4\pi r^2\sigma .$$

Bearing in mind that $n_l = 4\pi r^3/3v_l$ (v_l being the molecular volume of the liquid) one obtains

$$\Delta G(r) = -\frac{4}{3}\frac{\pi r^3}{v_l}\Delta\mu + 4\pi r^3\sigma , \qquad (2.1)$$

where $\Delta\mu = \mu_v - \mu_l$ is the supersaturation (Eq. 1.9). The $\Delta G(r)$ dependence is plotted in Fig. 2.1.

Thus in the simplest case of a droplet formation in vapor ΔG consists of two terms: a volume term $4\pi r^3\Delta\mu/3v_l = (4\pi r^3/3)(P_l - P_v)$ and a surface term $4\pi r^2\sigma$. The minus before the volume term reflects the fact that energy is gained when the liquid phase is thermodynamically stable ($\mu_l < \mu_v$). The increase of the thermodynamic potential of the system is due to the formation of a dividing surface. Then ΔG displays a maximum at some critical size r^* given by (Fig. 2.1)

$$r^* = \frac{2\sigma v_l}{\Delta\mu} . \qquad (2.2)$$

Equation (2.2) is in fact the equation of Thomson–Gibbs (1.18) and gives the condition for equilibrium of the nucleus with the ambient phase. Note, however, that this equilibrium is unstable. Indeed, if some more atoms join the critical nucleus its radius increases and in turn its equilibrium vapor pressure becomes smaller than the one available in the system (Eq. (1.19)). Then the probability of decay becomes smaller than the probability of

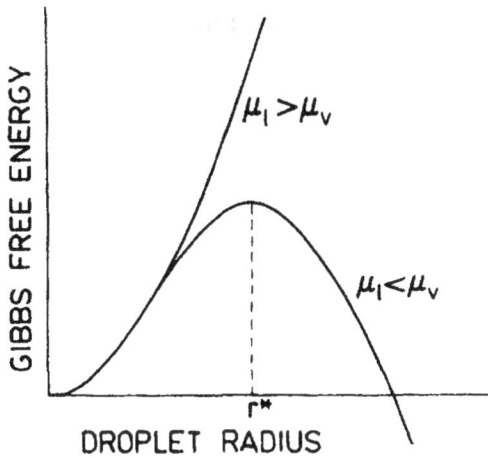

Fig. 2.1. Dependence on the droplet radius of the Gibbs free energy change ΔG connected with the formation of liquid nuclei from a supersaturated vapor phase. When the liquid phase is stable ($\mu_l < \mu_v$) ΔG displays a maximum at some critical radius $r = r^*$. Beyond this size, growth of the nucleus leads to a decrease of the Gibbs free energy of the system. The maximum Gibbs free energy ΔG^* is the work of formation of the critical nucleus. When the vapor phase is stable ($\mu_v < \mu_l$) both terms in Eq. (2.1) are positive and the formation of nuclei capable of unlimited growth is thermodynamically prohibited as it leads to infinite increase of the Gibbs free energy.

growth and the nucleus should grow further. In the opposite case its equilibrium vapor pressure becomes greater than that available in the system and the nucleus reveals a tendency for further decay. In other words, any infinitesimal deviation of the size of the nucleus from the critical one leads to a decrease of the thermodynamic potential of the system. In this sense a cluster of size r^* is a critical nucleus of the new phase.

The maximal value of ΔG which is obtained by the substitution of r^* into Eq. (2.1),

$$\Delta G^* = \frac{16\pi}{3} \frac{\sigma^3 v_l^2}{\Delta\mu^2} , \qquad (2.3)$$

gives the height of the energy barrier which should be overcome for condensation to take place. It is inversely proportional to the square of the supersaturation (a result which is obtained for the first time by Gibbs [1878]) and increases steeply near the phase equilibrium (i.e. at small supersaturations), thus imposing great difficulties for the phase transition to occur.

When the ambient phase is stable ($\mu_v < \mu_l$) both terms in Eq. (2.1) are positive and ΔG tends to infinity (see Fig. 2.1), thus reflecting the fact that the heterophase density fluctuations at undersaturation are thermodynamically unfavored.

Substituting the supersaturation $\Delta\mu$ into (2.3) by the radius r^* of the critical nucleus from the Thomson–Gibbs equation (2.2) gives

$$\Delta G^* = \frac{1}{3} 4\pi r^{*2}\sigma = \frac{1}{3}\sigma\Sigma \ . \tag{2.4}$$

As seen the Gibbs free energy required to form a critical nucleus of the new phase with equilibrium shape is precisely equal to one third of the surface energy $\sigma\Sigma$, a result obtained for the first time by Gibbs [1878].

Useful expressions for $\Delta G(r)$ are obtained if we substitute the supersaturation from the Thomson–Gibbs equation (2.2) into (2.1) in terms of the radius r^* of the critical nucleus:

$$\Delta G(r) = \Delta G^* \left[3\left(\frac{r}{r^*}\right)^2 - 2\left(\frac{r}{r^*}\right)^3 \right] \ , \tag{2.5}$$

or in terms of the number n^* of atoms in the nucleus (from $v_c n = 4\pi r^3/3$):

$$\Delta G(n) = \Delta G^* \left[3\left(\frac{n}{n^*}\right)^{2/3} - 2\left(\frac{n}{n^*}\right) \right] \ , \tag{2.6}$$

where ΔG is given by (2.4). We will use Eqs. (2.5) and (2.6) when deriving an expression for the rate of nucleation.

The result (2.4) is a universal one. It does not depend on the state of aggregation of the nucleus and can be easily obtained for crystalline nuclei in a general form. Indeed, in this case

$$\Delta G^* = -\frac{V^*}{v_c}\Delta\mu + \sum_n \sigma_n \Sigma_n \ ,$$

where V^* is the volume of the critical nucleus and v_c is the volume of one building unit in the crystal phase. Bearing in mind (see Chap. 1)

$$V^* = \frac{1}{3}\sum_n h_n \Sigma_n$$

and Eq. (1.27) for the equilibrium shape,

$$\frac{h_n}{\sigma_n} = \frac{2v_c}{\Delta\mu} \ ,$$

one obtains

$$\Delta G^* = \frac{1}{3}\sum_n \sigma_n \Sigma_n \ . \tag{2.7}$$

One can use this expression to obtain the work of formation of crystalline nuclei with arbitrary symmetry and radius of action of the interatomic forces. In the simplest case of a nucleus of a Kossel crystal with first neighbor interactions the (100) faces appear only in the equilibrium shape. Then

$$\Delta G^* = 2l^{*2}\sigma = \frac{32\sigma^3 v_c^2}{\Delta\mu^2} . \tag{2.8}$$

Under the same condition (first neighbor interactions) the equilibrium shape of a crystal with face-centred cubic (fcc) lattice has the form of a truncated octahedron and consists of six square faces (100) and eight hexagonal (111) faces with equal edge lengths [Markov and Kaischew 1976b]. Then

$$\Delta G^* = \frac{1}{3}\left(6l^{*2}\sigma_{100} + 8\frac{3\sqrt{3}}{2}l^{*2}\sigma_{111}\right) .$$

From the equilibrium shape condition $h_{111} : h_{100} = \sigma_{111} : \sigma_{100}$ with the first neighbor model relation $\sigma_{111}/\sigma_{100} = \sqrt{3}/2$ it follows

$$l^* = \frac{\sigma_{100}v_c}{\Delta\mu} = \frac{2\sigma_{111}v_c}{\Delta\mu\sqrt{3}}$$

and

$$\Delta G^* = \frac{1}{3}\frac{v_c^2}{\Delta\mu^2}\left(6\sigma_{100}^3 + 16\sqrt{3}\sigma_{111}^3\right) = \frac{8\sigma_{100}^3 v_c^2}{\Delta\mu^2} . \tag{2.9}$$

Equation (2.3), (2.8) or (2.9) is applicable for nucleation from any supersaturated (undercooled) phase (vapor, liquid or solution). For this aim the corresponding differences of the chemical potentials $\Delta\mu$ (Eq. (1.9), (1.10) or (1.12)) and the specific energies of the corresponding interfaces (crystal–vapor, crystal–melt or crystal–solution) should be taken into account.

2.1.2. Heterogeneous formation of 3D nuclei

The process of nucleation is stimulated by the presence of impurity particles, ions or foreign surfaces. Nuclei are usually formed on the walls of the reaction vessels. While these effects are usually undesirable, the process of nucleation on foreign substrates is essential for epitaxial deposition of thin films. We will illustrate this problem with the formation of a liquid droplet on the so-called structureless substrate [Volmer 1939]. But before doing that we have to clarify what "structureless substrate" means.

A single crystal substrate exerts on the particles of the mother phase a periodic potential which is characterized by a period equal to the interatomic spacing and the overall amplitude equal to the activation energy for surface diffusion [Frenkel and Kontorova 1938]. In the simplest case it can be presented by a sinusoid (Fig. 2.2(a)). If a nucleus is formed on such a surface it should be elastically strained to fit the substrate. Then the energy of the elastic strains should be added to the change of the thermodynamic potential. In order to simplify the problem we assume that the modulation of the periodic potential is equal to zero. The hypothetical structureless substrate (Fig. 2.2(b)) is the result which gives the possibility to study the effect of interatomic forces on the process of nucleation, neglecting the lattice mismatch as a first approximation. Then we can make the necessary correction for the latter.

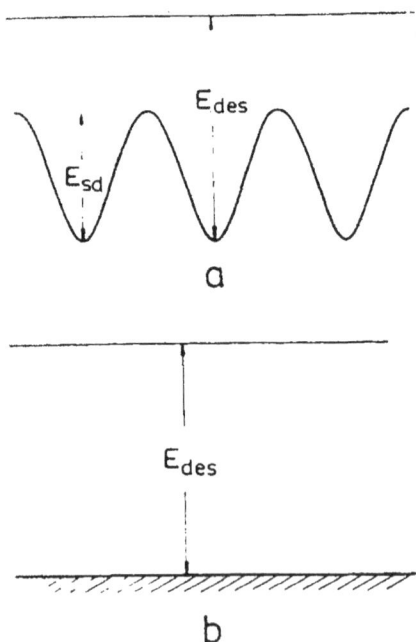

Fig. 2.2. For the determination of the concept of a structureless substrate which is very convenient for studying the catalytic potency of the substrate on the nucleus formation. (a) Shows the energetic profile of a single crystal substrate, where E_{sd} and E_{des} are the activation energies for surface diffusion and desorption, respectively. In (b) the surface potential is no longer a periodic function ($E_{sd} = 0$). This simplification excludes the effect of the lattice misfit but permits the study of the effect of $E_{des} \neq 0$.

When considering the homogeneous formation of liquid droplets from a vapor phase we assume that the equilibrium shape of the droplet is a sphere. We also assume that the shape of the crystalline nuclei is the equilibrium one in order to derive the expression for the Gibbs energy change of their formation. Obviously, we should first derive an expression for the equilibrium shape of a droplet on a foreign substrate.

We consider a liquid droplet on a smooth structureless substrate (Fig. 2.3). It represents a segment of a sphere with radius of curvature r and projected radius $r \sin \theta$, where θ is the so-called wetting angle. The latter characterizes the energetic influence of the substrate. We denote the specific surface energies of the free surfaces of the droplet and the substrate, and of the substrate–droplet interface by σ, σ_s and σ_i, respectively. Then the condition of equilibrium is expressed by the well-known relation of Young [1805] (see also Adam [1968]):

$$\sigma_s = \sigma_i + \sigma \cos \theta \,, \qquad (2.10)$$

which is an analog of the relation of Dupré [1869] (Eq. (1.28)) for the case of a liquid droplet on a solid surface.

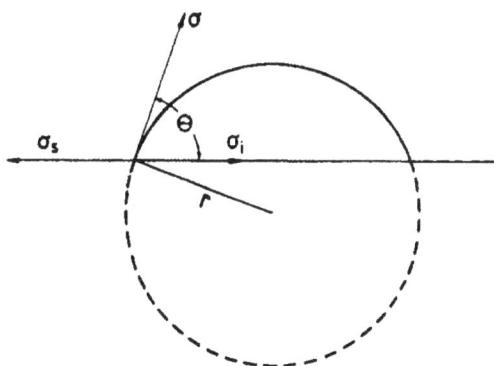

Fig. 2.3. Equilibrium shape of a liquid droplet with radius of curvature r on a structureless substrate. The latter is characterized by the wetting angle θ between the substrate plane and the tangent to the droplet surface. The wetting angle is determined through the Young equation from the specific surface energies of the interfaces between the liquid droplet and the solid substrate and the vapor phase, and between the substrate and the liquid droplet, denoted by σ, σ_s and σ_i, respectively.

The Young relation is easy to derive following the approach given in Chap. 1. We have to find as before the minimum of the surface energy Φ of the liquid droplet at constant volume V_l of the segment. Bearing in mind that the area of the free surface and the area of the contact are given by

$$\Sigma = 2\pi r^2 (1 - \cos\theta)$$

and

$$\Sigma_i = \pi r^2 \sin^2\theta \ ,$$

respectively, the surface energy Φ of the liquid droplet reads

$$\Phi = 2\pi r^2 (1 - \cos\theta)\sigma + \pi r^2 \sin^2\theta(\sigma_i - \sigma_s) \ . \tag{2.11}$$

The volume of the segment is

$$V_l = \frac{4}{3}\pi r^3 \frac{(1 - \cos\theta)^2(2 + \cos\theta)}{4} \ . \tag{2.12}$$

From $dV_l = 0$ we find

$$dr = -\frac{(1 + \cos\theta)\sin\theta}{(1 - \cos\theta)(2 + \cos\theta)} r\, d\theta \ .$$

Substituting it into $d\Phi = 0$ results in Eq. (2.10).

The change of the thermodynamic potential upon formation of the droplet is given by

$$\Delta G = -\frac{V_l}{v_l}\Delta\mu + \Phi \ ,$$

where V_l and Φ are given by (2.12) and (2.11), respectively.

Substituting $\sigma_i - \sigma_s = -\sigma\cos\theta$ from the equation of the equilibrium shape (2.10) into the expression for ΔG gives

$$\Delta G = \frac{4}{3}\pi r^3 \frac{(1 - \cos\theta)^2(2 + \cos\theta)}{4}\frac{\Delta\mu}{v_l} + 2\pi r^2\sigma(1 - \cos\theta) - \pi r^2\sigma\sin^2\theta\cos\theta \ .$$

Following the same procedure as before we find that the change of the thermodynamic potential reaches a maximum at a critical size

$$r^* = \frac{2\sigma v_l}{\Delta\mu}$$

which does not depend on the wetting angle. The latter is clear recalling that the equilibrium vapor pressure depends only on the curvature but not on whether the droplet is a complete sphere or not.

Then for the work of nucleus formation one obtains

$$\Delta G_{\text{het}}^* = \frac{16\pi}{3}\frac{\sigma^3 v_l^2}{\Delta\mu^2}\phi(\theta) = \Delta G_{\text{hom}}^*\phi(\theta) \ , \tag{2.13}$$

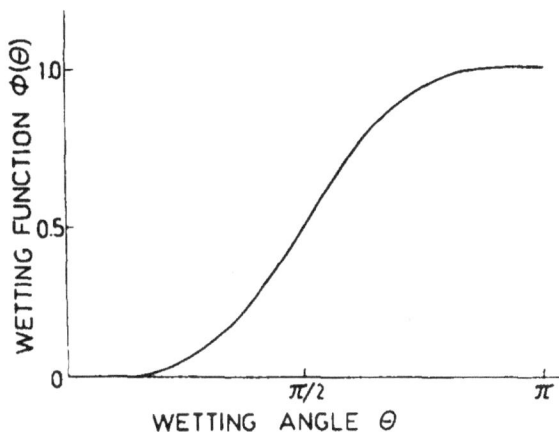

Fig. 2.4. Plot of the wetting function $\phi(\theta) = (1 - \cos \theta)^2 (2 + \cos \theta)/4$ vs the wetting angle θ. It varies from zero to unity when θ varies from 0 (complete wetting) to π (complete nonwetting).

where

$$\phi(\theta) = \frac{1}{4}(1 - \cos \theta)^2 (2 + \cos \theta) \tag{2.14}$$

is a function of the wetting angle and accounts for the catalytic potency of the substrate with respect to nucleus formation.

A graphic representation of $\phi(\theta)$ is given in Fig. 2.4. As seen it varies from 0 to 1 when θ varies from 0 to π. In other words, in the case of complete wetting $(\phi(\theta = 0) = 0)$ $\Delta G^*_{\text{het}} = 0$, the formation of 3D droplets is thermodynamically unfavored and the liquid has a tendency to cover the substrate as a continuous film. In the other extreme (complete nonwetting; $\phi(\theta = \pi) = 1$) $\Delta G^*_{\text{het}} = \Delta G^*_{\text{hom}}$ which means that the substrate does not exert any energetic influence on the nucleus formation and the nucleus has the shape of a complete sphere, i.e. we have in practice homogeneous nucleation. It should be noted, however, that in the case of complete wetting $(\theta = 0)$ the formation of a new phase still requires overcoming an energetic barrier which is connected with the formation of a two-dimensional nucleus. The increase of the Gibbs free energy in this case is due to the formation of a single height step surrounding the nucleus.

Comparing the wetting function $\phi(\theta)$ with the volume of the segment it becomes immediately clear that $\phi(\theta) = V^*/V_0$, where V_0 is the volume of the complete sphere. Then

$$\Delta G^*_{\text{het}} = \Delta G^*_{\text{hom}} \frac{V^*}{V_0} \; , \tag{2.15}$$

i.e. the ratio of the works for heterogeneous and homogeneous formation of nuclei at one and the same supersaturation is simply equal to the ratio of the corresponding volumes of the nuclei. As will be shown below this is also a universal result which depends neither on the state of aggregation of the nucleus nor its crystal lattice.

Indeed, the work of homogeneous formation of a crystalline nucleus in a general form is

$$\Delta G^*_{\text{hom}} = -\frac{V_0}{v_c} \Delta\mu + \sum_n \sigma_n \Sigma_n \; ,$$

where V_0 is the volume of the homogeneously formed critical nucleus.

Substituting σ_n from the Gibbs–Curie–Wulff theorem (Eq. 1.27) into the surface energy term gives

$$\Delta G^*_{\text{hom}} = \frac{1}{2} \frac{V^*_0}{v_c} \Delta\mu \; , \tag{2.16}$$

i.e. the work of nucleus formation is equal to one-half of the volume work which is required to transfer $n = V^*_0/v_c$ atoms from the parent to the new phase.

The work of heterogeneous formation of a 3D nucleus is

$$\Delta G^*_{\text{het}} = -\frac{V^*}{v_c} \Delta\mu + \sum_{n \neq m} \sigma_n \Sigma_n + (\sigma_m - \beta)\Sigma_i \; ,$$

where V^* is the volume of the critical nucleus formed on a foreign substrate and $\sigma_m - \beta = \sigma_i - \sigma_s$ follows from the relation of Dupré.

Substituting again σ_n and $\sigma_m - \beta$ from the Gibbs–Curie–Wulff theorem valid for the heterogeneous case (Eq. 1.30) gives

$$\Delta G^*_{\text{het}} = \frac{1}{2} \frac{V^*}{v_c} \Delta\mu \; . \tag{2.17}$$

Equation (2.15) follows from Eqs. (2.16) and (2.17).

Equations (2.16) and (2.17) can be written in the form

$$n^* = \frac{2\Delta G^*}{\Delta\mu} \; , \tag{2.18}$$

where $n^* = V^*/v_c$ is the number of atoms in the critical nucleus.

It follows that in all cases the number of atoms in the critical nucleus is equal to the doubled work for nucleus formation divided by the supersaturation. Bearing in mind that the work of nucleus formation is inversely proportional to the square of the supersaturation (Eq. (2.3)) we find that the number of atoms in the critical nucleus decreases with the cube of the supersaturation.

Equation (2.17) allows us to calculate easily the work for nucleus formation of, say, a cubic nucleus on foreign substrate (Fig. 1.18). With $V^* = l^{*2}h^*$, the condition for the equilibrium shape $h^*/l^* = (\sigma + \sigma_i - \sigma_s)/2\sigma = \Delta\sigma/2\sigma$ (Eq. 1.69) and the Thomson–Gibbs equation $l^* = 4\sigma v_c/\Delta\mu$ one obtains

$$\Delta G^*_{\text{het}} = \frac{32\sigma^3 v_c^2}{\Delta\mu^2}\frac{(\sigma + \sigma_i - \sigma_s)}{2\sigma} = \Delta G^*_{\text{hom}}\frac{\Delta\sigma}{2\sigma}\,, \qquad (2.19)$$

where $\Delta\sigma/2\sigma = h^*/l^* = h^* l^{*2}/l^{*3} = V^*/V_0^*$. Equation (2.19) can be easily obtained also combining Eqs. (2.8), (2.15) and (1.69).

It appears (Eqs. (1.69) and (2.14)) as if the foreign substrate cuts the homogeneous nucleus at a height determined by the equilibrium shape or, in other words, by the ratio of the interatomic forces. The latter is given by the wetting function $\phi(\theta)$ for liquid droplets or the equivalent expression $(\sigma + \sigma_i - \sigma_s)/2\sigma = 1 - \psi'/\psi$ valid for crystals.

2.1.3. Heterogeneous formation of elastically strained 3D nuclei

The problem of formation of nuclei on single crystal substrates is much more complicated and can be solved more or less approximately. The difficulties arise from the fact that the strains are anisotropic. There are lateral strains due to the tendency of the nuclei to fit the substrate and normal strains with opposite sign due to the Poisson effect. (The latter accounts for the transverse deformations of the crystal lattice which appear as a result of longitudinal strains.) This strain energy should be added to the change of the thermodynamic potential. The role of the elastic strain is, however, twofold. One problem arises in accounting for the strain dependence of the specific surface energies. As mentioned in Chap. 1 a new surface can be created by stretching out an old one (by stretching out the whole crystal). Thus if the crystal is laterally strained in both orthogonal directions parallel to the substrate surface the upper crystal surface will be "isotropically" strained in two directions. As a result its specific energy will be changed. On the other hand, the lateral deformation of the crystal leads to strongly anisotropic change of the specific energy of the side faces. Bearing in mind the Poisson effect the chemical bonds will be stretched out in directions parallel to the substrate surface but compressed in normal direction and vice versa. As a result the specific energy of the side crystal faces will be changed in a very peculiar way. Also, the elastic strains should affect the chemical bonds between the substrate and deposit atoms and in turn the specific energy of the interfacial boundary. The contribution of the strains to the specific surface energies should be added to the surface term of the thermodynamic potential of the system. A second problem is that the strains change the strength of the lateral chemical bonds between the atoms and thus the chemical potential of the crystal. This change should obviously be included in the volume term of the thermodynamic potential.

That is why in this section a more or less qualitative treatment of the problem will be given based on the atomistic approach developed by Kossel, Stranski and Kaischew (see Sec. 1.4). A Kossel crystal is adopted for simplicity. We will confine ourselves to first neighbor interactions and will neglect the Poisson effect. We assume further that the lattice misfit is accommodated completely by the homogeneous elastic strain, misfit dislocations being ruled out. The atomistic approach has in this case one important advantage. It permits one to account for the effect of the strains on the specific surface energies in an implicit way without entering into sophisticated details.

The change of the thermodynamic potential when a nucleus of the new phase is formed in a general form reads

$$\Delta G = -n\Delta\mu + \Phi , \qquad (2.20)$$

where n is the number of building units and Φ is the surface energy. According to the definition of Stranski [1936/7] the latter is given by

$$\Phi = n\varphi_{1/2} - U_n , \qquad (2.21)$$

where U_n is the energy of disintegration of the whole crystal into single atoms. In fact this quantity taken with negative sign, $-U_n$, is the potential (binding) energy of the chemical bonds of the crystal. Equation (2.21) can be easily understood. The first term in the right-hand side gives the energy of the bonds as if all the atoms are in the bulk of the crystal. The second term gives the energy of the bonds between the atoms of the cluster and hence the difference is simply the number of the unsaturated dangling bonds on the cluster "surface" multiplied by the energy required to break a bond. Note that Φ can be expressed in terms of the surface, edge and apex energies in the case of large enough crystals, but, as written above, is applicable for arbitrarily small clusters. It is also very important to note that whereas the first term is only a function of the crystal volume the second is additionally a function of the crystal shape.

As shown in Chap. 1 the equation of Thomson–Gibbs (1.62) in atomistic terms reads

$$\Delta\mu = \varphi_{1/2} - \bar{\varphi}_3 \;, \tag{2.22}$$

where $\bar{\varphi}_3$ (Eq. (1.60)) is the mean work of separation representing the energy of disintegration per atom of a whole uppermost lattice plane of the crystal and must have one and the same value for all crystal faces belonging to the equilibrium shape.

Substituting Eq. (2.22) into Eq. (2.21) gives the work of formation of the critical nucleus in the atomistic approach as

$$\Delta G^* = n^* \bar{\varphi}_3 - U_{n^*} \;. \tag{2.23}$$

We have to account now for the effect of the elastic strain on the energy of the first neighbor bonds. We consider the harmonic approximation of the interatomic potential which is usually represented by a Lennard–Jones or Morse potential. As seen in Fig. 2.5 the work necessary to disrupt a strained bond will be equal to $\psi - \varepsilon$, where ψ is the work to break an unstrained bond and ε is the strain energy of a bond. The latter is given by

$$\varepsilon = \frac{1}{2}\gamma(a - b)^2 \;,$$

where γ is the elastic constant of the first neighbor bond, and a and b are the natural interatomic spacings of the substrate and nucleus crystals.

The nucleus of our model is shown in Fig. 2.6. It can be imagined as consisting of blocks having the shape of square-based prisms (rather than cubes) with thicknesses smaller or larger than the lateral size, thus reflecting the fact that the lateral bonds are strained whereas the normal bonds preserve their length (the Poisson effect is neglected). Then the energy to break a lateral bond will be $\psi - \varepsilon$ and that for a normal one, ψ. The numbers of atoms in the lateral and normal edges are n_e and n'_e, respectively.

The mean works of separation calculated from the upper (subscript u) and side (subscript s) faces read

$$\bar{\varphi}_u = 3\psi - 2\varepsilon - 2\frac{\psi - \varepsilon}{n_e} \tag{2.24}$$

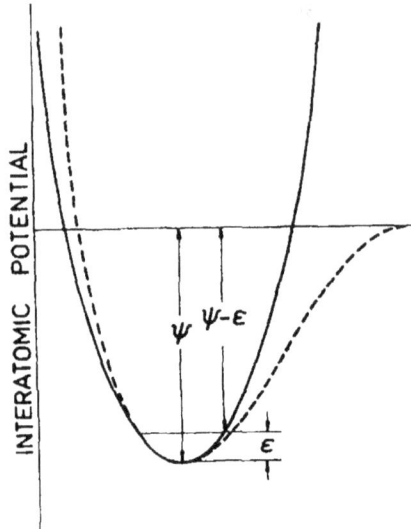

Fig. 2.5. Schematic representation of the change of the interatomic bond strength with
the bond strain. The solid line shows the elastic (Hookean) approximation to a pairwise
interatomic potential given by the dashed line. The work to break an unstrained bond is
ψ. The work to break an elastically strained bond is $\psi - \varepsilon$, where ε is the strain energy
per bond.

and

$$\bar{\varphi}_s = 3\psi - 2\varepsilon - \frac{\psi - \varepsilon}{n_e} - \frac{\psi - \psi'}{n'_e} \,, \tag{2.24'}$$

respectively. The condition $\bar{\varphi}_u = \bar{\varphi}_s$ gives the expression for the equilibrium
shape:

$$\frac{n'_e}{n_e} = \left(1 - \frac{\psi'}{\psi}\right)\left(1 - \frac{\varepsilon}{\psi}\right)^{-1} \,. \tag{2.25}$$

Bearing in mind that the condition for the equilibrium shape of unstrained nucleus is given by (Eq. 1.68)

$$\frac{n'}{n} = 1 - \frac{\psi'}{\psi} \,,$$

it follows that

$$\frac{n'_e}{n_e} = \frac{n'}{n}\left(1 - \frac{\varepsilon}{\psi}\right)^{-1} \,, \tag{2.25'}$$

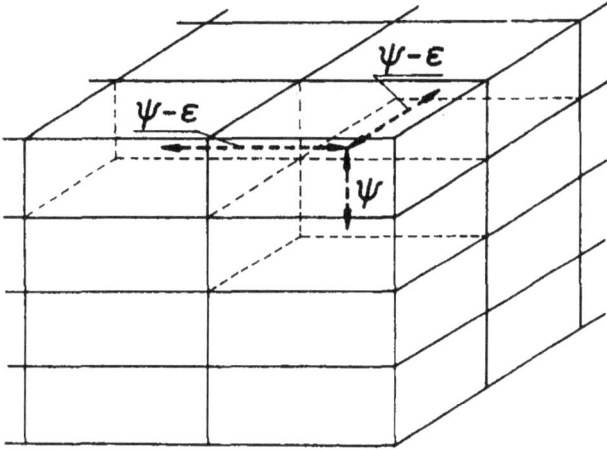

Fig. 2.6. Part of a nucleus of a Kossel crystal formed on a single crystal substrate with quadratic symmetry. The nucleus is homogeneously strained in both lateral directions to fit the substrate. The lateral bonds are equally strained and the work to break them is equal to $\psi - \varepsilon$. The normal bonds remain unchanged if the Poisson effect is neglected. Then the work to break a bond perpendicular to the substrate surface is equal to ψ.

i.e. an elastically strained nucleus consists of a larger number of lattice planes than an unstrained one.

From Eqs. (2.22) and (2.24), recalling that $\varphi_{1/2} = 3\psi$ for a Kossel crystal, for the supersaturation one obtains

$$\Delta\mu = 2\frac{\psi}{n_e^*} + 2\varepsilon\left(1 - \frac{1}{n_e^*}\right) , \qquad (2.26)$$

where n_e^* is the number of atoms in the lateral edge of the strained critical nucleus. Rearranging (2.26) gives for the latter

$$n_e^* = \frac{2(\psi - \varepsilon)}{\Delta\mu - 2\varepsilon} = n^*\left(1 - \frac{\varepsilon}{\psi}\right)\left(1 - n^*\frac{\varepsilon}{\psi}\right)^{-1} , \qquad (2.26')$$

where $n^* = 2\psi/\Delta\mu$ is the number of atoms on the edge of the unstrained critical nucleus ($\varepsilon = 0$) at the same value of the supersaturation. As seen n_e^* is inversely proportional to the difference $\Delta\mu_e = \Delta\mu - 2\varepsilon$, which reflects the increase of the chemical potential of the strained nucleus due to the strain energy per atom, 2ε. An inspection of Eq. (2.26') shows that $n_e^* > n^*$ and the difference increases sharply with the increase of the strain energy per bond, ε. Hence, we have both $n_e'^* > n'^*$ and $n_e^* > n^*$, i.e. the laterally strained nucleus is larger than the unstrained one in both height and width at one and the same supersaturation.

The equilibrium vapor pressure of an unstrained nucleus is $kT \ln(P/P_\infty)$ $= \Delta\mu = 2\psi/n^*$ and then the analogous quantity for strained nucleus reads

$$P_e = P \exp\left(\frac{2\varepsilon}{kT}\right) . \tag{2.27}$$

We have obtained the important result that the equilibrium vapor pressure of a laterally strained small crystal formed on a foreign single crystal substrate is higher than that of the unstrained one due to its increased chemical potential as a consequence of the elastic strain.

Making use of Eqs. (2.22), (2.25) and (2.26′) and counting the bonds between the atoms in the nucleus gives for the work of elastically strained nucleus

$$\Delta G_e^* = n_e^{*2}\psi\left(1 - \frac{\psi'}{\psi}\right) = \frac{4\psi(\psi - \varepsilon)^2}{(\Delta\mu - 2\varepsilon)^2}\left(1 - \frac{\psi'}{\psi}\right) . \tag{2.28}$$

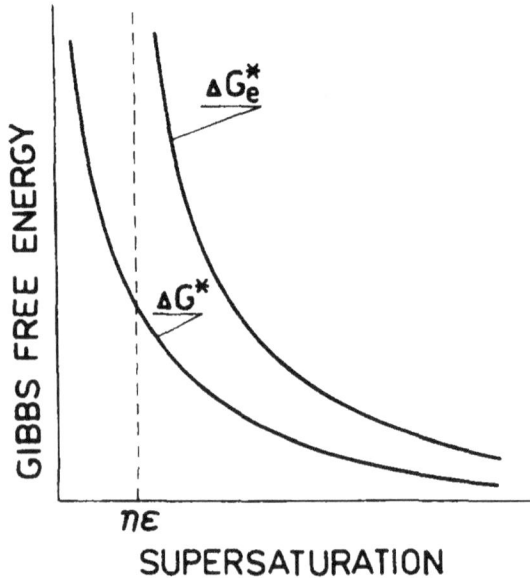

Fig. 2.7. Dependence of the Gibbs free energy ΔG_e^* of formation of an elastically strained 3D nucleus on a foreign single crystal substrate on the supersaturation. A critical supersaturation $\Delta\mu = n\varepsilon$ should be exceeded which is equal to the strain energy per atom. The Gibbs free energy ΔG^* of formation of elastically unstrained 3D nucleus is also given for comparison. As seen, ΔG_e^* is greater than ΔG^* and the formation of elastically strained nuclei requires higher supersaturation.

In other words

$$\Delta G_e^* = \Delta G^* \left(1 - \frac{\varepsilon}{\psi}\right)^2 \left(1 - \frac{2\varepsilon}{\Delta\mu}\right)^{-2} , \qquad (2.29)$$

where ΔG^* is the work of formation of the unstrained nucleus ($\varepsilon = 0$) given by Eq. (2.19).

ΔG_e^* and ΔG^* are plotted against supersaturation in Fig. 2.7. As seen ΔG_e^* is larger than ΔG^* and tends asymptotically to infinity when $\Delta\mu$ tends to 2ε, whereas ΔG^* still has a finite value. It follows that the formation of elastically strained nuclei requires supersaturation which is higher than the strain energy per bond.

2.1.4. *Formation of 2D nuclei*

It was Brandes [1927] who first considered the possibility of formation of 2D nuclei on the surface of a foreign substrate. He found that the Gibbs free energy of formation of such nuclei is precisely equal to one-half of their edge energy:

$$\Delta G_2^* = \frac{1}{2} \sum_n \varkappa_n l_n , \qquad (2.30)$$

where $\varkappa_n$ is the specific edge energy of the nth edge and l_n is its length. The similarity of Eqs. (2.30) and (2.7) is apparent.

Following the same procedure as before one can derive an expression for the work of formation of 2D nuclei. The more general case of nucleation on a foreign substrate will be treated first.

We consider a cluster with square equilibrium shape (in principle the equilibrium shape has to be determined beforehand making use of the Gibbs–Curie–Wulff theorem for the 2D case (Eq. (1.44))). In the simplest case of a Kossel crystal with first neighbor interactions only the equilibrium shape is a square with edge length l formed on the surface of a foreign structureless substrate. The variation of the Gibbs free energy reads

$$\Delta G = -\frac{l^2}{s_c}\Delta\mu + l^2(\sigma + \sigma_i - \sigma_s) + 4l\varkappa , \qquad (2.31)$$

where $n = l^2/s_c$ is the number of atoms in the cluster. The dependence of ΔG on l is similar to that shown in Fig. 2.1. It displays a maximum at a critical edge length

$$l^* = \frac{2\varkappa s_c}{\Delta\mu - s_c(\sigma + \sigma_i - \sigma_s)} . \qquad (2.32)$$

Then the work of formation of a critical 2D nucleus is

$$\Delta G_2^* = \frac{4\varkappa^2 s_c}{\Delta\mu - s_c(\sigma + \sigma_i - \sigma_s)} . \tag{2.33}$$

Substituting $\Delta\mu$ from the Thomson–Gibbs equation (2.32) into (2.33) one obtains $\Delta G_2^* = 2l^*\varkappa$ which is identical to (2.30). The latter can be easily obtained following the procedure applied for the 3D case.

Without going into details and following the same procedure as above one can treat the problem of formation of elastically strained 2D nuclei. For the corresponding work of formation one obtains [Markov *et al.* 1978]

$$\Delta G_{e2}^* = \frac{(\psi - \varepsilon)^2}{(\Delta\mu - 2\varepsilon) - (\psi - \psi')}$$

or

$$\Delta G_{e2}^* = \frac{4\varkappa_e^2 s_c}{(\Delta\mu - 2\varepsilon) - s_c(\sigma + \sigma_i - \sigma_s)} , \tag{2.34}$$

where now the specific edge energy $\varkappa_e = (\psi - \varepsilon)/2a$ accounts for the elastic strain and $s_c = a^2$ is the area occupied by a surface atom.

In the case of nucleation on the surface of the same crystal ($\varepsilon = 0, \sigma_i = 0, \sigma_s = \sigma$ and $\sigma + \sigma_i - \sigma_s = 0$),

$$l^* = \frac{2\varkappa s_c}{\Delta\mu} \tag{2.35}$$

and

$$\Delta G_2^* = \frac{4\varkappa^2 s_c}{\Delta\mu} . \tag{2.36}$$

Equations (2.33), (2.34) and (2.36) are valid for polygonized square nuclei without rounded regions as required by Herring's formula at finite temperatures, i.e. when the steps which appeared as a result of the nucleus formation are straight. If the steps are roughened to some extent below the roughening temperature T_r the equilibrium shape will be more or less circular, and instead of Eqs. (2.35) and (2.36) we have

$$r^* = \frac{\varkappa s_c}{\Delta\mu} \tag{2.37}$$

and

$$\Delta G_2^* = \frac{\pi\varkappa^2 s_c}{\Delta\mu} , \tag{2.38}$$

where r^* is the radius of the critical nucleus. Note that in this case $\varkappa$ will have a value different (smaller) from that in Eq. (2.36) which is valid for straight steps without kinks. For more details concerning the formation of 2D nuclei the reader is referred to the work of Burton, Cabrera and Frank [1951].

2.1.5. Mode of nucleation on a foreign substrate

We discuss in this section the question as to which mode of nucleation, 2D or 3D, is thermodynamically preferred as a function of the supersaturation depending on the difference of the cohesion and adhesion energies $\Delta\sigma = \sigma + \sigma_i - \sigma_s = 2\sigma - \beta = (\psi - \psi')/b^2$ on the one hand, and the strain energy ε on the other.

Let us consider first the case of $\Delta\sigma < 0$, i.e. when the attractive forces exerted by the substrate on the deposit atoms are stronger than the forces between the deposit atoms. As follows from the above considerations 3D nucleation is thermodynamically prohibited ($\Delta G_3^* = 0$) and only 2D nuclei can be formed. Under this condition the quantity in the denominator of Eq. (2.33) is positive and ΔG_2^* has a finite value at $\Delta\mu = 0$ and even at undersaturation $\Delta\mu < 0$. Thus in this case (Fig. 2.8) 2D nuclei of at least the first monolayer can be formed in an undersaturated system, an idea introduced in the theory of the epitaxial crystal growth by Stranski and Krastanov [1938]. Obviously, ΔG_2^* tends to infinity at an undersaturation $-\Delta\mu = s_c\Delta\sigma$ which determines the equilibrium vapor pressure of an adlayer under stronger forces across the interface.

We have practically the same case when $\Delta\sigma = 0$. In fact this condition means that the 2D nuclei are formed on the surface of the same crystal, only in this case σ_i is precisely equal to zero and $\sigma = \sigma_s$. 3D nucleation is again prohibited (the wetting is complete) and 2D nuclei can be formed only in a supersaturated system $\Delta\mu > 0$.

The case of a positive surface energy change, $\Delta\sigma > 0$, offers greater variety — 3D nuclei can be formed as well in addition to 2D nuclei. We consider first the case when the strain energy $\varepsilon = 0$. 3D nuclei can be formed only in a supersaturated system $\Delta\mu > 0$ and ΔG_3^* decreases with the square of the supersaturation (Eq. (2.16)). 2D nuclei, however, can be formed at positive supersaturations higher than a critical one:

$$\Delta\mu_0 = s_c(\sigma + \sigma_i - \sigma_s) , \qquad (2.39)$$

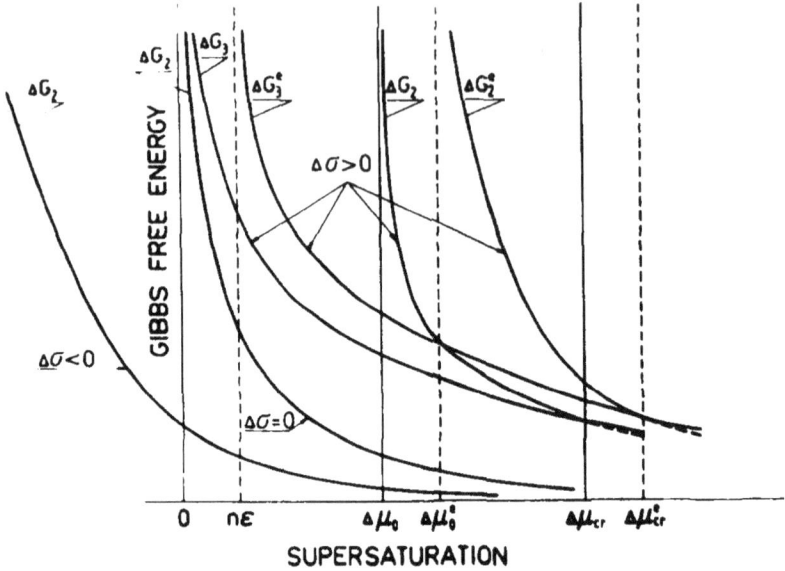

Fig. 2.8. Dependence of the Gibbs free energies of formation of 2D and 3D nuclei on the supersaturation for different values of the surface energy change $\Delta\sigma = \sigma + \sigma_i - \sigma_s = (\psi - \psi')/b^2$. 2D nucleation is only possible at complete wetting $\Delta\sigma < 0$ or $\Delta\sigma = 0$. In the case of $\Delta\sigma < 0$, 2D nucleation can take place even at undersaturation as discussed in Chap. 1, whereas at $\Delta\sigma = 0$, 2D nucleation always requires a supersaturation. In the case of incomplete wetting, $\Delta\sigma > 0$, 2D and 3D nucleation can occur, 3D nucleation being always more probable than 2D nucleation. 2D nucleation occurs at supersaturations higher than some value $\Delta\mu_0 = s_c\Delta\sigma = \psi - \psi'$ determined by the difference in bonding. At a critical supersaturation $\Delta\mu_{cr} = 2\Delta\mu_0$, the 3D nucleus transforms into 2D nucleus (Fig. 2.9) and 3D nucleation is no longer possible. This is the reason why the corresponding curves for ΔG_3 are given by dashed lines at $\Delta\mu > \Delta\mu_{cr}$. In this interval, only 2D nucleation is possible. The critical supersaturations $\Delta\mu_0^\bullet$ and $\Delta\mu_{cr}^\bullet$ for formation of strained nuclei are shifted to greater values by the strain energy per atom $n\varepsilon$ (the straight dashed lines).

which determines the equilibrium vapor pressure (or solubility) of the adlayer.

Beyond this value ΔG_2^* decreases with the supersaturation and at some critical supersaturation,

$$\Delta\mu_{cr} = 2s_c(\sigma + \sigma_i - \sigma_s) = 2\Delta\mu_0 \qquad (2.40)$$

becomes equal to ΔG_3^*. Obviously the condition $\Delta G_2^* = \Delta G_3^*$ means that the height of the 3D nucleus becomes equal to that of one monolayer or, in other words, the 3D nuclei turn into 2D ones (Fig. 2.9). This is easy to understand bearing in mind the assumption made at the beginning of this section that the nuclei retain the equilibrium shape. Therefore when the supersaturation increases the 3D nuclei preserve their height-to-width ratio h/l and as a result they turn into 2D nuclei at $\Delta\mu_{cr}$ [Lacmann 1961; Toschev, Paunov and Kaischew 1968].

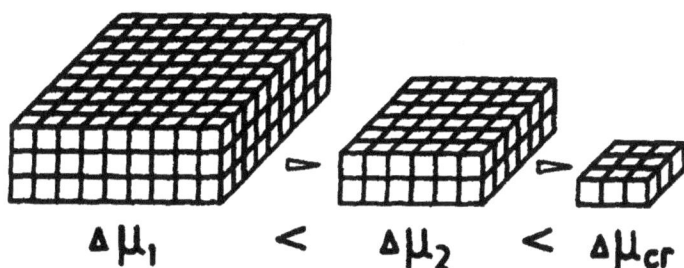

$$\Delta\mu_1 \quad < \quad \Delta\mu_2 \quad < \quad \Delta\mu_{cr}$$

Fig. 2.9. Transformation of a 3D nucleus into a 2D nucleus with increasing supersaturation assuming the equilibrium shape is preserved (after Toschev, Paunov and Kaischew [1968]).

In the same way one can consider the case of elastically strained ($\varepsilon \neq 0$) 2D and 3D nuclei. 3D nuclei can be formed at supersaturations $\Delta\mu$ higher than the strain energy per atom, $n\varepsilon$, where n is the number of lateral bonds per atom in the nucleus ($n = 2$ and $n = 3$ for square and hexagonal meshes of the substrate surface, respectively). 2D nuclei can be formed at supersaturations higher than

$$\Delta\mu_0^e = n\varepsilon + s_c(\sigma + \sigma_i - \sigma_s) \, . \tag{2.41}$$

The critical supersaturation $\Delta\mu_{cr}^e$ at which the 3D nuclei turn into 2D nuclei is shifted by $n\varepsilon$ with respect to $\Delta\mu_{cr}$ and reads

$$\Delta\mu_{cr}^e = n\varepsilon + 2s_c(\sigma + \sigma_i - \sigma_s) = \Delta\mu_{cr} + n\varepsilon \, . \tag{2.42}$$

It follows that at supersaturations in the interval $\Delta\mu_0$ to $\Delta\mu_{cr}$ or, corrrespondingly, $\Delta\mu_0^e$ to $\Delta\mu_{cr}^e$, 3D nucleation is thermodynamically favored although 2D nucleation is in principle also possible. Beyond $\Delta\mu_{cr}$, or corrrespondingly, $\Delta\mu_{cr}^e$, only 2D nucleation is possible.

Rearranging (2.42) gives the following criterion for the mode of nucleation on a foreign substrate [Markov and Kaischew 1976a, 1976b]. 3D nucleation is energetically favored when

$$\sigma_s < \sigma + \sigma_i - \frac{\Delta\mu - n\varepsilon}{2s_c} \ . \tag{2.43}$$

2D nucleation will take place when

$$\sigma_s > \sigma + \sigma_i - \frac{\Delta\mu - n\varepsilon}{2s_c} \ . \tag{2.44}$$

The above criterion can be easily generalized for crystals with other lattices and orientations [Markov and Kaischew 1976a, 1976b].

We can rewrite Eqs. (2.43) and (2.44) in the form

$$\sigma_s \lessgtr \sigma + \sigma_i^* - \frac{\Delta\mu}{2s_c} \ , \tag{2.45}$$

where $\sigma_i^* = \sigma_i + n\varepsilon/2s_c$ is the specific energy of the interface accounting not only for the different interaction energies but also for the homogeneous strains at the interface (for more details see Chap. 4).

One concludes that the mode of nucleation in one and the same system can be varied by changing the supersaturation. Thus at high enough supersaturations 2D nucleation is only possible from the point of view of classical thermodynamics, whereas at lower supersaturations 3D nucleation prevails.

2.2. Rate of Nucleation

In this chapter the classical (capillary) theory of nucleation will be considered first, treating consecutively the cases of homogeneous and heterogeneous nucleation and the formation of 3D and 2D nuclei in the latter case. Then we will treat in some detail the atomistic theory of heterogeneous nucleation which plays an important role in the deposition of thin films at high supersaturations. At the end of this chapter the nonsteady state effects in nucleation as well as the saturation of the nucleus density will be briefly considered. The Ostwald step rule for the case when nuclei of thermodynamically less stable phase are initially formed will be briefly discussed.

2.2.1. *General formulation*

As mentioned at the beginning of the previous section the nuclei of the new phase appear in the bulk of the ambient one as heterophase fluctuations of the density, i.e. the nucleation is a random process. The number of nuclei formed in a fixed interval of time is a random quantity and is subject to statistical laws [Toschev 1973]. The average values, however, can be calculated and are subject to the kinetic theory of nucleation. Thus the aim of this chapter is to calculate the rate of nucleation or, in other words, the average number of nuclei formed per unit time and volume (or unit area of the substrate in the heterogeneous case) of the ambient phase.

We will follow an approach developed by Becker and Döring [1935]. In fact the first treatment of the problem was given by Volmer and Weber [1926]. The latter was further elaborated by Farkas [1927], Stranski and Kaischew [1934] and Frenkel [1939] (for a review see Christian [1981]).

We treat first the rate of formation of liquid droplets in vapors. We consider a vessel with volume V containing supersaturated vapors with pressure P and temperature T. The following simplifying assumptions are adopted:

1. The growing clusters preserve a constant geometrical shape (spherical in the particular case) which coincides with the equilibrium one. As mentioned in the previous chapter it ensures minimal free energy.
2. Clusters consisting of N atoms (N being sufficiently greater than the critical number n^*) are removed from the system and replaced by an equivalent number of single atoms, thus ensuring a constant supersaturation in the system.
3. The nucleation process is considered as a series of consecutive bimolecular reactions (a scheme proposed by Leo Scillard) (see Benson [1960])

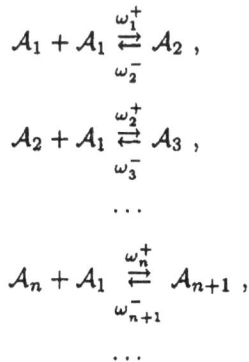

$$\mathcal{A}_1 + \mathcal{A}_1 \underset{\omega_2^-}{\overset{\omega_1^+}{\rightleftarrows}} \mathcal{A}_2 \, ,$$

$$\mathcal{A}_2 + \mathcal{A}_1 \underset{\omega_3^-}{\overset{\omega_2^+}{\rightleftarrows}} \mathcal{A}_3 \, ,$$

$$\cdots$$

$$\mathcal{A}_n + \mathcal{A}_1 \underset{\omega_{n+1}^-}{\overset{\omega_n^+}{\rightleftarrows}} \mathcal{A}_{n+1} \, ,$$

$$\cdots$$

in which the growth and decay of the clusters take place by attachment and detachment of single atoms. Triple and multiple collisions are ruled out as less probable. ω_n^+ and ω_n^- denote the rate constants of the direct and reverse reactions, respectively. Here $\mathcal{A}$ is used as a chemical symbol.

Clusters consisting of n atoms (to be called clusters of class n) are formed by the growth of clusters of class $n-1$ and decay of clusters of class $n+1$ (birth processes), and disappear by the growth and decay into clusters of classes $n+1$ and $n-1$ (death processes), respectively. Then the change of the concentration $Z_n(t)$ of clusters of class n with time is given by

$$\frac{dZ_n(t)}{dt} = \omega_{n-1}^+ Z_{n-1}(t) - \omega_n^- Z_n(t) - \omega_n^+ Z_n(t) + \omega_{n+1}^- Z_{n+1}(t) . \quad (2.46)$$

Introducing the net flux of clusters through the size n,

$$J_n(t) = \omega_{n-1}^+ Z_{n-1}(t) - \omega_n^- Z_n(t) , \quad (2.47)$$

turns Eq. (2.46) into

$$\frac{dZ_n(t)}{dt} = J_n(t) - J_{n+1}(t) . \quad (2.48)$$

In the steady state, $dZ_n(t)/dt = 0$ and

$$J_n(t) = J_{n+1}(t) = J_0 , \quad (2.49)$$

where we denote by J_0 the steady state rate or the frequency of formation of clusters of any class which obviously does not depend on the cluster size n. Hence J_0 is also equal to the rate of formation of clusters with critical size n^*. In other words, in the steady state,

$$J_0 = \omega_1^+ Z_1 - w_2^- Z_2 ,$$
$$J_0 = \omega_2^+ Z_2 - w_3^- Z_3 ,$$
$$\cdots$$
$$J_0 = \omega_n^+ Z_n - w_{n+1}^- Z_{n+1} , \quad (2.49')$$
$$\cdots$$
$$J_0 = \omega_{N-1}^+ Z_{N-1} ,$$

where $Z_N = 0$ as accepted at point 2.

Following Becker and Döring we multiply each equation by a ratio of the rate constants. The first equation is multiplied by $1/\omega_1^+$, the second by $\omega_2^-/\omega_1^+ \omega_2^+$, the nth by $\omega_2^- \omega_3^- \cdots \omega_n^-/\omega_1^+ \omega_2^+ \cdots \omega_n^+$, etc. Then we sum up the equations and obtain after rearrangement (Becker and Döring [1935])

$$J_0 = \frac{Z_1}{\displaystyle\sum_{n=1}^{N-1} \left(\frac{1}{\omega_n^+} \frac{\omega_2^- \omega_3^- \cdots \omega_n^-}{\omega_1^+ \omega_2^+ \cdots \omega_{n-1}^+} \right)} . \quad (2.50)$$

This is the general expression for the steady state rate of nucleation. It is applicable to any case of nucleation (homogeneous or heterogeneous, 3D or 2D, or even 1D) taking the appropriate expressions for the rate constants ω_n^+ and ω_n^-. Moreover, it allows the derivation of equations for the classical as well as the atomistic nucleation rate at small and high supersaturations as limiting cases. We will consider in more detail the classical (capillary) theory of homogeneous nucleation. The corresponding equations for heterogeneous formation of 2D and 3D nuclei will then be written down by analogy without derivation. Finally, the atomistic theory of nucleation will be treated on the basis of Eq. (2.50). 1D nucleation which takes place when smooth single height steps propagate will be considered in Chap. 3.

2.2.2. The equilibrium state

Before going further it is instructive to consider the equilibrium state. In an undersaturated system the equilibrium state gives the equilibrium distribution of the homophase density fluctuations [Frenkel 1955]. In a supersaturated system the equilibrium state can be realized near enough to the phase equilibrium. Obviously the latter will be a metastable equilibrium and will give the equilibrium distribution of the heterophase fluctuations. Far from the line of phase equilibrium ($\mu_\alpha = \mu_\beta$) an equilibrium state cannot be realized. However, we can write an expression even in this case which will serve as a convenient reference.

Under the condition $J_0 = 0$, from (2.47) follows

$$\omega_{n-1}^+ N_{n-1} = \omega_n^- N_n \;, \tag{2.51}$$

where N_n denote now the equilibrium concentrations of clusters of class n in the absence of molecular flux in the system.

Equation (2.51) is known as the equation of detailed balance. It can be rewritten in the form

$$\frac{N_n}{N_{n-1}} = \frac{\omega_{n-1}^+}{\omega_n^-} \;.$$

Multiplying the ratios N_n/N_{n-1} from $n = 2$ to n gives

$$\frac{N_n}{N_1} = \prod_{i=2}^{n} \left(\frac{\omega_{i-1}^+}{\omega_i^-} \right) = \left(\frac{\omega_2^- \omega_3^- \cdots \omega_n^-}{\omega_1^+ \omega_2^+ \cdots \omega_{n-1}^+} \right)^{-1} \;. \tag{2.52}$$

As seen the rate constant ratio on the right-hand side appears in the expressions for both the equilibrium concentration of clusters N_n of class

n and the steady state rate of nucleation (2.50). Obviously, the problem is reduced to finding expressions for the rate constants ω_n^+ and ω_n^-. We will do that for the simplest case of homogeneous formation of liquid nuclei in supersaturated vapors adopting the idea of Gibbs that the nuclei represent small liquid droplets.

The rate constant of the growth reactions, ω_n^+, is given by the number of collisions of atoms from the vapor phase on the surface of the droplets. Then for the rate constant of growth of a cluster of class $n - 1$ to form a cluster of class n we have

$$\omega_{n-1}^+ = \frac{P}{(2\pi m k T)^{1/2}}\Sigma_{n-1} , \qquad (2.53)$$

where $P/(2\pi m k T)^{1/2}$ is the number of collisions per unit area, P being the vapor pressure available in the system, and Σ_{n-1} is the surface area of a cluster of class $n - 1$.

The rate constant of the reverse reaction of decay of a cluster of class n to form a cluster of class $n - 1$ can be evaluated as follows [Volmer 1939]. In equilibrium with the vapor phase the number of atoms leaving the cluster in a fixed interval of time is equal to the number of atoms arriving at its surface. Hence the flux of atoms leaving the cluster is equal to the equilibrium flux of atoms arriving at its surface. On the other hand, condensation takes place when the center of mass of the molecule joining the droplet crosses the sphere of action of the interatomic forces (Fig. 2.10). The radius of the droplet increases just after that event. The evaporation of a molecule is the reverse process. When it leaves the droplet its center of mass should cross the same sphere of action and at that moment the droplet shrinks and its surface area becomes equal to Σ_{n-1}. This area, namely, should be taken into account when considering the rate constant of the reverse reaction and

$$\omega_n^- = \frac{P_n}{(2\pi m k T)^{1/2}}\Sigma_{n-1} , \qquad (2.54)$$

where P_n is the equilibrium vapor pressure of the cluster of class n.

The pressures P and P_n can be expressed through the Thomson–Gibbs equation

$$kT \ln \frac{P}{P_\infty} = \frac{2\sigma v_1}{r^*}$$

and

$$kT \ln \frac{P_n}{P_\infty} = \frac{2\sigma v_1}{r_n} ,$$

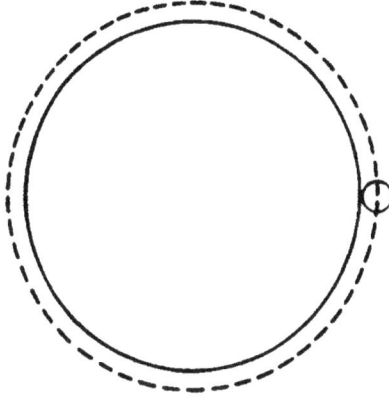

Fig. 2.10. For the determination of the surface area of a liquid droplet containing n molecules upon detachment of a single molecule shown by the small circle. The molecule leaves the droplet when its center of mass crosses the surface of action of the intermolecular forces given by the dashed circle. Precisely at that moment the surface area of the droplet is Σ_{n-1} as given by solid circle (after Volmer [1939]).

where r_n is the radius of a droplet consisting of n molecules. Then

$$\frac{\omega_n^-}{\omega_{n-1}^+} = \frac{P_n}{P} = \exp\left[\frac{2\sigma v_1}{kT}\left(\frac{1}{r_n} - \frac{1}{r^*}\right)\right] \ .$$

We replace then the radii r_n and r^* by the number of atoms through $n v_1 = 4\pi r^3/3$ and for every term in (2.52) (and also in the sum of (2.50)) one obtains

$$\frac{\omega_2^- \omega_3^- \cdots \omega_n^-}{\omega_1^+ \omega_2^+ \cdots \omega_{n-1}^+} = \exp\left[\frac{2\sigma}{kT}\left(\frac{4\pi v_1^2}{3}\right)^{1/3}\sum_1^n\left(\frac{1}{n^{1/3}} - \frac{1}{n^{*1/3}}\right)\right] \ .$$

Assuming $n^* \gg 1$ (the capillary approximation) we replace the sum by an integral and carrying out the integration yields

$$\frac{\omega_2^- \omega_3^- \cdots \omega_n^-}{\omega_1^+ \omega_2^+ \cdots \omega_{n-1}^+} = \exp\left\{\frac{\sigma}{kT}\left(\frac{4\pi v_1^2 n^{*2}}{3}\right)^{1/3}\left[3\left(\frac{n}{n^*}\right)^{2/3} - 2\left(\frac{n}{n^*}\right)\right]\right\} \ .$$

A comparison of the above equation with Eqs. (2.6) and (2.4) shows immediately that the expression in the curly brackets in the right-hand side is simply the function $\Delta G(n)/kT$. Hence

$$\frac{\omega_2^- \omega_3^- \cdots \omega_n^-}{\omega_1^+ \omega_2^+ \cdots \omega_{n-1}^+} = \exp\left(\frac{\Delta G(n)}{kT}\right) \ , \tag{2.55}$$

or

$$\frac{\omega_2^- \omega_3^- \cdots \omega_n^-}{\omega_1^+ \omega_2^+ \cdots \omega_{n-1}^+} = \exp\left\{\frac{\Delta G^*}{kT}\left[3\left(\frac{n}{n^*}\right)^{2/3} - 2\left(\frac{n}{n^*}\right)\right]\right\} , \qquad (2.55')$$

i.e. every term in the sum in the denominator of (2.50) represents one point of the dependence $\exp[\Delta G(n)/kT]$.

Substituting (2.55) into (2.52) gives for the equilibrium concentration of clusters of class n

$$N_n = N_1 \exp\left(-\frac{\Delta G(n)}{kT}\right) . \qquad (2.56)$$

The reader can find a more rigorous derivation of (2.56) in the monograph of Frenkel [1955] (see also Toschev [1973]).

2.2.3. *Steady state nucleation rate*

Replacing the sum in the denominator of Eq. (2.50) by an integral gives

$$\sum_{n=1}^{N-1}\left(\frac{1}{\omega_n^+}\frac{\omega_2^- \omega_3^- \cdots \omega_n^-}{\omega_1^+ \omega_2^+ \cdots \omega_{n-1}^+}\right)$$

$$\cong \int_1^N \frac{1}{\omega_n^+}\exp\left\{\frac{\Delta G^*}{kT}\left[3\left(\frac{n}{n^*}\right)^{2/3} - 2\left(\frac{n}{n^*}\right)\right]\right\} dn .$$

The function in the exponent in the right-hand side displays a maximum at $n = n^*$ (Fig. 2.1) and can be expanded in Taylor series in the vicinity of the maximum:

$$\Delta G(n) = \Delta G^*\left[3\left(\frac{n}{n^*}\right)^{2/3} - 2\left(\frac{n}{n^*}\right)\right] \cong \Delta G^*\left(1 - \frac{1}{3n^{*2}}(n - n^*)^2\right) . \qquad (2.57)$$

Then the sum attains the form

$$\exp\left(\frac{\Delta G^*}{kT}\right)\int_1^N \frac{1}{\omega_n^+}\exp\left(-\frac{\Delta G^*}{kT}\frac{1}{3n^{*2}}(n - n^*)^2\right) dn .$$

In order to carry out the integration we make the following approximations. As shown in Fig. 2.11 the exponent under the integral displays a sharp maximum in the vicinity of n^* and the limits of integration can be extended to $-\infty$ and $+\infty$ without making a significant error. Also, the

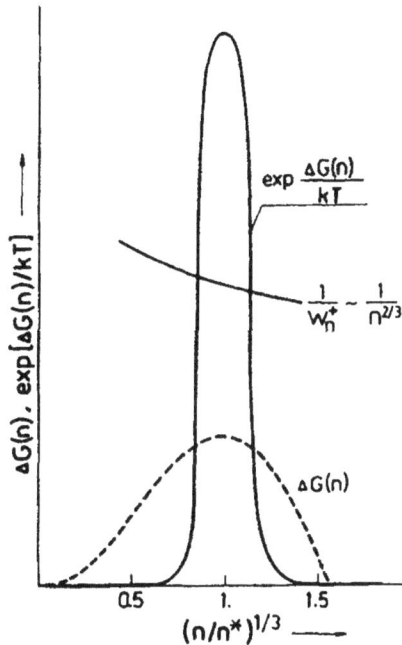

Fig. 2.11. Dependence of the Gibbs free energy change $\Delta G(n)/kT$ (dashed line), $\exp[\Delta G(n)/kT]$ and the reciprocal of the rate constant of the forward reaction on the cluster size. As seen, $\exp[\Delta G(n)/kT]$ displays a sharp maximum and its width is confined to the near vicinity of the critical size. The reciprocal of the rate constant $1/\omega_n^+$ is a weak function of n and can be taken as a constant at the critical size.

rate constant ω_n^+ is not a sensitive function of n and can be replaced by $w_{n^*}^+ \equiv \omega^* = $ const and then taken out before the integral.

Finally, carrying out the integration from minus infinity to plus infinity, one obtains for the steady state nucleation rate

$$J_0 = \omega^* \Gamma Z_1 \exp\left(-\frac{\Delta G^*}{kT}\right) , \qquad (2.58)$$

where Z_1 is the steady state concentration of single molecules in the vapor phase and ω^* is the frequency of the attachment of molecules to the critical nucleus.

The parameter Γ in the expression (2.58) for the steady state nucleation rate

$$\Gamma = \left(\frac{\Delta G^*}{3\pi kT n^{*2}}\right)^{1/2} \qquad (2.59)$$

is known in the literature as a factor of Zeldovich [1943] (see also Frenkel [1955]). In fact it was derived for the first time by Farkas [1927]. Its physical meaning can be revealed if one inspects more closely the definition of the steady state nucleation rate (2.50). Zeldovich [1943] argued that the steady state distribution function Z_n deviates perceptibly from the equilibrium one, N_n, only in the vicinity of the critical size n^*. In other words, the processes taking place in an interval $\Delta n^* = n_r - n_l$ (Fig. 2.12) around the critical size determine the overall rate of nucleation. According to Zeldovich the width of this interval is determined by the condition that the free energy change $\Delta G(n)$ varies by kT around the maximum (at $n < n_l$, $Z_n \cong N_n$ and at $n \geq n_r = N$, $Z_N = 0$), i.e.

$$\Delta G^* - \Delta G(n = n_l, n_r) = kT \ .$$

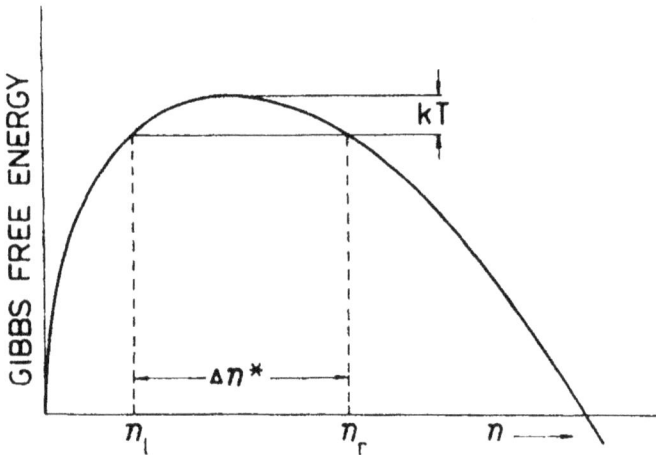

Fig. 2.12. For the determination of the interval $\Delta n^* = n_r - n_l$ around the critical size. According to Zeldovich [1943] the processes which take place in this interval determine the overall rate of the nucleation. The width of the interval is in fact the reciprocal to the nonequilibrium Zeldovich factor Γ.

Bearing in mind (2.57) it follows that the width of the interval Δn^* reads

$$\Delta n^* = 2 \left(\frac{3kTn^{*2}}{\Delta G^*} \right)^{1/2} = \frac{2}{\Gamma\sqrt{\pi}} \ . \tag{2.60}$$

In other words, the factor of Zeldovich is simply a reciprocal of Δn^* and thus accounts for the deviation of the system from the equilibrium state. A more rigorous treatment of this problem is given by Kashchiev [1969].

Bearing in mind Eqs. (2.2) and (2.3) it turns out that Γ is directly proportional to the square of the supersaturation, $\Gamma = \Delta\mu^2/8\pi v_l(\sigma^3 kT)^{1/2}$. In other words, $\Gamma = 0$ at the phase equilibrium ($\Delta\mu = 0$) and increases steeply when deviating from it. In the case of water condensation from the vapor ($\sigma = 70$ erg/cm^2) in the range of P/P_∞ from 2 to 6, n^* varies from 470 to 30 and correspondingly Γ increases from 0.004 to 0.054 [Toschev 1973]. Thus the Zeldovich factor is usually of the order of 1×10^{-2} in the case of homogeneous nucleation.

The steady state rate of nucleation (2.58) can be rewritten in the form ($Z_1 \cong N_1$)

$$J_0 = \omega^* \Gamma N^* , \qquad (2.61)$$

that is, it is a product of the equilibrium concentration of critical nuclei,

$$N^* = N_1 \exp\left(-\frac{\Delta G^*}{kT}\right) , \qquad (2.62)$$

the Zeldovich factor Γ and the frequency of attachment of building units to the critical nucleus, ω^*. It can be easily proved that this is a general expression valid for all possible cases of nucleation and can be used in any particular case.

2.2.4. Nucleation of liquids from vapors

In this particular case the surface of the nucleus is given by $4\pi r^{*2}$ and

$$\omega^* = P(2\pi m kT)^{-1/2} 4\pi r^{*2} . \qquad (2.63)$$

Then

$$J_0 = \frac{P}{\sqrt{2\pi m kT}} 4\pi r^{*2} \frac{\Delta\mu^2}{8\pi v_l \sigma \sqrt{\sigma kT}} \frac{P}{kT} \exp\left(-\frac{16\pi\sigma^3 v_l^2}{3kT\Delta\mu^2}\right) , \qquad (2.64)$$

where $P/kT = N_1$ assuming the vapor phase behaves as an ideal gas. Recalling the Thomson–Gibbs equation (2.2) for the critical radius, (2.64) turns into

$$J_0 = \frac{P^2}{(kT)^2} \frac{2\sigma^{1/2} v_l}{\sqrt{2\pi m}} \exp\left(-\frac{16\pi\sigma^3 v_l^2}{3(kT)^3[\ln(P/P_\infty)]^2}\right) . \qquad (2.65)$$

A closer inspection of Eq. (2.65) shows that the pre-exponent $K_1 = \omega^* \Gamma Z_1$ is not very sensitive to the supersaturation in comparison with $\exp(-\Delta G^*/kT) = \exp(-K_2/\Delta\mu^2)$. Therefore one can accept that K_1 is approximately a constant, i.e.

$$J_0 = K_1 \exp\left(-\frac{K_2}{\Delta\mu^2}\right) , \tag{2.66}$$

where $K_2 = 16\pi\sigma^3 v_l^2/3kT$. With typical values of the quantities involved one finds that K_1 is of order of 1×10^{25} cm^{-3}sec^{-1}.

Equation (2.66) is demonstrated in Fig. 2.13. As seen there exists a critical supersaturation $\Delta\mu_c$ below which the nucleation rate is practically equal to zero and increases steeply beyond it. The critical supersaturation can be determined by the condition $J_0 = 1$ cm^{-3}sec^{-1}. After taking the logarithm of J_0 in (2.66) we find

$$\Delta\mu_c = \left(\frac{K_2}{\ln K_1}\right)^{1/2} . \tag{2.67}$$

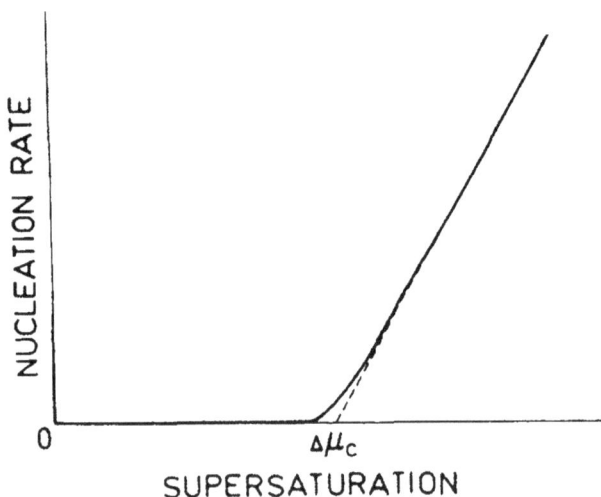

Fig. 2.13. Plot of the nucleation rate versus the supersaturation. The nucleation rate is practically equal to zero up to a critical supersaturation $\Delta\mu_c$. Beyond this value the rate of nucleation increases sharply by many orders of magnitude. This is the reason why the nucleation rate can be measured in a very narrow interval of supersaturations.

In the case of nucleation in water vapor at a temperature $T = 275$ K, $K_1 \cong 1 \times 10^{25}$ cm^{-3}sec^{-1}, $\sigma = 75.2$ erg/cm^2, $v_l = 3 \times 10^{-23}$ cm^{-3}, $\Delta\mu_c = 5.42 \times 10^{-14}$ erg or $P_c/P_\infty = 4.16$, which is in excellent agreement with the value 4.21 found experimentally by Volmer and Flood [Volmer 1939].

The existence of critical supersaturation leads to the conclusion that condensation of vapors will be experimentally observed only at $P > P_c$. In the opposite case the vapors will be in a metastable state, i.e. due to kinetic reasons no condensation will take place for the time of the experiment. Thus we identify the critical supersaturation with the *limit of metastability* of the ambient phase. Recalling the phase diagram (Fig. 1.1) of a one-component system the existence of such a limit of metastability means that observable phase transitions will not occur when the lines of the phase equilibrium (determined by the equality of the chemical potentials) are crossed. For this purpose the limits of metastability should be crossed and they lie to the left of the phase equilibrium lines. It should be pointed out that the metastability limits are very sensitive to the values of the specific surface energies of the corresponding phase boundaries. In general, the metastability limits will be smaller for liquid–crystal transition than for vapor–liquid transition due to the smaller surface energy between the condensed phases. The presence of impurity particles, ions or foreign substrates will reduce the metastability limit with the square root of the wetting function. Besides, the presence of surfactants (substances which once adsorbed at the phase boundaries change drastically the surface energies) will also reduce the values of the specific surface energies and in turn the metastability limit.

2.2.5. Statistical contributions

Later theoretical studies showed, however, that the above agreement between theory and experiment is apparent. Lothe and Pound [1962] (see also Dunning [1969]) noted that the Gibbs free energy of formation of liquid nuclei as given by Eq. (2.1) is valid for the state of rest. As discussed by Christian [1981] it gives in fact the free energy of formation of a liquid droplet confined in a liquid rather than a vapor. When formed in a vapor phase the droplet must acquire the gas-like translational and rotational degrees of freedom and must lose six internal or liquid-like degrees of freedom. Then instead of (2.1) one should write

$$\Delta G(r) = -\frac{4}{3}\frac{\pi r^3}{v_l}\Delta\mu + 4\pi r^2\sigma + \Delta G_{\mathrm{tr}} + \Delta G_{\mathrm{rot}} + \Delta G_{\mathrm{rep}} ,$$

where ΔG_{tr} and ΔG_{rot} must be positive and ΔG_{rep} negative.

The corresponding expressions are [Lothe and Pound 1962] respectively

$$\Delta G_{tr} = -kT \ln \left(\frac{(2\pi n^* mkT)^{3/2} kT}{Ph^3} \right) ,$$

$$\Delta G_{rot} = -kT \ln \left(\frac{(2kT)^{3/2} (\pi I^3)^{1/2}}{\hbar^3} \right)$$

and

$$\Delta G_{rep} = \frac{1}{2} kT \ln(2\pi n^*) + Ts ,$$

where $I \cong 2n^* mr^{*2}$ is the moment of inertia of the nucleus, s is the entropy of the liquid, h is Plank's constant and $\hbar = h/2\pi$. Then for water at $T = 300$ K, $n^* = 100$, $I = 8.6 \times 10^{-36}$ g cm^2, $P = 0.075$ atm, $m = 3 \times 10^{-23}$ g and $s = 70$ J K^{-1}mole^{-1}, $\Delta G_{tr} = -24.4$ kT, $\Delta G_{rot} = -20.6$ kT and $\Delta G_s = 11.5$ kT. As these terms for the free energy depend very weakly on the nucleus size they contribute primarily to the pre-exponent with a total of 1×10^{17}. Then for the critical supersaturation one obtains $P_c/P_\infty = 3.09$, in marked disagreement with the experimentally found value of 4.21. The interested reader can find more information in Feder *et al.* [1966], Lothe and Pound [1969], Reiss [1977], Nishioka and Pound [1977], Kikuchi [1977] and Christian [1981].

2.2.6. *Nucleation from solutions and melts*

In the first case the flux of building units to the critical nuclei is determined by diffusion in the bulk of the solution. On the other hand, the molecules of the solute should break the bonds with the molecules of the solvent before being attached to the nucleus. In other words, an energy barrier for desolvation should be overcome. The frequency ω^* of attachment of the molecules to the critical nucleus will be proportional to the concentration of the solute C and thus instead of (2.63) we have

$$\omega^* = 4\pi r^{*2} C\nu\lambda \exp(-\Delta U/kT) , \qquad (2.68)$$

where ν is a frequency factor, ΔU is the energy of desolvation and λ is the mean free path of particles in the liquid which is approximately equal to the atomic diameter. Then for the nucleation rate one obtains [Walton A G 1969]

$$J_0 = 4\pi r^{*2} C^2 \nu\lambda \exp(-\Delta U/kT)\Gamma \exp(-\Delta G^*/kT) , \qquad (2.69)$$

in which the concentration C is expressed in the number of molecules per unit volume and the supersaturation is expressed by (1.10).

The main problem in the estimation of the pre-exponent is the lack of knowledge concerning the value of ΔU. We may assume that it is of the order of the molecular interactions, i.e. 10–20 kcal/mole. Measurements of the critical supersaturation in the nucleation of a series of salts in aqueous solutions [Nielsen 1967; Walton 1967] gave reasonable values for the specific surface energies of the crystal–solution interface ($\cong$ 100 erg/cm^2) and for the pre-exponential factors ($\approx 10^{24}$–10^{25} cm^{-3}sec^{-1}). The values of ΔU calculated from these data vary from 7 kcal/mole for BaSO$_4$ to 14.5 kcal/mole for PbSO$_4$. Laudise determined the value 20 kcal/mole for quartz in aqueous solutions of NaOH from measurements of the rate of hydrothermal growth of the (0001) face in the temperature interval 570–660 K [Laudise 1959; see also Laudise 1970].

Solidification of liquids does not differ too much from crystallization in solutions. In this case the activation energy barrier ΔU originates from rearrangement of the molecules in the liquids when crossing the crystal–melt boundary to occupy precise positions in the crystal lattice, i.e. from the replacement of the long range disorder in the liquid by long range order in the crystal. That is the reason why ΔU is usually identified with the activation energy for viscous flow. The latter varies from 1 to 6 kcal/mole for metals [Grosse 1963], 10 kcal/mole for organic melts to 50–150 kcal/mole in glass forming melts (SiO$_2$, GeO$_2$) [Mackenzie 1960; Gutzow 1975; Oqui 1990]. On the other hand, there are no transport difficulties in melts and the concentration C in (2.68) and (2.69) should be replaced by the number of atoms in a unit volume $1/v$. Then

$$\omega^* = 4\pi r^{*2} \frac{1}{v_1} \nu \lambda \exp(-\Delta U/kT) \tag{2.70}$$

and

$$J_0 = 4\pi r^{*2} \frac{\nu \lambda}{v_1^2} \exp(-\Delta U/kT)\Gamma \exp(-\Delta G^*/kT) , \tag{2.71}$$

where the supersaturation is given by (1.12)

$$\Delta \mu = \Delta h_{\mathrm{m}} \frac{\Delta T}{T_{\mathrm{m}}} = \Delta s_{\mathrm{m}} \Delta T .$$

Assuming spherical symmetry and isotropic interfacial tension,

$$\Delta G^* = \frac{16\pi\sigma^3 v_c^2 T_m^2}{3\Delta h_m^2 (\Delta T)^2} \ , \tag{2.72}$$

$$\Gamma = \frac{(\Delta h_m \Delta T)^2}{8\pi v_c (\sigma^3 kT)^{1/2} T_m^2} \tag{2.73}$$

and

$$r^* = \frac{2\sigma v_c T_m}{\Delta h_m \Delta T} \tag{2.74}$$

for the pre-exponent, one obtains

$$K_1 = \frac{2\sigma^{1/2} v_c}{(kT)^{1/2}} \frac{\nu\lambda}{v_l^2} \exp(-\Delta U/kT) \ . \tag{2.75}$$

Thus for homogeneous nucleation in metal melts, say Ag [Turnbull and Sech 1950] with $1/v_l \cong 5 \times 10^{22}$ ($v_c \cong v_l$), $T \cong 1000$ K, $\sigma \cong 150$ erg/cm^2, $\nu \cong 2 \times 10^{13}$ sec^{-1}, $\lambda \cong 3 \times 10^{-8}$ cm and $\Delta U/kT \cong 3$ the typical value of 1×10^{35} for K_1 is obtained.

Considering Eq. (2.71) more closely shows that the nucleation rate depends not only on the undercooling $\Delta T = T_m - T$ but also on the absolute value of the temperature. Hence the temperature dependence of the nucleation rate should display a maximum as decrease of the temperature leads to an increase of undercooling. It is easy to find by differentiation of (2.71) that $T_{max} > T_m/3$ ($T_{max} = T_m/3$ when $\Delta G^* \gg \Delta U$). The physical reason for this behavior of the nucleation rate in melts, which is uncharacteristic for nucleation in vapors, consists in the competition between the inhibition of the transport processes in the melt (higher viscosity) and increasing the thermodynamic driving force for nucleation to occur. At the lower temperature side of the maximum the viscosity becomes so large that the melt glassifies before crystallization takes place. The above behavior has been established experimentally for the first time by Tamman [1933] in the case of glycerine and piperine. Typical nucleation rate versus temperature plots are shown in Fig. 2.14 for lithium disilicate melts in the interval 425–527°C [James 1974].

The main difficulty in the experimental verification of nucleation theory in melts lies in the purification of the melts to avoid heterogeneous nucleation on impurity particles. Turnbull and Sech [1950] (see also Hollomon and Turnbull [1951]) measured the critical undercooling for homogeneous nucleation to take place applying the following method. The bulk melt sample was dispersed into small droplets in an inert matrix to outnumber the impurity particles. Then the maximum undercooling measured is taken

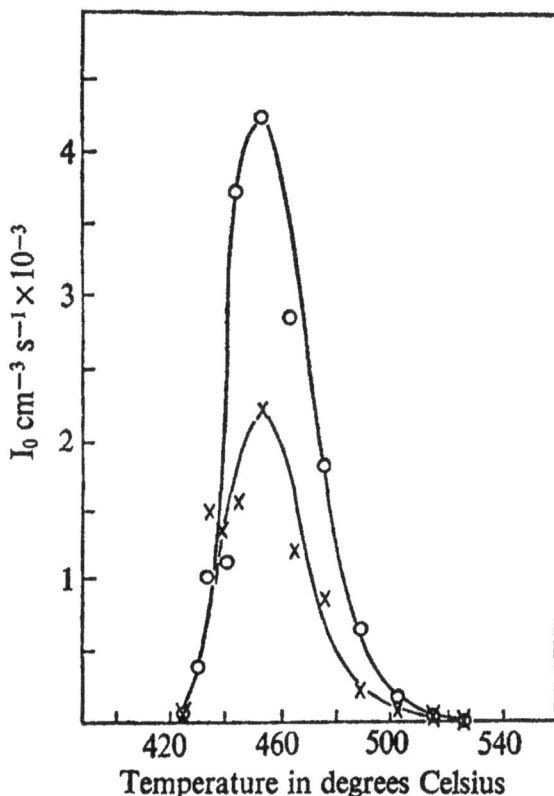

Fig. 2.14. Experimentally measured rate of nucleation in lithium disilicate melt versus temperature. The high temperature branch is determined by the activation energy for nucleus formation. The low temperature branch is determined by transport processes in the melt. (P. F. James, *Phys. Chem. Glasses* **15**, 95 (1974). By permission of Society of Glass Technology and courtesy of P. F. James.)

as the one for the homogeneous nucleation. Their data for a series of 17 metals are compiled in Table 2.1 [Strickland-Constable 1968; Toschev 1973]. An updated table to include the later results of Skripov *et al.* [1970], Powell and Hogan [1968] and others is given by Chernov [1984]. Turnbull [1950] correlated the experimental results with (2.71) together with (2.72) and (2.73) and found that the maximum relative undercoolings $\Delta T_{max}/T_m$ are nearly constant for all metals investigated ($\Delta T_{max}/T_m \cong 0.183$). The same behavior shows the ratio of the gramatomic surface energy σ_m of

Table 2.1. Homogenous nucleation of metals after Turnbull and Cech [1950]. ΔT_{max} is the maximum supercooling, T_m is the melting point, $\Delta T_{max}/T_m$ is the maximum supercooling relative to the melting point, σ is the specific surface energy at the solid-liquid boundary, σ_m is the molar surface energy and Δh_m is the molar enthalpy of melting.

Metal	ΔT_{max}	$\dfrac{\Delta T_{max}}{T_m}$	σ erg cm^{-2}	σ_m cal g atom^{-1}	$\dfrac{\sigma_m}{\Delta h_m}$
Mercury	58	0.247	24.4	296	0.530
Gallium	76	0.250	55.9	581	0.436
Tin	105	0.208	54.5	720	0.418
Bismuth	90	0.166	54.4	825	0.330
Lead	80	0.133	33.3	479	0.386
Antimony	135	0.150	101	1430	0.302
Aluminium	130	0.140	93	932	0.364
Germanium	227	0.184	181	2120	0.348
Silver	227	0.184	126	1240	0.457
Gold	230	0.172	132	1320	0.436
Copper	236	0.174	177	1360	0.439
Manganese	308	0.206	206	1660	0.480
Nickel	319	0.185	255	1860	0.444
Cobalt	330	0.187	234	1800	0.490
Iron	295	0.164	204	1580	0.445
Palladium	332	0.182	209	1850	0.450
Platium	370	0.181	240	2140	0.455

the crystal–melt interface and the molar enthalpy of melting Δh_m. The gramatomic surface energy is defined as $\sigma_m = \sigma N_A v_c^{2/3}$ where N_A is Avogadro's number $(6.023 \times 10^{23} \ \text{mol}^{-1})$ and v_c is the volume of one building unit in the crystal. It was found that for most of the metals studied the ratio $\sigma_m/\Delta h_m \cong 0.5$. Figure 2.15 demonstrates the plot of σ_m vs Δh_m. As seen most of the metals with the exception of Ge, Bi and Sb follow the straight line with a slope of 0.46. The above result becomes immediately clear bearing in mind that the enthalpy of melting is connected with the separation work from the half-crystal position, i.e. it is proportional to $Z/2$ times the energy per bond. On the other hand, $v_c^{2/3}$ represents the area occupied by such a unit. Then σ_m is proportional to the work to separate two nearest neighbor atoms [Stranski, Kaischew and Krastanov 1933].

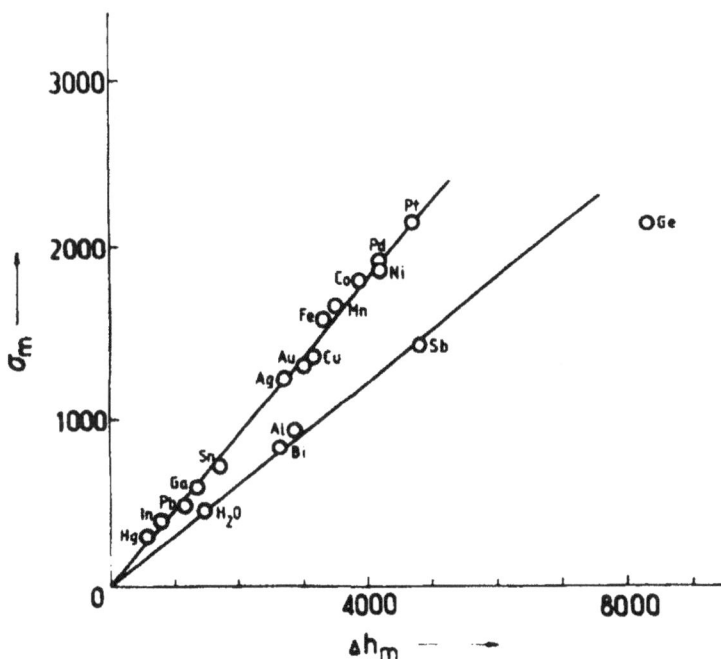

Fig. 2.15. Dependence of the molar surface energy $\sigma_m = \sigma N_A v_c^{2/3}$ on the molar enthalpy of fusion Δh_m for a series of metals. As seen, for most of the metals the ratio $\sigma_m/\Delta h_m = $ const $\cong 0.5$ (after Turnbull [1950]).

Analogous results have also been obtained for nucleation for melts of alkali halides [Buckle and Ubbelohde 1960, 1961], organic melts [Thomas and Staveley 1952; Nordwall and Staveley 1954] and polymers [Koutsky 1966; Cormia, Price and Turnbull 1962]. They are reviewed by Jackson [1966] and critically analysed by Walton A G [1969].

2.2.7. Rate of heterogeneous nucleation

As mentioned in the previous section one of the main difficulties in the experimental verification of the theory of the homogeneous nucleation is the preferred nucleation on the walls of the reaction vessels, foreign particles, ions, etc. The rate of heterogeneous nucleation should obviously be much greater than that of the homogeneous one under the same conditions due to the wetting from the substrate. At the same time when depositing a substance on a foreign substrate the latter is never homogeneous in the sense that there are always some defect sites with higher chemical potentials such as emerging points of screw dislocations, embedded foreign atoms, etc., which are more active in comparison with the remaining part with respect to crystal nucleation. On the other hand, in the case of vapor condensation the frequency of attachment of adsorbed atoms through diffusion of the latter on the substrate surface to the periphery of the growing clusters is usually much greater in comparison with volume diffusion in the vapor phase [Pound, Simnad and Yang 1954; Hirth and Pound 1963]. We have the opposite case when deposition takes place from solutions and melts.

We treat first the case of vapor condensation bearing in mind the general expression (2.58) for the steady state nucleation rate. Atoms arrive from the vapor phase on the substrate surface and after a thermal accommodation period, which is of the order of magnitude of several atomic vibrations [Hirth and Pound 1963], begin to migrate on the surface. Then they collide with each other to produce clusters of different sizes thus giving rise to critical nuclei. The concentration of atoms, Z_1, is now identified with the adatom concentration n_s. The latter is determined by the adsorption-desorption equilibrium and is equal to the product of the adsorption flux

$$R = \frac{P}{(2\pi mkT)^{1/2}} \tag{2.76}$$

and the mean residence time

$$\tau_s = \frac{1}{\nu_\perp} \exp\left(\frac{E_{des}}{kT}\right) , \tag{2.77}$$

which lapses before re-evaporation takes place. E_{des} denotes the activation energy for desorption and $\nu_\perp$ the vibrational frequency of the adatoms in a direction normal to the surface plane. Then

$$n_s = \frac{P}{(2\pi mkT)^{1/2}} \frac{1}{\nu_\perp} \exp\left(\frac{E_{des}}{kT}\right) . \tag{2.78}$$

The flux of adatoms towards the critical nuclei along the substrate surface is

$$j_s = D_s \, \mathrm{grad}\, n_s \cong D_s \frac{n_s}{a} , \tag{2.79}$$

where

$$D_s = a^2 \nu_= \exp\left(-\frac{E_{sd}}{kT}\right) \tag{2.80}$$

is the surface diffusion coefficient, E_{sd} is the activation energy for surface diffusion, $\nu_=$ is the vibrational frequency of the adatoms in a direction parallel to the surface plane and a is the length of a diffusion jump.

Bearing in mind that the periphery of the critical nucleus, assuming a semispherical shape, is $2\pi r^* \sin\theta$ and $\nu_\perp \cong \nu_= = \nu$ for the frequency of attachment of atoms to the latter, one obtains

$$\omega^* = 2\pi r^* \sin\theta \frac{P}{(2\pi mkT)^{1/2}} a \exp\left(\frac{E_{des} - E_{sd}}{kT}\right) . \qquad (2.81)$$

Finally, for the steady state nucleation rate one obtains

$$J_0 = 2\pi r^* \sin\theta \frac{R^2 a}{\nu} \Gamma \exp\left(\frac{2E_{des} - E_{sd}}{kT}\right) \exp\left(-\frac{\Delta G^*}{kT}\right) . \qquad (2.82)$$

This equation is valid for nucleation of liquid droplets. The Zeldovich factor $\Gamma = \Delta\mu^2/8\pi v_l[\sigma^3 \phi(\theta)kT]^{1/2}$ now includes the wetting function $\phi(\theta)$. In the case of crystal nucleation the periphery of the nucleus $4l^*$ should be inserted instead of $2\pi r^* \sin\theta$ and for the ΔG^* the corresponding expression (2.16) must be taken.

In the case of heterogeneous nucleation there is a statistical contribution to the work of nucleus formation, which is independent of the nucleus size [Lothe and Pound 1962; see also Sigsbee 1969], and which accounts for the distribution of the clusters and the single adatoms among the adsorption sites of density N_0 ($\cong 1 \times 10^{15}$ cm^{-2}). Assuming the density of clusters is negligible compared with the adatom concentration,

$$\Delta G_{conf} \cong -kT \ln\left(\frac{N_0}{n_s}\right) . \qquad (2.83)$$

As a result the adatom concentration n_s is replaced by the density of the adsorption sites N_0 and the following expression for the nucleation rate is obtained:

$$J_0 = 2\pi r^* \sin\theta Ra\Gamma N_0 \exp\left(\frac{E_{des} - E_{sd}}{kT}\right) \exp\left(-\frac{\Delta G^*}{kT}\right) . \qquad (2.84)$$

In the case of nucleation in condensed phases (solutions or melts) expressions for the steady state rate are easy to obtain taking the frequency of attachment of atoms ω^* as that given by Eq. (2.62) or (2.64) and replacing the surface area of the critical nucleus $4\pi r^{*2}$ by the area of the segment $4\pi r^{*2}(1 - \cos\theta)$.

2.2.8. *Rate of 2D nucleation*

The 2D nucleation kinetics can be treated in the same way as the 3D one. Following the procedure outlined in Chap. 2.1 we find

$$\frac{\omega_2^- \omega_3^- \cdots \omega_n^-}{\omega_1^+ \omega_2^+ \cdots \omega_{n-1}^+} = \exp\left\{ \frac{\Delta G_2^*}{kT} \left[2\left(\frac{n}{n^*}\right)^{1/2} - \frac{n}{n^*} \right] \right\} .$$

After expanding the function in the exponent as a Taylor series up to the parabolic term and replacing the sum in (2.50) by an integral with limits of integration extended to $-\infty$ and $+\infty$ we obtain

$$\sum_{n=1}^{N-1} \left(\frac{1}{\omega_n^+} \frac{\omega_2^- \omega_3^- \cdots \omega_n^-}{\omega_1^+ \omega_2^+ \cdots \omega_{n-1}^+} \right) \cong \frac{1}{\omega^*} \left(\frac{4\pi kT n^{*2}}{\Delta G_2^*} \right)^{1/2} \exp\left(\frac{\Delta G_2^*}{kT} \right) .$$

Then the expression (2.61) results, in which the Zeldovich factor is given by

$$\Gamma = \left(\frac{\Delta G_2^*}{4\pi kT n^{*2}} \right)^{1/2} . \tag{2.85}$$

2.2.8.1. *Rate of 2D nucleation from vapors*

The frequency of attachment of single atoms to the periphery of the critical nucleus to produce a cluster with a supercritical size is given by

$$\omega^* = 4l^* \frac{P}{(2\pi mkT)^{1/2}} a \exp\left(\frac{E_{des} - E_{sd}}{kT} \right) ,$$

and for the nucleation rate one obtains

$$J_0(2D) = 4l^* Ra\Gamma N_0 \exp\left(\frac{E_{des} - E_{sd}}{kT} \right) \exp\left(-\frac{\Delta G_2^*}{kT} \right) , \tag{2.86}$$

where Γ, l^* and ΔG_2^* are given by (2.85), (2.32) and (2.33), respectively. The same expression is valid for the case of 2D nucleation on the same substrate ($\Delta\sigma = 0$) by taking the expressions (2.35) and (2.36) for l^* and ΔG_2^*, respectively. Also the values for the activation energies for surface self-diffusion and desorption from the same substrate should be taken.

2.2.8.2. *Rate of 2D nucleation from solutions*

When considering the solution and melt growth of crystals through the formation and lateral propagation of 2D nuclei we will need expressions for the rate of 2D nucleation. In the first case the flux of arrival of molecules of the solute $(\mathrm{cm}^{-2}\mathrm{sec}^{-1})$ is given by $a\nu C\exp(-\Delta U/kT)$ (Eq. (2.68)) and the frequency of arrival of molecules per molecular site will be

$$j_+ = a^3\nu C\exp\left(-\frac{\Delta U}{kT}\right) = \nu Cv_c\exp\left(-\frac{\Delta U}{kT}\right) . \qquad (2.87)$$

The number of sites available for attachment of molecules along the perimeter of the 2D nucleus is $2\pi r^*/a$ and

$$\omega^* = 2\pi\frac{r^*}{a}Cv_c\nu\exp\left(-\frac{\Delta U}{kT}\right) . \qquad (2.88)$$

Bearing in mind (2.37), (2.38), (2.61), (2.85) and (2.88), the rate of 2D nucleation reads

$$J_0(2D) = \nu Cv_c\left(\frac{\Delta\mu}{kT}\right)^{1/2}\exp\left(-\frac{\Delta U}{kT}\right)N_0\exp\left(-\frac{\pi\varkappa^2 a^2}{kT\Delta\mu}\right) , \qquad (2.89)$$

where $\Delta\mu$ is given by Eq. (1.10). For typical values of the quantities involved, $\nu \cong 1\times 10^{13}\ \mathrm{sec}^{-1}$, $Cv_c \cong 0.1$ (10% concentration of the solute), $\Delta\mu/kT \cong 0.04$, $\Delta U = 10$ kcal mole^{-1}, $N_0 = 1\times 10^{15}\ \mathrm{cm}^{-2}$ and $T = 300$ K, the pre-exponential factor has a value $K_1 \cong 1\times 10^{19}\ \mathrm{cm}^{-2}\mathrm{sec}^{-1}$.

2.2.8.3. Rate of 2D nucleation in melts

In order to calculate the rate of 2D nucleation in melts our first task is to find an expression for the frequency of attachment of building units to the critical nucleus. The flux of atoms which cross the phase boundary to be incorporated into the crystal lattice is given by

$$j_+ = k_+\exp\left(-\frac{\Delta U}{kT}\right) . \qquad (2.90)$$

The corresponding reverse flux is given by

$$j_- = k_-\exp\left(-\frac{\Delta h_m + \Delta U}{kT}\right) , \qquad (2.91)$$

where k_+ and k_- are rate constants.

At phase equilibrium $T = T_m$ both fluxes are equal and

$$k_+ = k_-\exp\left(-\frac{\Delta h_m}{kT_m}\right) = k_-\exp\left(-\frac{\Delta s_m}{k}\right) . \qquad (2.92)$$

The reverse rate constant can be identified with the vibration frequency of the surface atoms, $k_- = \nu$, and

$$j_+ = \nu \exp\left(-\frac{\Delta s_m}{k}\right) \exp\left(-\frac{\Delta U}{kT}\right) , \qquad (2.93)$$

$$j_- = \nu \exp\left(-\frac{\Delta h_m + \Delta U}{kT}\right) . \qquad (2.94)$$

Making use of (2.93) gives

$$\omega^* = 2\pi \frac{r^*}{a} \nu \exp\left(-\frac{\Delta s_m}{k}\right) \exp\left(-\frac{\Delta U}{kT}\right) \qquad (2.95)$$

and

$$J_0 = \nu N_0 \left(\frac{\Delta s_m \Delta T}{kT}\right)^{1/2}$$

$$\times \exp\left(-\frac{\Delta s_m}{k}\right) \exp\left(-\frac{\Delta U}{kT}\right) \exp\left(-\frac{\pi \varkappa^2 a^2}{\Delta s_m \Delta T kT}\right) . \qquad (2.96)$$

An estimate in the case of formation of 2D nuclei on the (111) face of Si growing at a temperature which is 1 K under the melting point $T_m = 1685$ K, with $\nu = 3 \times 10^{13}$ sec^{-1}, $\Delta s_m/k = 3.6$, $\Delta U/kT = 3$ and $N_0 \cong 1 \times 10^{15}$ cm^{-2}, gives for the pre-exponent the value $K_1 = 2 \times 10^{22}$ cm^{-2}sec^{-1}.

2.2.9. Atomistic theory of nucleation

Experimental investigations of the heterogeneous nucleation showed that the number of atoms constituting the critical nucleus is very small [Robinson and Robins 1970, 1974; Paunov and Harsdorff 1974; Toschev and Markov 1969]. It does not exceed several atoms and in some particular cases of nucleation on active sites this number is equal to zero. This means physically that the adatom is so strongly bound to the active site that the combination active site–atom is a stable configuration. It is thus obvious that the quantities used by the phenomenological thermodynamics such as specific surface energies, equilibrium shape, even state of aggregation (we cannot say whether a cluster of 4 to 5 atoms is solid or liquid as we do not know the long range order) cannot be defined. That is why an atomistic approach which excludes the use of such quantities has been developed [Walton D 1962, 1969].

In order to understand the atomistic approach we should establish the limits of validity of the classical theory of nucleation. For this aim we should consider the equilibrium of small particles of the new phase or, in other words, we should go back to the equation of Thomson–Gibbs (1.19) represented graphically in Fig. 1.4.

We replace the radius of the particle in the equation of Thomson–Gibbs by the number of atoms in it and obtain

$$\frac{P_n}{P_\infty} = \exp\left(\frac{2\sigma b v_l^{2/3}}{n^{1/3}}\right) \ ,$$

where P_n is the equilibrium vapor pressure of a cluster consisting of n atoms or molecules and b is a geometrical factor equal to $(4\pi/3)^{1/3}$ for a spherical droplet.

It is immediately seen that the left-hand side of the equation (the ratio of the equilibrium vapor pressures) is a continuous quantity whereas the right-hand side is a discrete function of the number of molecules. In other words, a fixed value of the vapor pressure or of the chemical potential corresponds to each number n of the molecules. At the same time there are intermediate values of the vapor pressure to which correspond values of the number of molecules which are not integers. This situation is represented graphically in Fig. 2.16 [Milchev and Malinowski 1985]. The value P_2 of the vapor pressure corresponds to two-atom clusters, P_3 to three-atom clusters, etc. If the actual vapor pressure $P = P_2$, a pair of atoms is precisely in equilibrium with the vapor phase (the critical nucleus) and the three-atom cluster is stable, as the vapor phase is supersaturated with respect to it, and it can grow further. If $P = P_3$, the three-atom cluster is the critical nucleus and the four-atom cluster is stable. However, if $P_3 < P < P_2$ the pair of atoms becomes unstable as the vapor phase is undersaturated with respect to it, and it should decay. At the same time the three-atom cluster is still stable and will remain stable as long as the vapor pressure is higher than P_3. It follows that contrary to the classical concept a cluster with a fixed size is stable in an interval of supersaturation which is as larger as the cluster size is smaller. An increase of the cluster size leads to a sharp decrease of the width of the intervals and the discrete dependence can be approximated by a smooth curve. In other words, the classical approach becomes applicable. For small clusters, however, the latter is a very rough approximation although the tendency remains the same. Thus the fundamental difference between the classical and atomistic

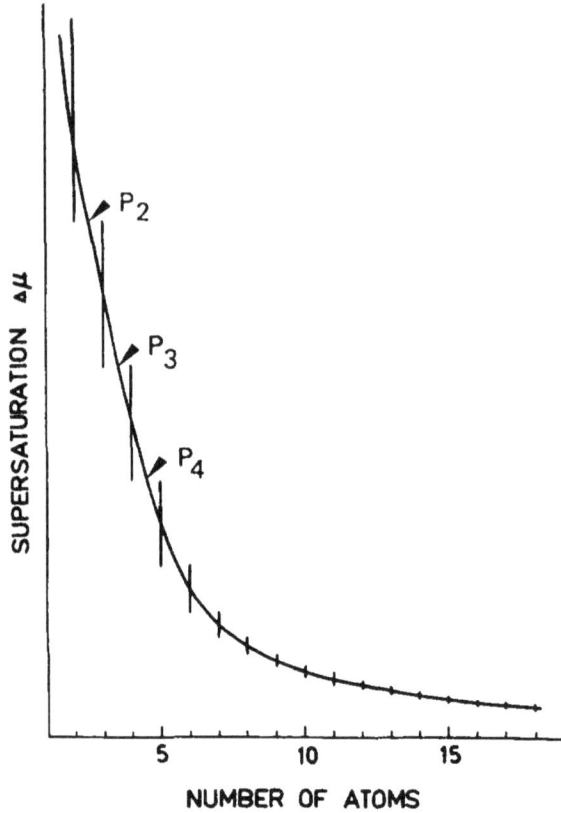

Fig. 2.16. Dependence of the equilibrium vapor pressure of atomic clusters on the number of atoms in the clusters. Clusters of a given size are stable in intervals of supersaturation given by the bars. The classical Thomson–Gibbs equation is plotted as a solid line. As seen, it is a good approximation for large clusters (after Milchev and Malinowski [1985]).

considerations is that a single nucleus size should be operative over a range of temperature or atom arrival rate.

In order to calculate the Gibbs free energy change of formation of small clusters we will use the atomistic approach introduced by Stranski and Kaischew (see Sec. 2.1.3) which has the advantage that it avoids the use of macroscopic quantities. In terms of this approach, by making use of Eqs. (2.20) and (2.21) the Gibbs free energy change of formation of a cluster consisting of n atoms reads

$$\Delta G(n) = n(\varphi_{1/2} - \Delta\mu) - U_n \ . \tag{2.97}$$

At small enough supersaturations n is large and the $\Delta G(n)$ dependence resembles very much the classical one shown in Fig. 2.1 given in terms of n instead of r. This is not the case, however, when the supersaturation is large. As shown in Fig. 2.17(a), $\Delta G(n)$ is a discrete function of n and displays a highest value at certain value of $n = n^*$. In the particular case of $\Delta\mu/\psi = 3.25$ the cluster with a highest ΔG value consists of two atoms and the equilibrium vapor pressure is determined by the breaking of one lateral bond. The equilibrium vapor pressure of the three-atom cluster is lower than that of the atom pair as it is determined by the breaking of two bonds per atom. Hence, the latter is more stable than the pair which thus plays the role of the critical nucleus. Analogously, in the case of $\Delta\mu/\psi = 2.75$ the cluster with maximum value of ΔG consists of six atoms and the equilibrium vapor pressure is determined by the breaking of two bonds per atom, whereas the equilibrium vapor pressure of seven-atom cluster is determined by the breaking of three bonds per atom.

It follows from the above that when the size of the critical nucleus is small its geometrical shape does not remain constant as adopted by the classical theory. No analytical expression can be derived for n^* and its structure should be determined by a trial-and-error procedure estimating the binding energy of each configuration. It turns out for example [Stoyanov 1979] that small clusters with a fivefold symmetry have lower potential energy than clusters with the normal fcc lattice with (111) orientation.

In order to calculate the steady state rate of nucleation we make use of the general expression derived by Becker and Döring (2.50) [Stoyanov 1973]. As was shown each term in the denominator of (2.50) represents one point of the dependence $\exp[\Delta G(n)/kT]$ (Eq. 2.55). For n sufficiently large, $\exp[\Delta G(n)/kT]$ is more or less a smooth function of n, which justifies the replacement of the sum by an integral (see Sec. 2.1; Fig. 2.17(b)). At small values of n this procedure is obviously unapplicable. The $\exp[\Delta G(n)/kT]$ vs n dependence, normalized to $\exp[\Delta G(n^*)/kT]$, is shown in Fig. 2.17(b) for two different values of the supersaturation. As seen $\exp[\Delta G(n)/kT]$ displays a sharp maximum at $n = n^*$, all other terms in the sum of the denominator being negligible. Obviously, the term corresponding to the critical size gives the main contribution to the sum. The latter constitutes the main difference between the classical and the atomistic approach to nucleation. In the former case we have to sum up (or integrate) over a large number of terms whereas in the latter case we just take one of them corresponding to the critical nucleus and neglect all the others.

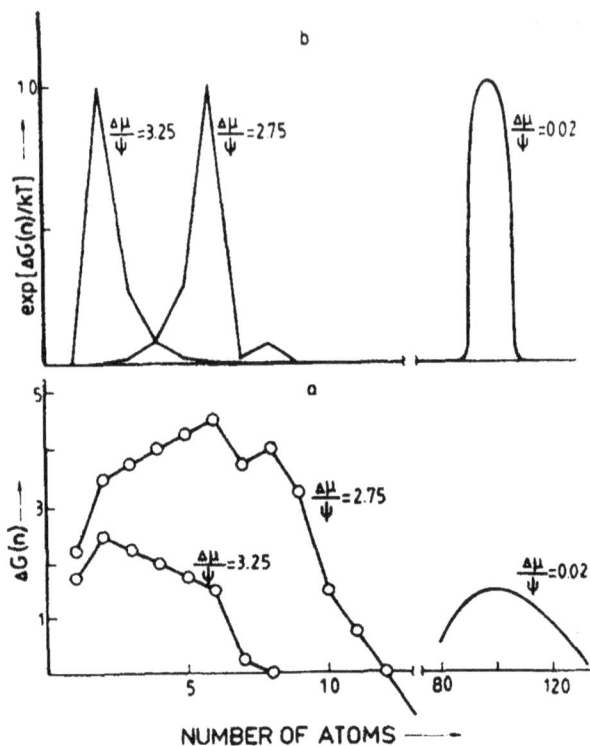

Fig. 2.17. Dependence of (a) the Gibbs free energy change $\Delta G(n)/\psi$ in units of the work ψ to break a first neighbor bond and (b) $\exp[\Delta G(n)/kT]$ on the number of atoms n in the cluster at different values of the supersaturation. At small supersaturations ($\Delta\mu = 0.02\psi$) the number of atoms in the critical nucleus is large ($\cong 100$) and it can be described in thermodynamic terms. The corresponding curves are fluent and the summation in Eq. (2.50) can be replaced by integration. At extremely small supersaturations ($\Delta\mu = 3.25\psi$ and $\Delta\mu = 2.75\psi$) the critical nuclei consist of 2 and 6 atoms, respectively. The $\Delta G(n)/\psi$ and $\exp[\Delta G(n)/kT]$ dependencies are represented by broken lines and integration of $\exp[\Delta G(n)/kT]$ is no longer possible. Instead, the value $\exp[\Delta G(n^*)/kT]$ is taken with the contribution of the remaining terms being neglected. In fact the contribution of the remaining terms gives the nonequilibrium Zeldovich factor which is close to unity in this particular case.

Hence

$$\sum_{n=1}^{N-1} \left(\frac{1}{\omega_n^+} \frac{\omega_2^- \omega_3^- \cdots \omega_n^-}{\omega_1^+ \omega_2^+ \cdots \omega_{n-1}^+} \right) \cong \frac{1}{\omega^*} \gamma \exp\left(\frac{\Delta G^*}{kT} \right) , \qquad (2.98)$$

where $\Gamma = 1/\gamma$ is the Zeldovich factor which accounts for the remaining smaller terms in the sum and in this case is of the order of unity. $\omega_n^+ \cong \omega^* =$ const as before.

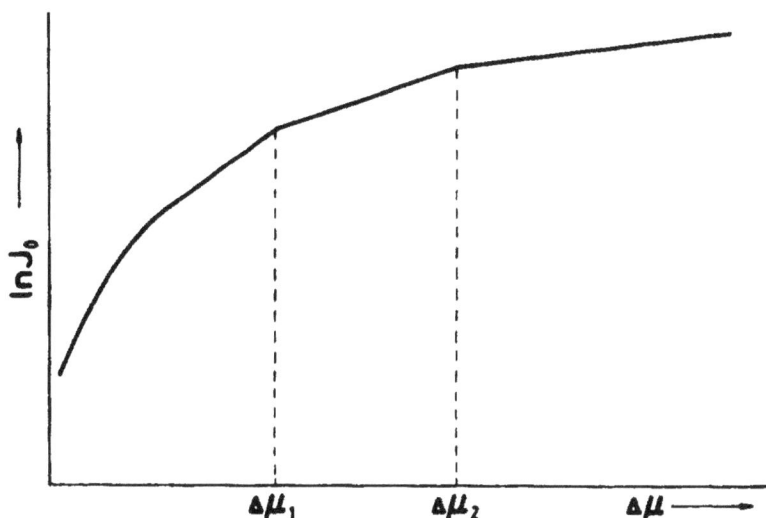

Fig. 2.18. Dependence of the logarithm of the nucleation rate on the supersaturation $\Delta\mu$. At high supersaturations the dependence is represented by a broken line reflecting the real dependence of the supersaturation on the cluster size. $\Delta\mu_1$ and $\Delta\mu_2$ denote the critical supersaturations at which the number of atoms in the critical nucleus changes from one integer value to another. At small supersaturations the broken line turns gradually into a fluent curve due to the decrease of the widths of the cluster stability intervals (see Fig. 2.16). Thus the classical nucleation theory appears to be a good approximation at small supersaturations (after Milchev *et al.* [1974]).

Then bearing in mind (2.20) with $n = n^*$

$$\Delta G^* = -n^*\Delta\mu + \Phi ,$$

and for the nucleation rate one obtains

$$J_0 = \omega^*\Gamma n_s \exp\left(-\frac{\Phi}{kT}\right)\exp\left(\frac{\Delta\mu}{kT}n^*\right) . \tag{2.99}$$

The logarithm of the steady state nucleation rate is plotted against the supersaturation in Fig. 2.18. It represents a broken line when the experimental data cover more than one supersaturation interval. This is easy to understand recalling that the size of the critical nucleus remains constant in more or less wide intervals of the supersaturation [Stoyanov 1979] and so is its geometrical shape and in turn its "surface energy" Φ. The slopes of the straight lines give directly the number of the atoms in the critical nuclei which can be evaluated from a comparison with experimental data. Figure 2.19 represents experimental data for the nucleation rate in electrodeposition of mercury on platinum single crystal spheres [Toschev and Markov 1969] interpreted in the terms of the atomistic theory by Milchev and Stoyanov [1976]. The values for $n^* = 6$ and 10 have been found from the slopes of both parts of the plot. The same values ($n^* = 2$ and 5) have been obtained also in electrolytic nucleation of silver on platinum single crystal spheres in a solution of $AgNO_3$ in fused salts at high temperatures [Toschev *et al.* 1969].

The expression (2.99) does not give explicitly the dependence of the nucleation rate on the atom arrival rate and the temperature of deposition from the vapor, and in this sense is not suitable for the interpretation of the experimental data in this particular case. For this purpose we have to derive expressions for the growth and decay frequencies specific for the case under consideration and to insert them into Eq. (2.50).

In the capillary approach,

$$\omega_n^+ = P_n D_s \text{ grad } n_s \cong P_n D_s \frac{n_s}{a} = \frac{P_n}{a} D_s n_s \ ,$$

where P_n is the perimeter of the cluster and P_n/a is in fact the number of the lateral unsaturated bonds.

In the atomistic approach [Stoyanov 1973],

$$\omega_n^+ = \alpha_n D_s n_s \ , \tag{2.100}$$

where α_n, in complete analogy with the capillary model (P_n/a), gives the number of ways of formation of a cluster of size $n+1$ by joining an adatom to a cluster of size n or, in other words, the number of the adsorption sites neighboring a cluster of n atoms.

The decay constant reads

$$\omega_n^- = \beta_n \nu \exp\left(-\frac{U_n - U_{n-1} + E_{sd}}{kT}\right) \ , \tag{2.101}$$

where U_n is the energy required to disintegrate a cluster of size n into single atoms. The difference $U_n - U_{n-1}$ gives the energy of detachment of an atom from a cluster of size n. E_{sd} is the activation energy for surface diffusion and β_n is the number of ways of detachment of single atoms from a cluster of size n. A one-to-one correspondence exists between each decay process $n+1 \rightarrow n$ and each growth process $n \rightarrow n+1$, and hence

$$\alpha_n = \beta_{n+1} \ . \tag{2.102}$$

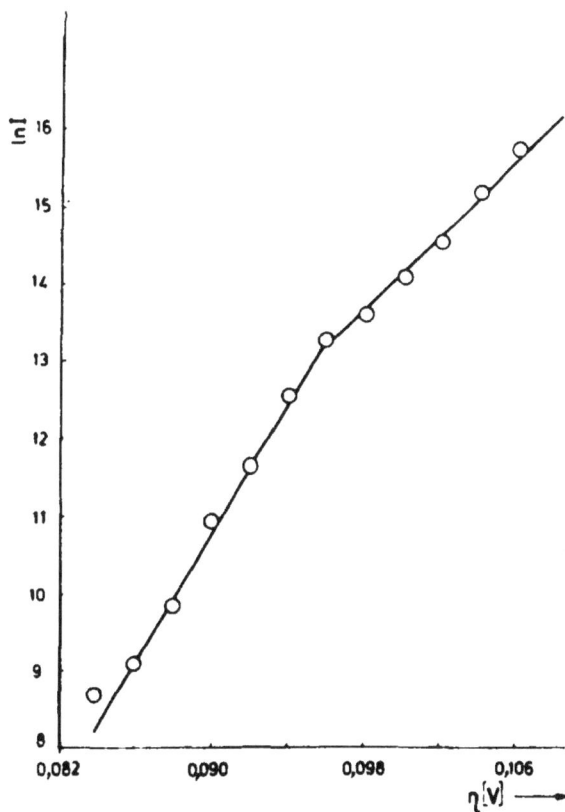

Fig. 2.19. Experimental data for the nucleation rate as a function of the overpotential η in the case of electrochemical nucleation of mercury on platinum single crystals [Toschev and Markov 1969]. The data are plotted in "atomistic" coordinates ln J_0 vs η. As seen the number of atoms in the critical nucleus changes at about 0.096 V. (A. Milchev and S. Stoyanov, *J. Electroanal. Chem.* **72**, 33 (1976). By permission of Elsevier Sequoia S.A. and courtesy of A. Milchev.)

Recalling the expression for the surface diffusion coefficient Eq. (2.101) can be rewritten in the form

$$\omega_n^- = \beta_n D_s N_0 \exp\left(-\frac{U_n - U_{n-1}}{kT}\right) , \tag{2.103}$$

where $N_0 \cong a^{-2}$ is the density of the adsorption sites on the substrate surface.

Equation (2.50) can be rewritten in the form

$$J_0 = \omega_1^+ n_s \left(1 + \frac{\omega_2^-}{\omega_2^+} + \frac{\omega_2^- \omega_3^-}{\omega_2^+ \omega_3^+} + \frac{\omega_2^- \omega_3^- \omega_4^-}{\omega_2^+ \omega_3^+ \omega_4^+} + \cdots \right)^{-1} . \qquad (2.104)$$

Assuming $n^* = 1$ (all terms in the denominator are much smaller than unity and $U_1 = 0$), with (2.100) one obtains

$$J_0 = \alpha_1 D_s n_s^2 = \alpha_1 \frac{R^2}{\nu N_0} \exp\left(\frac{2E_{\text{des}} - E_{\text{sd}}}{kT} \right) . \qquad (2.105)$$

For $n^* = 2$ ($\omega_2^- / \omega_2^+ \gg 1$ and $\omega_2^- / \omega_2^+ \gg \omega_2^- \omega_3^- / \omega_2^+ \omega_3^+$),

$$J_0 = \alpha_2 D_s^2 n_s^3 \nu^{-1} \exp\left(\frac{U_2 + E_{\text{sd}}}{kT} \right)$$

or

$$J_0 = \alpha_2 \frac{R^3}{\nu^2 N_0^2} \exp\left(\frac{3E_{\text{des}} - E_{\text{sd}}}{kT} \right) \exp\left(\frac{U_2}{kT} \right) . \qquad (2.106)$$

Respectively, for $n^* = 6$,

$$J_0 = \alpha_6 D_s^6 n_s^7 \nu^{-5} \exp\left(\frac{U_6 + 5E_{\text{sd}}}{kT} \right)$$

or

$$J_0 = \alpha_6 \frac{R^7}{\nu^6 N_0^6} \exp\left(\frac{7E_{\text{des}} - E_{\text{sd}}}{kT} \right) \exp\left(\frac{U_6}{kT} \right) . \qquad (2.107)$$

Then in the general case,

$$J_0 = \alpha^* D_s^{n^*} n_s^{n^*+1} \nu^{-(n^*-1)} \exp\left(\frac{U^* + (n^* - 1)E_{\text{sd}}}{kT} \right)$$

or

$$J_0 = \alpha^* R \left(\frac{R}{\nu N_0} \right)^{n^*} \exp\left(\frac{U^* + (n^* + 1)E_{\text{des}} - E_{\text{sd}}}{kT} \right) , \qquad (2.108)$$

where $\alpha^* \equiv \alpha(n^*)$ and $U^* \equiv U(n^*)$.

Equation (2.108) can be written also in terms of the adatom concentration in the form

$$J_0 = \alpha^* D_s \frac{n_s^{n^*+1}}{N_0^{n^*-1}} \exp\left(\frac{U^*}{kT} \right) , \qquad (2.109)$$

which will be used when considering the 2D nucleation growth of crystals from vapors in Chap. 3.

The critical nucleus size can change by an integer number of atoms and a single nucleus size is operative over a temperature interval. The latter can be defined as follows. Let us consider for example nucleation on Ag (111) surface. At extremely low temperatures a single atom will be a critical nucleus. Above some temperature T_1 the critical nucleus will be a pair of atoms, and the cluster of three atoms, each one situated at the apex of a triangle, will be a stable configuration so that two chemical bonds correspond to each atom. Above some other critical temperature $T_2 > T_1$, the six-atom cluster will be a critical nucleus and the seven-atom cluster with three bonds per atom will be stable. We can easily calculate T_2 and T_1 and determine the interval of stability of the two-atom nucleus. The left-hand limit of the interval T_1 is determined from the condition $w_2^- / w_2^+ = 1$:

$$T_1 = \frac{U_2 + E_{\text{des}}}{k \ln(\beta_2 N_0 \nu / \alpha_2 R)} \ .$$

In the same way we find

$$T_2 = \frac{U_6 - U_2 + 4E_{\text{des}}}{k \ln(\beta_3 N_0^4 \nu^4 / \alpha_6 R^4)} \ .$$

Then for Ag (111) with $\Delta H_e = 60720$ cal/mole, $U_2 = \psi = \Delta H_e/6 = 10120$ cal/mole, $U_6 = 9\psi$, $E_{\text{des}} = 3\psi$, $N_0 \cong 1 \times 10^{15}$ cm^{-2}, $\nu \cong 1 \times 10^{13}$ sec^{-1}, $R = 1 \times 10^{14}$ cm^{-2}sec^{-1}, and neglecting the coefficients α_n and β_n, for $\Delta T = T_2 - T_1$ one obtains 160 K. This means that one and the same critical size will be operative under the conditions of the experiment except for the case when we work in a temperature region around the critical temperature. This is the case in the electrolytic nucleation of mercury shown in Fig. 2.19.

Critical nuclei consisting of 0, 1 and 2 atoms are found in the electrolytic nucleation of silver on glassy carbon [Milchev 1983]. The critical size of 0 atom is interpreted as nucleation on active sites whose binding energy $-U_0$ to the adatoms is strong enough so that the mean residence time of the atom on the site is longer than the mean time elapsed between the arrivals of two successive atoms to the site. Under such a condition the couple adatom–active-site is considered as a stable cluster because the probability of its growth is larger than the probability of its decay. If this is not the case the pair adatom–active-site is no longer a stable cluster and nuclei can form also randomly on defect-free surface. Thus random nucleation on defect-free surface and selective nucleation on active sites can take place simultaneously. Depending on the density of the active sites and their binding energy, one or the other mechanism predominates. In the limiting case of nucleation on active sites the following expression is obtained [Stoyanov 1974]:

$$J_0 = \alpha_1 \frac{Z_0}{N_0} \frac{R^2}{\nu N_0} \exp\left(\frac{U_0 + 2E_{\text{des}} - E_{\text{sd}}}{kT}\right) , \qquad (2.110)$$

where z_0 is the density of the active sites.

As seen in both cases of critical sizes $n^* = 0$ and 1 (Eq. 2.105) the nucleation rate is directly proportional to the square of the atom arrival rate. In that sense both cases are practically undistinguishable. Actually, the pre-exponent in (2.110) is smaller than that in (2.105) by $Z_0/N_0 \cong 10^{-2}$–10^{-4} but $\exp(U_0/kT)$ overcompensates this effect.

2.2.10. *Nonsteady state nucleation*

One can assume that at the initial moment after "switching on" of the supersaturation the concentration of the homophase fluctuations given by Eq. (2.56) is negligible and $J_0 = 0$. It follows that some time from that initial moment should pass in order that a steady state molecular flow in the system be established or, in other words, the concentrations of the clusters attain their steady state values Z_n. The solution of this problem attracted the attention of many authors [Zeldovich 1943; Kantrowitz 1951; Probstein 1951; Farley 1952; Wakeshima 1954; Collins 1955; Chakraverty 1966; Kashchiev 1969] and the problem has been experimentally studied in detail for different systems [Toschev and Markov 1968, 1969; James 1974] and reviewed extensively [Lyubov and Roitburd 1958; Toschev and Gutzow 1972; Toschev 1973]. We will show in this section how the problem of transient states can be treated and will evaluate the nonsteady state effects for different supersaturated (undercooled) systems.

As mentioned in Sec. 2.2.1 the steady state is determined by the condition $dZ_n(t)/dt = 0$. We have now to solve the general problem as given by Eq. (2.48). In other words, as follows from Eq. (2.47) the rates of formation of clusters of different sizes are no longer equal in the time-dependent case. This is the case at least at the beginning of the process after the supersaturation is "switched on."

We mentioned that the formation of a cluster with a critical size appears as a result of fluctuations of the density in the parent phase. Let us consider this process more closely. Imagine we have a cluster with a size

n smaller than the critical one n^*. When a single atom joins the cluster it becomes of a size $n + 1$. When an atom is detached from the cluster the latter goes to a size $n - 1$. The attachment and detachment of atoms are random processes and hence the cluster increases or decreases in size randomly. In other words, the cluster performs random walk back and forth on the size axis up to the moment it reaches the critical size. Then any further growth is connected with the fall of the thermodynamic potential and loses its random character. We can easily prove this following the approach developed by Frenkel and Zeldovich [Frenkel 1955]. Considering n as a continuous variable Zeldovich and Frenkel replaced Eq. (2.48) by the differential equation

$$\frac{dZ(n,t)}{dt} = -\frac{dJ(n,t)}{dn} , \qquad (2.111)$$

where n is now not an index to denote that it is no more a discrete variable.

The nucleation rate $J(n,t)$ is defined in the continuous case as (compare with Eq. (2.47))

$$J(n,t) = \omega^+_{n-1} Z(n-1,t) - \omega^-_n Z(n,t) . \qquad (2.112)$$

The expression of the detailed balance (2.51) in the absence of a molecular flux through the system now reads

$$\omega^+(n-1)N(n-1) = \omega^-(n)N(n) . \qquad (2.113)$$

Eliminating the decay constant $\omega^-(n)$ from (2.112) and (2.113) gives

$$J(n,t) = \omega^+(n-1)N(n-1) \left(\frac{Z(n-1,t)}{N(n-1)} - \frac{Z(n,t)}{N(n)} \right)$$

or

$$J(n,t) \cong -w^+(n)N(n)\frac{d}{dn}\left(\frac{Z(n,t)}{N(n)} \right) , \qquad (2.114)$$

where the approximation $\omega^+(n-1)N(n-1) \cong \omega^+(n)N(n)$ has been used. Combining (2.111) and (2.114) gives

$$\frac{dZ(n,t)}{dt} = \frac{d}{dn}\left[\omega^+(n)N(n)\frac{d}{dn}\left(\frac{Z(n,t)}{N(n)} \right) \right] . \qquad (2.115)$$

Recalling the equilibrium distribution of clusters of class n (2.56) we carry out the differentiation and obtain

$$\frac{dZ(n,t)}{N(n)dt}$$

$$= \omega^+(n)\frac{d^2}{dn^2}\left(\frac{Z(n,t)}{N(n)}\right) + \left(\frac{d\omega^+(n)}{dn} - \frac{\omega^+(n)}{kT}\frac{d\Delta G(n)}{dn}\right)\frac{d}{dn}\left(\frac{Z(n,t)}{N(n)}\right).$$

This equation is valid within the whole range of n from 1 to N. In the vicinity of the critical size it is simplified to

$$\frac{d}{dt}\left(\frac{Z(n,t)}{N(n)}\right) = \omega^+(n^*)\frac{d^2}{dn^2}\left(\frac{Z(n,t)}{N(n)}\right), \qquad (2.116)$$

assuming as above that the growth coefficient is nearly a constant, $\omega^+(n) \cong \omega^+(n^*) = $ const, and bearing in mind that $\Delta G(n)$ displays a maximum at $n = n^*$, i.e. $[d\Delta G(n)/dn]_{n=n^*} = 0$.

As seen the time and size dependence of the steady state concentration $Z(n,t)$ is governed by a partial differential equation of second order. In fact this is the familiar diffusion equation in which the diffusion coefficient is replaced by the growth rate constant $w^+(n^*)$ and which reflects the random character of the growth process in the critical region Δn^*. We can thus envisage the growth of the clusters as "diffusion" in the space of the size n.

Even without solving the governing Eq. (2.116) we can draw some qualitative conclusions by considering a simple analogy with diffusion process towards some boundary. At the initial moment the "concentration" is one and the same all over the system, which is equivalent to homogeneous ambient phase without clusters of any size. Once we have a supersaturation which is equivalent to the appearance of a diffusion gradient the "concentration" in the near vicinity of the boundary decreases and we have a "concentration" profile which changes with time up to reaching steady state. The same happens to the supersaturated medium. Clustering begins and the concentration of clusters of a given class n gradually increases with time up to the moment it reaches the steady state value. This picture has been directly verified by Courtney [1962] who computed the time dependence of the clusters concentrations.

To proceed further it is necessary to define the boundary conditions. As mentioned earlier, Zeldovich and Frenkel argued that rate determining are the processes confined to a small region around the critical size. A detailed mathematical analysis of the problem carried out by Kashchiev [1969] has shown that this is a very good approximation. It follows that $Z(n,t)$ and $N(n)$ differ only in this region Δn^*, i.e. at $n < n_l$ (see Fig. 2.12),

$Z(n,t) = N(n)$ and at $n > n_r$, $Z(n,t) = 0$ at any time t. Thus, the natural initial condition ($t = 0$) is that immediately after "switching on" the supersaturation clusters of any size are absent in the system. In other words, at the initial moment only single molecules (monomers) are present in the system. The boundary conditions arise from point 2 in Sec. 2.2.1. At any moment the concentration of clusters of class $N \gg n^*$ is equal to zero and the steady state concentration of the monomer is equal to the equilibrium concentration. In other words,

$$Z(1,0) = N(1), \qquad Z(n \geq 2,0) = 0 ,$$
$$Z(1,t) = N(1), \qquad Z(N,t) = 0 .$$

The solution of Eq. (2.116) subject to the above boundary conditions reads [Kashchiev 1969]

$$\frac{Z(n,t)}{N(n)} = \frac{1}{2} - \frac{n - n^*}{\Delta n^*} - \frac{2}{\pi} \sum_{i=1}^{\infty} \frac{1}{i} \sin\left(i\pi \frac{n - n^*}{\Delta n^*} + \frac{i\pi}{2} \right) \exp\left(-i^2 \frac{\pi \omega^* t}{16 \Delta n^{*2}} \right) .$$

It is immediately seen that at the steady state ($t \to \infty$) the sum vanishes and $Z(n^*) = N(n^*)/2$.

Substituting the above solution into (2.114) (the latter being taken for the rate of formation of critical nuclei) gives [Kashchiev 1969]

$$J(t) = J_0 \left[1 + 2 \sum_{i=1}^{\infty} (-1)^i \exp\left(-i^2 \frac{t}{\tau} \right) \right] , \qquad (2.117)$$

where the parameter

$$\tau = \frac{4(\Delta n^*)^2}{\pi \omega^*} \qquad (2.118)$$

is the so-called induction period.

Bearing in mind that the number of the nuclei versus time is given by the integral

$$N(t) = \int_0^t J(t)dt , \qquad (2.119)$$

the integration of (2.117) gives [Kashchiev 1969]

$$N(t) = J_0 \tau \left[\frac{t}{\tau} - \frac{\pi^2}{6} - 2 \sum_{i=1}^{\infty} \frac{(-1)^i}{i^2} \exp\left(-i^2 \frac{t}{\tau} \right) \right] , \qquad (2.120)$$

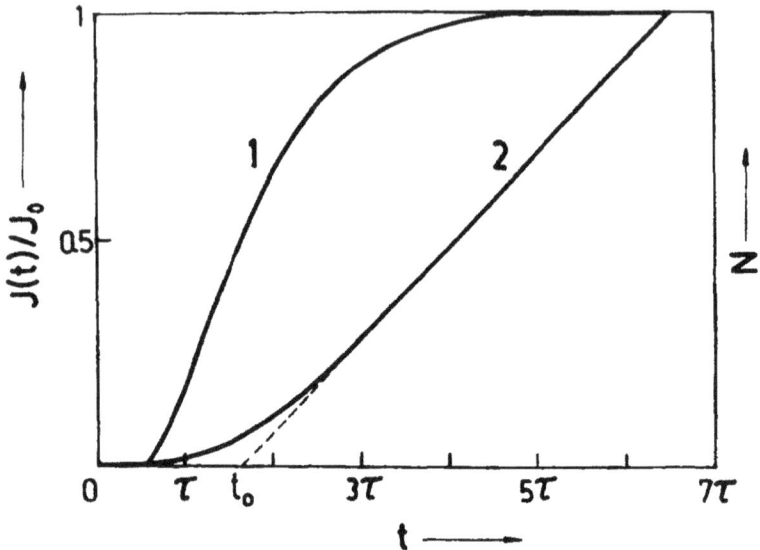

Fig. 2.20. Dependences of the nonsteady state nucleation rate $J(t)$ relative to the steady state nucleation rate J_0 (curve 1, Eq. (2.117)) and the number of nuclei (curve 2, Eq. (2.120)) on time in units of the time lag τ. As seen the steady state rate is reached after approximately 5τ. The incubation period t_0 of the $N(t)$ curve is referred to τ by $t_0 = \pi^2\tau/6 = 1.64\tau$.

where J_0 is given by any of the expressions derived in the previous section. As follows from (2.117) (Fig. 2.20) the steady state should be reached after an interval of approximately 5τ.

We can now evaluate the induction period τ for different cases of nucleation. In the case of formation of spherical nuclei from vapors from (2.118), (2.59), (2.60) and (1.9), one obtains

$$\tau = \frac{16}{\pi}\left(\frac{m}{kT}\right)^{1/2}\frac{\sigma}{P[\ln(P/P_\infty)]^2} \, . \qquad (2.121)$$

Then for homogeneous nucleation of water vapor at room temperature $T = 300$ K with $\sigma = 75.2$ erg/cm^2, $P_\infty \cong 20$ Torr $= 2.66 \times 10^4$ dyne/cm^2, $P/P_\infty = 4$, $m \cong 3 \times 10^{-23}$ g, $\tau \cong 1 \times 10^{-8}$ sec, i.e the induction period is practically negligible.

The problem differs considerably when condensed phases are involved. Making use of (2.68) for ω^* yields

$$\tau = \frac{16}{\pi} \frac{\sigma}{kT[\ln(C/C_\infty)]^2 C\nu\lambda} \exp\left(\frac{\Delta U}{kT}\right) \qquad (2.122)$$

for nucleation in solutions where the concentration must be taken in number of molecules per cubic centimeter.

Evaluation of τ shows that it varies by orders of magnitude due to the different values of the activation energy of desolvation. In any case the value of the induction period is several orders of magnitude greater than that from vapors. Thus for nucleation of $BaSO_4$ in aqueous solutions (assuming for simplicity a spherical shape) with $\sigma = 116$ erg/cm^2, $C = 1 \times 10^{-5}$ mole/l $= 6 \times 10^{15}$ molecules per cm^3, $C/C_\infty = 1000$, $\nu = 3 \times 10^{13}$ sec^{-1}, $a \cong 4 \times 10^{-8}$ cm and $\Delta U = 7$ kcal/mole ($\Delta U/kT = 11.67$), $\tau \cong 5 \times 10^{-3}$ sec. At the same time for $PbSO_4$ ($\sigma = 100$ erg/cm^2, $C = 8.5 \times 10^{16}$cm^{-3}, $C/C_\infty = 28$ and $\Delta U = 14.5$ kcal/mole) $\tau = 5 \times 10^2$ sec, i.e. the induction period is five orders of magnitude longer.

In the case of nucleation in melts with (2.70)

$$\tau = \frac{16}{\pi} \frac{\sigma kT v_c}{\Delta s_m^2 \Delta T^2 \nu a} \exp\left(\frac{\Delta U}{kT}\right) . \qquad (2.123)$$

The induction period is also negligible in homogeneous nucleation in simple metal melts. Thus in solidification of Ag with $\sigma = 150$ erg cm^{-2}, $T = 1230$ K, $\Delta T = 5$ K, $v_c = 5 \times 10^{-22}$ cm^3, $\Delta s_m = 2.19$ cal K^{-1}mole^{-1} $= 1.52 \times 10^{-16}$ erg K^{-1}, $\nu = 2 \times 10^{13}$ sec^{-1} and $\Delta U/kT \cong 2.5$, $\tau \cong 2 \times 10^{-6}$ sec. At the same time for typical glass-forming melts like SiO_2 and GeO_2 ($\Delta U/kT \cong 30 - 40$), $\tau \cong 1 \times 10^5$ sec, i.e. the induction period is as long as a day and night. This means that the process of phase transition can be completed before a steady state nucleation rate is reached. In other words, the whole crystallization process takes place in a transient regime.

Induction times of the order of tens of minutes have been observed in nucleation of polydecamethylenterephthalate [Sharples 1962]. In the case of crystallization of Graham glass ($NaPO_3$) on artificially introduced gold and iridium particles, Toschev and Gutzow [1972] found induction periods as long as 10 and 5 h respectively at about 300°C. Moreover, the induction time is a strong function of temperature. James [1974] found that the induction periods in crystallization of lithium disilicate melts vary from nearly 51 h at 425°C to 7 min at 489°C, i.e. more than two orders of magnitude in a temperature interval of 64°C. A value of 105 kcal/mole or, in other words, $\Delta U/kT \cong 72$, for the activation energy for viscous flow has been estimated from the data.

An expression for the induction period τ for heterogeneous nucleation from the vapor phase is easy to obtain from (2.118) and (2.81):

$$\tau = 17 \left(\frac{m}{k^3 T^3} \right)^{1/2} \frac{\phi(\theta)}{\sin \theta} \frac{\sigma^2 v_l}{aP[\ln(P/P_\infty)]^3} \exp \left(-\frac{E_{\text{des}} - E_{\text{sd}}}{kT} \right) . \quad (2.124)$$

Comparing (2.124) with (2.121) gives

$$\frac{\tau(\text{het})}{\tau(\text{hom})} \cong \frac{\pi}{2} \frac{r^*}{a} \frac{\phi(\phi)}{\sin \theta} \exp \left(-\frac{E_{\text{des}} - E_{\text{sd}}}{kT} \right) .$$

The pre-exponent is of the order of unity, with the exception of the cases of extremely high values of the wetting angle and the ratio $\tau(\text{het})/\tau(\text{hom})$, and is determined primarily by the activation energies for desorption and surface diffusion. One can conclude that in the case of heterogeneous nucleation the induction time is even smaller than in the homogeneous case.

Bearing in mind that the Zeldovich factor is inversely proportional to the square root of the wetting function $\phi(\theta)$ we find that in the case of nucleation in condensed phases (solutions or melts) the induction period τ for heterogeneous nucleation in condensed phases should be given by (2.122) or (2.123) but multiplied by the quantity $\phi(\theta)/(1 - \cos \theta) = (1 - \cos \theta)(2 + \cos \theta)/4$. This function displays a maximum of 0.5625 at $\theta = 120°$. Hence $\tau(\text{het})$ will again be smaller than $\tau(\text{hom})$ by this function.

A straightforward test of nonsteady state effects in the theory of heterogeneous nucleation requires measurements of either the time of appearance of the first nucleus or the dependence of the number of nuclei on time at a constant supersaturation. Precise measurements of the number of nuclei versus time have been performed in the case of electrolytic formation of metal nuclei on foreign metal substrates and the results have been compared with Eq. (2.120) [Toschev and Markov 1969]. In order to separate the process of nucleation from that of growth of the nuclei the so-called double pulse potentiostatic technique has been used. Π-shaped electric pulses (Fig. 2.21) are imposed on an electrolytic cell consisting of two electrodes immersed in an electrolytic solution of metal ions. The height of the first pulse of duration t is a measure of the overpotential $\eta = E - E_0$ (Eq. (1.13)). The height of the second pulse is sufficiently low (in fact lower than the critical supersaturation) so that no nuclei can be formed for a long enough time. The nuclei formed during the first pulse grow to visible sizes during the second pulse and are counted under microscope. A platinum single crystal sphere sealed into glass capillary served as a cathode while the

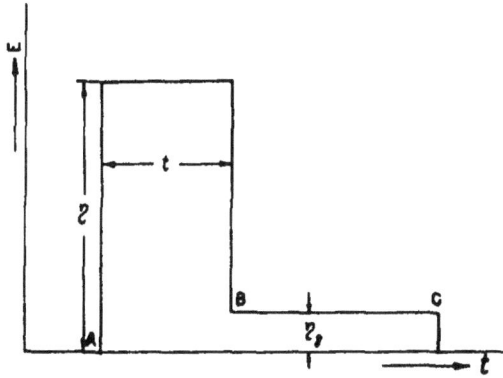

Fig. 2.21. Double pulse technique for investigation of the number of nuclei versus time when used to study the electrochemical nucleation of metals on foreign metal substrates. Nuclei are formed during the first pulse AB of height η. They grow to sizes visible under microscope during the second pulse BC of height η_g. The latter is lower than the critical overpotential required for nucleation to occur with a significant rate. The number of nuclei formed during the first pulse averaged from a large number of measurements is then plotted versus the pulse duration t at a constant pulse height [Toschev and Markov 1969].

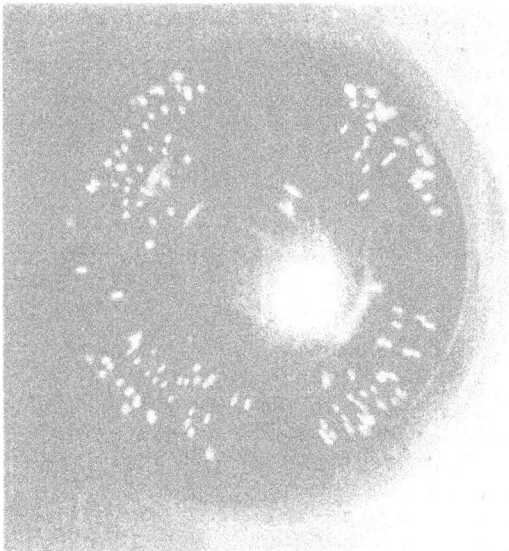

Fig. 2.22. Micrograph showing mercury droplets (bright points) electrodeposited on a platinum single crystal sphere. Every droplet is reflected from the mirror smooth surface of the electrode and looks like a bar. The large bright spot near the middle of the electrode is a reflection of the lamp. The nuclei are formed preferably around the (111) poles of the sphere [Toschev and Markov 1969].

anode was a sheet or a wire (or a pool of mercury in this particular case) of the metal whose ions are present in the solution. Figure 2.22 shows a typical picture of mercury droplets formed on the platinum single crystal sphere. As seen they are preferably formed around the (111) poles of the sphere.

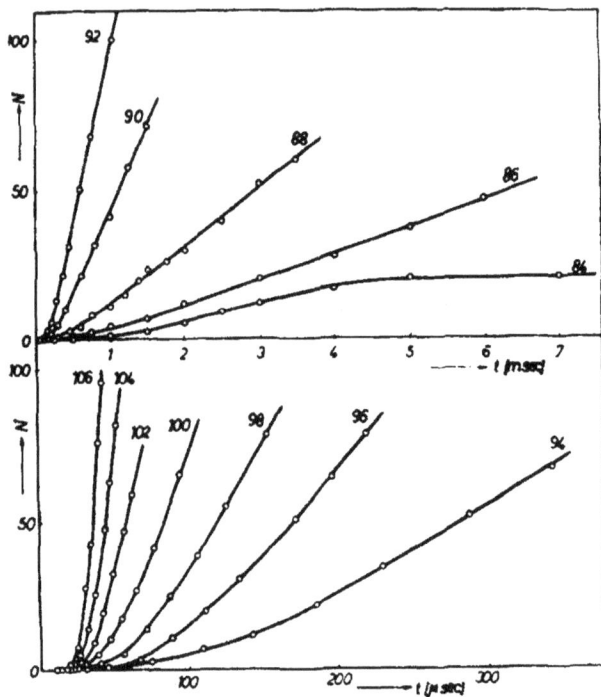

Fig. 2.23. A series of number of nuclei versus time curves in the case of electrochemical nucleation of mercury on platinum single crystal spheres at different overpotentials, denoted by figures in mV at the curves. The transient behavior of the nucleation process is clearly demonstrated. The curve obtained at $\eta = 84$ mV shows a saturation which is due to overlapping of nucleation exclusion zones [Toschev and Markov 1969].

Figure 2.23 shows a series of number of nuclei versus time curves for electrodeposition of mercury nuclei at different overpotentials denoted by the figures at the curves. The transient behavior required by the nonsteady state theory (Fig. 2.20) is clearly demonstrated. Induction periods of the order of milliseconds are observed. The validity of Eq. (2.120) is tested in Fig. 2.24. As seen the points fall close to the theoretical $N(t)$ curve

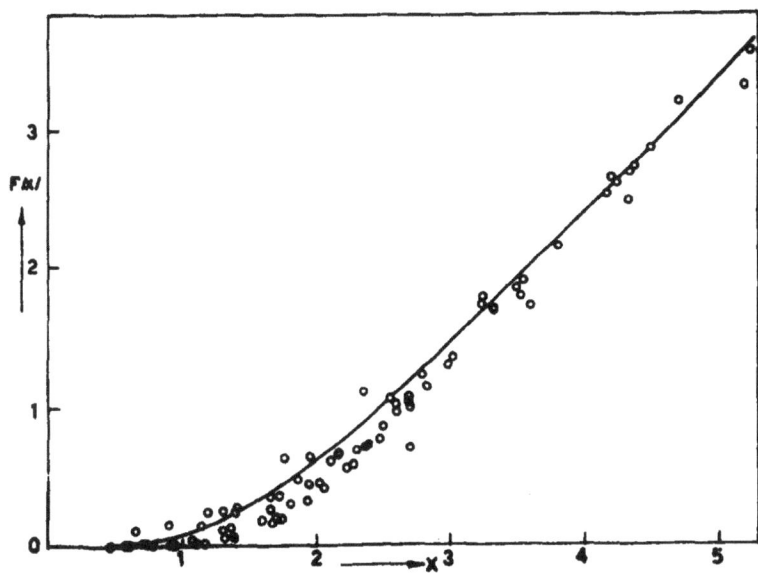

Fig. 2.24. Plot of the number of nuclei versus time from Fig. 2.23 in dimensionless coordinates $F(x) = N(t)/J_0\tau$ and $x = t/\tau$. The solid line represents the theoretical curve (2.125) [Toschev and Markov 1969].

$$f(x) = x - \frac{1}{6}\pi^2 - 2\sum_{n=1}^{\infty} \frac{(-1)^n}{n^2} \exp(-n^2 x) \qquad (2.125)$$

in dimensionless coordinates $f(x) = N/J_0\tau$ and $x = t/\tau$.

The slopes of the linear parts of the curves in Fig. 2.23 give a straight line in logarithmic coordinates $\log J_0$ vs $1/\eta^2$ (Fig. 2.25), thus confirming qualitatively the validity of the capillary model (2.66). The number of atoms constituting the critical nucleus can be evaluated as a function of the overpotential from the slope of the straight line, and the values from 3 to 8 have been obtained in this particular case. Obviously, the capillary model of heterogeneous nucleation which makes use of such phenomenological quantities as the bulk surface specific energies, etc., cannot be used for a quantitative description of the process at very high supersaturations. This is the reason the data of Fig. 2.23 have been interpreted in terms of the atomistic model of nucleation (Fig. 2.19).

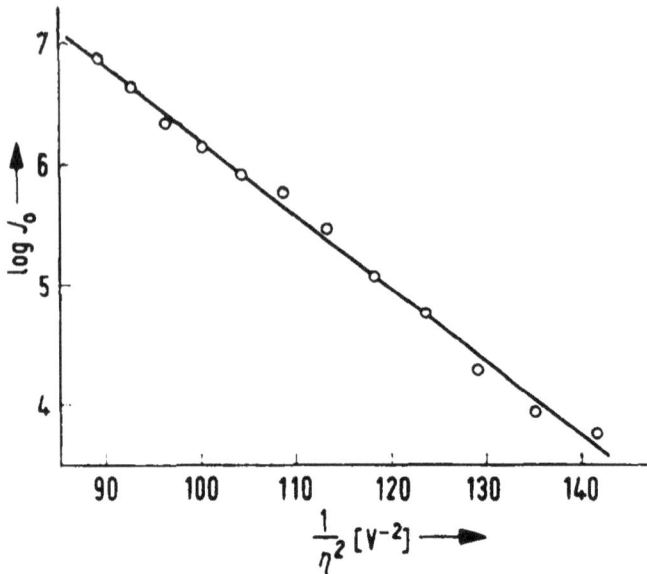

Fig. 2.25. Plot of the logarithm of the steady state rate of electrolytic nucleation of mercury on platinum single crystal spheres versus the reciprocal square of the overpotential as required by the classical theory of 3D nucleation. The rates represent in fact the slopes of the linear parts of the $N(t)$ curves shown in Fig. 2.23. The straight line demonstrates the qualitative validity of the classical nucleation theory although the number of atoms in the critical nuclei is about 8 [Toschev and Markov 1969].

2.2.11. *Saturation nucleus density*

As was mentioned above the main problem in the investigation of the nucleation kinetics in melts and solutions is the presence of impurity particles which stimulate the process due to the wetting. However, the existence of impurity particles leads, in addition, to another phenomenon. It is reasonable to assume that the particles have different activities with respect to crystal nucleation (or different wetting angles or adhesion forces) or, in other words, a different critical supersaturation, according to Eq. (2.67), in which the wetting function $\phi(\theta)$ should enter in the case of heterogeneous nucleation. Then, the active particles will take part in the process only when the supersaturation in the system is higher than their critical supersaturation. In such a case the number of the nuclei formed will be equal to the number of the particles whose critical supersaturation is lower than the actual value in the system. A saturation phenomenon will be observed in the sense that the number of nuclei will reach a constant value. As the particles have different critical supersaturations the increase of the current supersaturation in the system will lead to involvement of new particles with higher critical supersaturations, and the saturation nucleus density will grow. This process will continue up to the moment when the supersaturation exceeds the critical supersaturation for homogeneous nucleation. Then a large number of homogeneously formed nuclei will appear. Such a saturation behavior is clearly established in the case of nucleation of polymethylenterephthalate [Sharples 1962]. The author explains the phenomenon with the catalytic action of foreign particles of finite numbers and different activities in the melt.

The same picture holds for nucleation on foreign substrates. Defect sites on the substrate surface with different critical supersaturation play the role of the foreign particles in this particular case [Robins and Rhodin 1964; Kaischew and Mutaftschiev 1965].

An expression for the time dependence of the number of nuclei can be easily derived under the simplifying assumption of equal activity (or critical supersaturation) of the defect sites [Robins and Rhodin 1964]. Let us denote by N_d (cm^{-2}) the density of the defect sites and by J_0' (sec^{-1}) the frequency of nucleation per active site. Then the change of the number N of nuclei with time will be given by

$$\frac{dN}{dt} = J_0'(N_d - N) ,\qquad (2.126)$$

where $N_d - N$ is the number of the free active sites on which nuclei are not yet formed.

The integration of (2.126) subject to the initial condition $N(t = 0) = 0$ gives

$$N = N_d[1 - \exp(-J_0't)] .\qquad (2.127)$$

As seen, at $t \to \infty$, $N \to N_d = $ const. The saturation is reached in practice when $t > 5/J_0'$, where $1/J_0'$ is the time constant of the process.

This equation can be easily generalized to the case of active sites with different critical supersaturations to give [Kaischew and Mutaftschiev 1965]

$$N = \int_0^{\Delta\mu} N_d(\Delta\mu_c)\{1 - \exp[-J_0'(\Delta\mu_c)t]\}d\Delta\mu_c ,\qquad (2.128)$$

where $N_d(\Delta\mu_c)$ and $J_0'(\Delta\mu_c)$ are now functions of the activity or the critical supersaturation $\Delta\mu_c$. Obviously, Eq. (2.128) holds for nucleation in 3D systems (melts) as well as in 2D systems (surfaces). The main problem of such a treatment, however, is that the distributions $N_d(\Delta\mu_c)$ and $J_0'(\Delta\mu_c)$ are usually unknown.

There could be, however, other explanations of the saturation phenomenon. Thus, if the heat conductivity of the melt is low the temperature of the growing crystallites and their near vicinity increases as a result of accumulation of the latent heat of crystallization. As a result, zones with reduced undercooling appear around the growing crystallites in which nucleation is more or less prohibited. When these zones overlap and fill up the whole volume of the melt the nucleation process ceases and saturation of the nucleus density is reached. Then new nuclei do not form and the existing ones grow to complete solidification of the melt. In high heat conductive melts the zones with reduced undercooling can be reduced to the growing crystallites themselves. The final result, however, should be qualitatively the same.

We have an analogous picture in deposition on substrates. Assuming that the surface diffusion of adatoms is the process which determines the rate of growth of the clusters from the gas phase we have to account for the fact that the adatom concentration in the near vicinity of the growing nuclei is reduced and the system is *locally undersaturated*. As a result zones with reduced and even zero nucleation rate appear which grow together with the clusters [Lewis and Campbell 1967; Halpern 1969; Stowell 1970; Markov 1971]. Sigsbee and Pound [1967] (see also Sigsbee [1969]) coined the term "nucleation exclusion zones." When the zones overlap and cover the whole substrate surface the nucleation ceases and the saturation nucleus density is reached. The saturation phenomenon has been observed in the case of deposition of gold on amorphous carbon films [Paunov and Harsdorff 1974], of gold on cleaved surfaces of KCl and NaF at low temperatures [Robinson and Robins 1970], etc. In the same system (Au/(100) KCl, NaF but at high temperatures) Robins and Donohoe [1972] observed the appearance of maxima instead of plateaus. They explained this by the coalescence of crystallites which dominate at higher temperatures. Note that the first two reasons, namely, the presence of defect sites and the nucleation exclusion zones, lead to appearance of saturation whereas coalescence leads to well-pronounced maximum.

The nucleation exclusion zones are easily visualized in the case of electrolytic nucleation of metals on inert substrates. In this case the reduction of the supersaturation around the growing particles of the new phase is due predominantly to ohmic drop particularly in concentrated

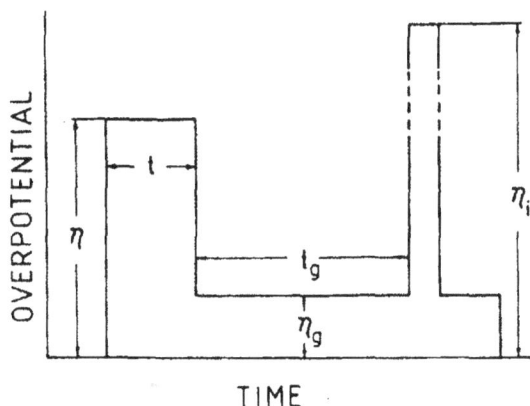

Fig. 2.26. Pulse train for visualization and investigation of the nucleation exclusion zones in electrodeposition of metals. The height and duration of the first pulse are chosen in such a way that only one nucleus is formed. It grows to a predetermined size during the second pulse. The height of the third pulse is chosen in such a way that the whole substrate surface is covered with metal nuclei except only the zone around the initial droplet or small crystal. The metal coating which serves to outline the nucleation exclusion zone grows to become visible under the microscope during the forth pulse. The sizes of the nucleation exclusion zones and the initial crystallites are then measured as functions of the duration and height of the second and third pulses. (I. Markov, A. Boynov and S. Toschev, *Electrochim. Acta* **18**, 377 (1973). By permission of Pergamon Press Ltd.)

solutions of the electrolyte, but bulk diffusion towards the growing particles also plays a part particularly in dilute solutions. A triple pulse train as shown in Fig. 2.26 [Markov, Boynov and Toschev 1973; Markov and Toschev 1975] is imposed on the cell consisting of a platinum single crystal hemispherical cathode and an anode of the metal to be deposited, both immersed in a solution of the electrolyte. The first pulse produces a single nucleus which grows to visible size during the second pulse. The third pulse is high enough to ensure a complete coverage of the platinum sphere with the metal except for a "prohibited" area around the initial particle where the actual overpotential is insufficient to cause nucleation, thus visualizing the nucleation exclusion zones. Typical pictures of this phenomenon are shown in Fig. 2.27 for the cases of electrodeposition of mercury and silver.

A mathematical treatment of the process of overlapping of nucleation exclusion zones irrespective of the dimensionality of the system is usually carried out on the basis of either the geometrical approach of Avrami [1939, 1940, 1941] or the probabilistic formalism developed by Kolmogorov [1937].

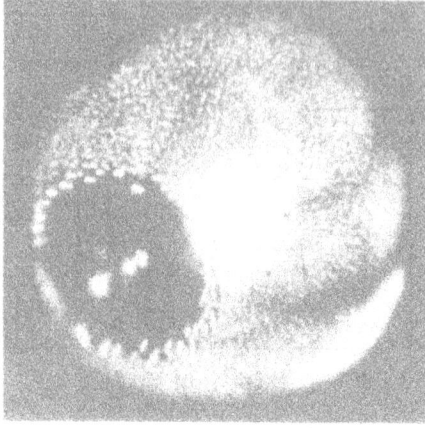

(a)

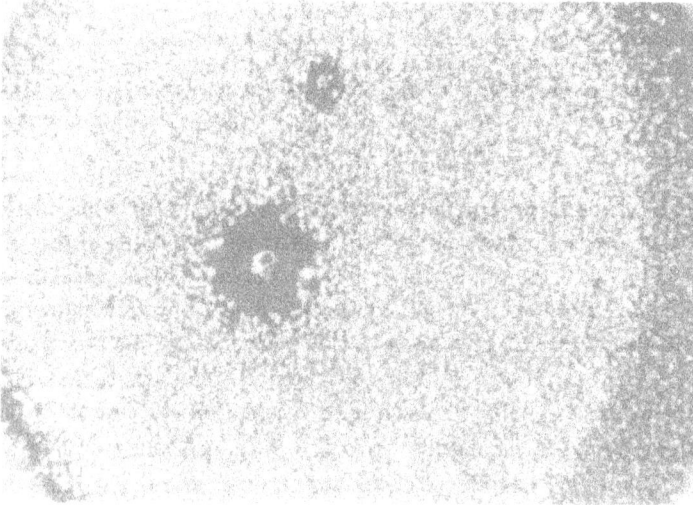

(b)

Fig. 2.27. Nucleation exclusion zones around (a) mercury droplet and (b) silver crystallites. It is clearly seen that the zones grow together with the crystallites. ((a) I. Markov, A. Boynov and S. Toschev, *Electrochim. Acta* **18**, 377 (1973). By permission of Pergamon Press Ltd. (b) By courtesy of A. Milchev.)

Both approaches give one and the same result. The reason is clearly seen in Fig. 2.28. The problem is reduced to calculation of the hatched areas (or volumes) covered simultaneously by two or more circular (or spherical) regions. It can be solved geometrically [Avrami 1939, 1940, 1941], but as is well known the area covered simultaneously by several regions is just equal to the probability of finding an arbitrary point simultaneously in all regions. That is why the probabilistic approach is much simpler and permits an easy generalization to account for nucleation on active sites. This is the reason why we shall follow the probabilistic approach of Kolmogorov. We will give a detailed derivation of Kolmogorov's formula as we shall need it further when considering the 2D growth of crystals and epitaxial films.

Fig. 2.28. A drawing illustrating the mathematical approaches of Avrami [1939] and Kolmogorov [1937]. In order to calculate the part of the volume (or the substrate surface) filled with circles representing the nucleation exclusion zones one has to subtract the shadowed regions covered simultaneously by two or more circles. This could be done geometrically (Avrami) or by using a probabilistic theory (Kolmogorov). It is clear that the fraction of the volume, or the surface covered by the shadowed regions, is equal to the probability of finding an arbitrary point simultaneously in two or more circles.

We consider a supersaturated phase of volume V. Nuclei are formed with a rate $J_0 = const$. A nucleation exclusion zone with a volume V' appears and spreads around each growing nucleus. The rate of growth $v(\mathbf{n}, t)$ of the zone is a function of the direction $\mathbf{n}$ and time t and can be expressed in the form

$$v(\mathbf{n}, t) = c(\mathbf{n})k(t) , \qquad (2.129)$$

assuming the rate of growth $v(\mathbf{n}, t)$ follows one and the same law of growth, $k(t)$, irrespective of the direction.

We introduce the quantity

$$c^3 = \frac{1}{4\pi} \int_\Sigma c^3(\mathbf{n})d\sigma \ ,$$

which has the sense of an average with respect to the direction and where the integration is carried out along the surface Σ of a sphere with center at the origin of the coordinate system. Then the volume of the zone growing around a nucleus formed at a time t' will be at a moment $t > t'$:

$$V'(t',t) = \frac{4\pi}{3} c^3 \left(\int_{t'}^{t} k(\tau - t')d\tau \right)^3 . \qquad (2.130)$$

The density of nuclei N as a function of time is given by

$$\frac{dN}{dt} = J_0 \Theta(t) \ , \qquad (2.131)$$

where $\Theta(t)$ is the fraction of the volume of the system uncovered by nucleation exclusion zones. Integration of (2.131) subject to the initial condition $N(t = 0) = 0$ yields

$$N = J_0 \int_0^t \Theta(\tau)d\tau \ . \qquad (2.132)$$

The fraction $\Theta(t)$ is equal to the probability that an arbitrarily chosen point P is at a moment t to be outside a nucleation exclusion zone (Fig. 2.29). The necessary and sufficient condition for the point P to be in a nucleation exclusion zone at a moment t is the formation of a nucleus at a moment $t' < t$ at another point P' which is spaced from P at a distance smaller than

$$r = c \int_{t'}^t k(\tau - t')d\tau \ ,$$

or, in other words, the point P has to be in the volume V' given by Eq. (2.130).

The probability that at least one nucleus will form in the time interval $\Delta t'$ in a volume V' with accuracy to infinitesimals of second and higher orders is

$$J_0 V'(t',t)\Delta t' \ .$$

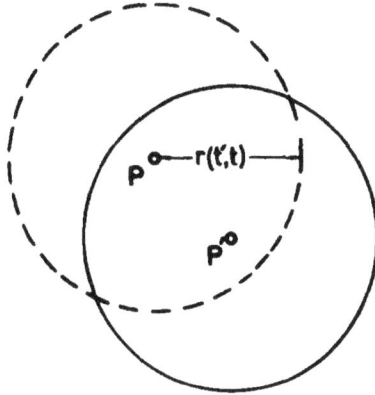

Fig. 2.29. For the calculation of the probability of finding an arbitrarily chosen point P in a nucleation exclusion zone or in a crystallized volume. The necessary and sufficient condition for the point P to be in a nucleation exclusion zone at a moment t is that a nucleus is formed at a moment $t' < t$ at point P' which is spaced from P at a distance smaller than $r(t', t)$, i.e. in the volume outlined by the dashed line. Then the point P will be in the nucleation exclusion zone outlined by the solid line.

The probability that a nucleus will not form in the time interval $\Delta t'$ in the volume V' is

$$1 - J_0 V'(t', t)\Delta t' .$$

The probability for the point P to be outside a nucleation exclusion zone at a moment t from the beginning of the process is

$$\Theta(t) = \prod_{i=1}^{s}[1 - J_0 V'(t_i)\Delta t'] , \qquad (2.133)$$

where $t = s\Delta t'$ and $t_i = i\Delta t'$. Taking the logarithm of (2.133) yields

$$\ln \Theta(t) = \sum_{i=1}^{s} \ln[1 - J_0 V'(t_i)\Delta t'] \cong -\sum_{i=1}^{s} J_0 V'(t_i)\Delta t'$$

$$\cong -J_0 \int_0^t V'(t')dt' .$$

Then the part of the volume V which is uncovered by nucleation exclusion zones is

$$\Theta(t) = \exp\left(-J_0 \int_0^t V'(t')dt'\right)$$

and

$$N = J_0 \int\limits_0^t \exp\left(-J_0 \int\limits_0^\tau V'(t')dt'\right) d\tau \, , \qquad (2.134)$$

where $V'(t')$ is given by (2.130).

The saturation nucleus density N_s obtained from the condition $t \to \infty$ is

$$N_s = J_0 \int\limits_0^\infty \exp\left(-J_0 \int\limits_0^\tau V'(t')dt'\right) d\tau \, . \qquad (2.135)$$

Without loss of generality we can represent the growth of nucleation exclusion zones by a power law

$$v(t) = \frac{dr}{dt} = c\alpha t^{\alpha-1} \, . \qquad (2.136)$$

Assuming a constant rate of growth $k(t) = 1$ $(\alpha = 1)$,

$$V' = \frac{4}{3}\pi c^3 (t - t')^3$$

and [Kolmogorov 1937]

$$N(t) = J_0 \int\limits_0^t \exp\left(-\frac{\pi}{3} J_0 c^3 t^4\right) dt \, .$$

Under the condition $t \to \infty$ for the saturation nucleus density one obtains

$$N_s = 0.9 \left(\frac{J_0}{c}\right)^{3/4} \, .$$

The volume of the melt covered by nucleation exclusion zones (or by growing crystallites), V_c, is given by

$$\frac{V_c(t)}{V} = 1 - \exp\left(-J_0 \int\limits_0^t V'(t')dt'\right) = 1 - \exp\left(-\frac{\pi}{3} J_0 c^3 t^4\right) \, .$$

For the particular case where the nucleation process takes place in a short interval at the beginning of crystallization so that a number of nuclei, N_s, is formed at the initial moment $t = 0$, we have one integration less and, instead of the above equation, we get

$$\frac{V_c(t)}{V} = 1 - \exp\left[-N_s V'(t')\right] = 1 - \exp\left(-\frac{4}{3}\pi N_s c^3 t^3\right) \, .$$

In the case of nucleation on the surface of a foreign substrate we have a two-dimensional system and Eq. (2.130) has to be replaced by

$$S'(t',t) = \pi c^2 \left(\int_{t'}^{t} k(\tau - t')d\tau \right)^2 . \qquad (2.137)$$

Assuming that the growth of the nuclei is governed by surface diffusion the nucleation exclusion zones will be determined by a decrease of the adatom concentration. The diffusion problems usually lead to square root of time dependence of the growth rate, $\alpha = 1/2$ and $c = k\sqrt{D_s}$, where k is a dimensionality constant of the order of unity [Markov 1970]. Then

$$N(t) = J_0 \int_{0}^{t} \exp\left(-\frac{\pi}{2}J_0 c^2 t'^2\right) dt' , \qquad (2.138)$$

from which, with $t \to \infty$,

$$N_s = \left(\frac{J_0}{2c^2}\right)^{1/2} \qquad (2.139)$$

or

$$N_s \cong \left(\frac{J_0}{D_s}\right)^{1/2} . \qquad (2.140)$$

The deposition of thin films takes place as a rule under conditions of either complete condensation (CC) or incomplete condensation (IC). The former case is characterized by negligible desorption flux and far from the growing nuclei the adatom concentration is a linear function of time, $n_s = Rt$. Then the steady state nucleation rate is a function of time through the adatom concentration. In the IC case the adatom concentration is determined by adsorption–desorption equilibrium and $n_s = R\tau_s$, and the nucleation rate is constant with respect to time.

Making use of Eqs. (2.78) and (2.108) for the IC case one obtains

$$N_s = \alpha^{*1/2} N_0 \left(\frac{R}{\nu N_0}\right)^{(n^*+1)/2} \exp\left(\frac{U^* + (n^*+1)E_{\text{des}}}{2kT}\right)$$

which reduces at high supersaturations, where $n^* = 1$ ($U^* = 0$), to

$$N_s = \alpha^{*1/2} \frac{R}{\nu} \exp\left(\frac{E_{\text{des}}}{kT}\right) .$$

In the CC case the nucleation rate has the form (2.109) and from (2.135) and (2.137) one obtains the expression first derived by Stowell [1970]:

$$N_s = q\alpha^{*1/(n^*+3)} \left(\frac{R}{D_s N_0^2}\right)^{\frac{n^*+1}{n^*+3}} N_0 \exp\left(\frac{U^*}{(n^* + 3)kT}\right) ,$$

where q is a dimensionless constant of the order of unity.

One can consider by the same way the coverage of a crystal face by laterally growing 2D nuclei. Then the 2D islands themselves will play the role of nucleation exclusion zones and with $k(t) = 1$ the surface coverage at a moment t will be given by

$$\frac{S_c}{S} = 1 - \exp\left(-\frac{\pi}{3}J_0 c^2 t^3\right) , \qquad (2.141)$$

where c is the rate of growth of the 2D nuclei and S is the surface area of the crystal face.

The active sites and nucleation exclusion zones usually influence simultaneously the nucleation kinetics. A general solution can be obtained by following the above procedure. We will consider for simplicity the case of equal activity of the nucleation centers with steady state nucleation rate.

Nuclei are formed with a frequency J_0 (sec^{-1}) on active centers whose number is N_d. To solve the problem we have to find the fraction of free active centers at a moment t. We consider as free those centers on which nuclei have not yet formed and which are not captured by nucleation exclusion zones. The latter means that a center on which nucleus has not yet formed can be covered by a zone originated by a nucleus growing in the near vicinity. Then the supersaturation in its vicinity can become lower than its critical supersaturation and a nucleus cannot form anymore on it. In that sense the center can be deactivated.

The probability of formation of at least one nucleus in the volume V' (Eq. (2.130)) within the time interval $\Delta t'$ is now given by

$$J_0 \bar{N}_d[V'(t')]\Delta t' ,$$

where

$$\bar{N}_d[V'(t')] = 1 + (N_d - 1)\frac{V'(t')}{V}$$

is the average number of active sites in the volume $V'(t')$. This equation accounts for the fact that in the volume $V'(t')$ there is one center with a probability of unity (the center P) and the remaining $N_d - 1$ centers with a probability of $V'(t')/V$. Then the fraction $\Theta(t)$ of the free centers on which nuclei are not yet formed and which are not covered by nucleation exclusion zones ($N_d \gg 1$) will be

$$\Theta(t) = \exp\left(-J_0 t - J_0 N_d \int\limits_0^t \frac{V'(t')}{V} dt'\right) . \tag{2.142}$$

The time dependence of the number N of nuclei formed up to time t is given by the definition equation

$$N(t) = J_0 N_d \int\limits_0^t \Theta(\tau) d\tau$$

or

$$N(t) = J_0 N_d \int\limits_0^t \exp\left(-J_0 \tau - J_0 N_d \int\limits_0^\tau \frac{V'(t')}{V} dt'\right) d\tau . \tag{2.143}$$

It is immediately seen that when the zone growth rate $c = 0$, i.e. $V'(t') = 0$, Eq. (2.143) turns into Eq. (2.127). In the other extreme when the number of active centers is large enough or the rate of growth of the nucleation exclusion zones is sufficiently high so that the second term in the exponent is much greater than $J_0 t$, Kolmogorov's formula (2.134) results from (2.143). The physical meaning of this result is that the major part of the active centers are deactivated by nucleation exclusion zones and the latter govern the nucleation kinetics. Generalization for time-dependent nucleation and activity distribution of the centers is easy to carry out and the interested reader is referred to the original papers [Markov and Kashchiev 1972a, 1972b, 1973].

As mentioned at the beginning of this section the coalescence of growing crystallites can also lead to limitation of the nucleus density. The reader is referred to the numerous review papers and monographs [Stoyanov and Kashchiev 1981; Lewis and Anderson 1978] and the references therein.

2.2.12. Ostwald's step rule

It was found long ago that when the new phase has several (at least two) modifications, one of which is thermodynamically stable and the others are metastable, the formation of one or more metastable phases is often (but not always) observed first. A typical example is the crystallization of zeolites (for a review see Barrer [1988]). It appears that the first zeolite which crystallizes is not stable when it is left for some time in the reaction vessel in contact with the solution at the temperature of growth. After some time it dissolves in the mother solution and a new, more stable type of zeolite crystallizes at the expense of the first one. The second type can also dissolve, and a third type of zeolite nucleates and grows. Thus, for example, the first type of zeolite (faujasite, pore size 7.4 Å) is replaced by mazzite (ZSM-4) which is more dense (pore size 5.8 Å). At about 100°C the faujasite displays a maximum yield after approximately 20 h of crystallization time. The mazzite first appears at the time of the maximum of the faujasite and reaches a maximum yield after 40 h more [Rollmann 1979]. If the first zeolite is isolated from the mother solution it usually remains stable for quite a long time, which is an indication that the transformation occurs through dissolution and crystallization in the mother phase. A similar step-like behavior shows the crystallization of amorphous Si–Ti alloy upon annealing at 500°C [Wang and Chen 1992]. Ti_5Si_3, Ti_5Si_4, TiSi and $TiSi_2$ nucleate and grow consecutively. After sufficiently long annealing the thermodynamically most stable phases $TiSi_2$ and TiSi only are present.

It was Wilhelm Ostwald [1897] who first compiled the available observations and gave his famous empirical rule according to which the thermodynamically metastable phase should nucleate first. Then at a later stage the metastable phase should transform into the phase which is thermodynamically stable under the given conditions (temperature and pressure). Thus the formation of the new stable phase should take place by consecutive steps from one phase to another with increasing thermodynamic stability. The first theoretical interpretation of this phenomenon, which is known as Ostwald's step rule, was given by Stranski and Totomanow [1933] in terms of the steady state nucleation rate. They showed that more often the metastable phase should have higher nucleation rate provided the system has not been transferred very far below the transformation point. We will repeat here in more detail their considerations.

We consider for simplicity the phase diagram given in Fig. 1.1. We know that the liquid can be undercooled to a considerable temperature without visible crystallization taking place. This means that the liquid phase can in principle nucleate and grow from the vapor phase when the system is supersaturated (undercooled) below the triple point, i.e. along the line AA' or AA''. The liquid phase will be metastable and should solidify at a later stage. We have to compare now the steady state rates of nucleation of the metastable liquid droplets and stable crystallites. The considerations are valid for any crystallization process which includes more than one new phase. We will assume for simplicity that the pre-exponential factors K_1 (see Eq. 2.65) are equal. Bearing in mind Eq. (2.65) for the ratio of the nucleation rates the following holds:

$$\ln\left(\frac{J_{om}}{J_{os}}\right) = \frac{b_s \sigma_s^3 v_s^2}{kT\Delta\mu_s^2} - \frac{b_m \sigma_m^3 v_m^2}{kT\Delta\mu_m^2} \ ,$$

where the subscripts s and m refer to the stable and metastable phases, respectively, and b_s and b_m are geometric factors. It follows that the nucleation rate of the metastable phase will be higher or, in other words, Ostwald's step rule will be valid, when the first term on the right-hand side is greater than the second one.

The stable phase is usually more dense so $v_s < v_m$. On the other hand, the more dense phase has greater specific surface energy than the less dense phase, i.e. $\sigma_s > \sigma_m$, as the density of unsaturated dangling bonds on the surface is greater. Near the triple point, or near any point of transformation, for which the equilibrium vapor pressures (or solubilities) are equal, i.e. $P_{os} = P_{om}$ (or $C_{os} = C_{om}$), the supersaturations $\Delta\mu_s = kT\ln(P/P_{os})$ and $\Delta\mu_m = kT\ln(P/P_{om})$ are equal. Taking into account the third power of the surface energy and the second power of the molecular volume one could anticipate that the specific surface energy will overcompensate the influence of the molecular volume and the geometric factors and

$$b_s \sigma_s^3 v_s^2 > b_m \sigma_m^3 v_m^2 \ , \tag{2.144}$$

which is the necessary and sufficient condition for Ostwald's step rule to hold.

The situation becomes more complicated far from the point of transformation. As seen from Fig. 2.30, the equilibrium vapor pressure P_{om} of the metastable phase will be greater than that (P_{os}) of the stable phase. Then $\Delta\mu_s > \Delta\mu_m$ at one and the same pressure in the system. Then in order that Ostwald's step rule holds (2.144) should be replaced by

$$\left(\frac{\Delta\mu_s}{\Delta\mu_m}\right)^2 < \frac{b_s}{b_m}\left(\frac{\sigma_s}{\sigma_m}\right)^3\left(\frac{v_s}{v_m}\right)^2 \ , \tag{2.145}$$

where

$$\frac{\Delta\mu_s}{\Delta\mu_m} = \frac{\ln\left(\frac{P}{P_{os}}\right)}{\ln\left(\frac{P}{P_{om}}\right)} = \frac{\ln\left(\frac{P}{P_{om}}\right) + \ln\left(\frac{P_{om}}{P_{os}}\right)}{\ln\left(\frac{P}{P_{om}}\right)} > 1 \ . \tag{2.146}$$

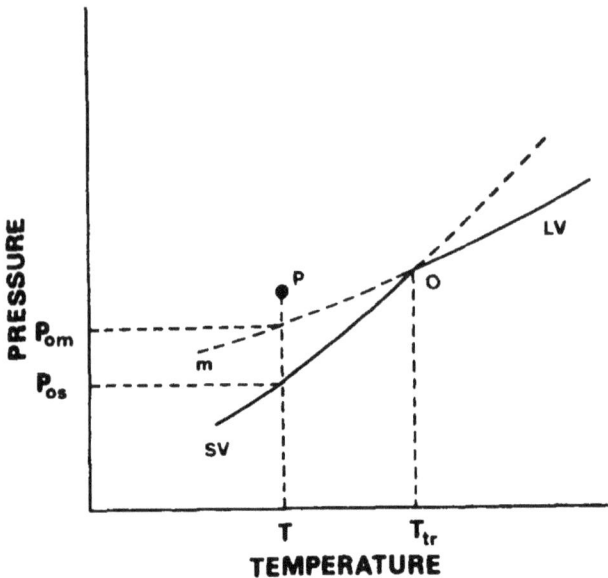

Fig. 2.30. Part of the phase diagram shown in Fig. 1.1 near the triple point O. The solid lines denoted by SV and LV give the solid–vapor and liquid–vapor equilibrium, respectively. The dashed line denoted by m is a continuation of the LV line in the region of stability of the crystal phase and gives the equilibrium between the vapor phase and the metastable liquid phase. As seen, the equilibrium vapor pressure P_{om} of the metastable liquid is higher than the equilibrium vapor pressure P_{os} of the stable crystal phase. It follows that at any temperature $T < T_{tr}$ the supersaturation $\Delta\mu_s = kT \ln(P/P_{os})$ with respect to the stable phase will be higher than the supersaturation $\Delta\mu_m = kT \ln(P/P_{om})$ with respect to the metastable phase.

The ratio $\Delta\mu_s/\Delta\mu_m$ is always greater than unity as P_{om} is always higher than P_{os} by definition.

The physical meaning of (2.145) is immediately seen from Fig. 2.31 where the corresponding works of nucleus formation are plotted versus the actual vapor pressure in the system. The metastable phase begins to nucleate at a higher vapor pressure P_{om} and initially the work of formation of nuclei of the metastable phase is greater than that of the stable one. Beyond some critical pressure P_{cr} the two curves intersect each other and nucleation of the metastable phase becomes thermodynamically favored. When the vapor pressure is lower than P_{cr} but higher than P_{om}, $\Delta\mu_m$ becomes very small and the ratio $\Delta\mu_s/\Delta\mu_m$ can become large. The sign in (2.145) is changed and the stable phase will nucleate first. At very

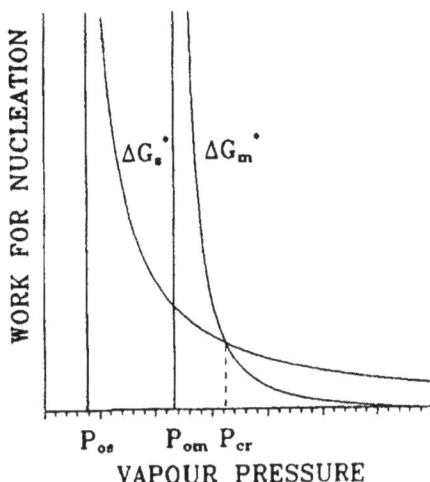

Fig. 2.31. Plot of the works of formation of nuclei of the stable, ΔG_s^*, and metastable, ΔG_m^*, phases versus the vapor pressure. P_{os} and P_{om} denote the corresponding equilibrium vapor pressures. Both curves intersect at some critical vapor pressure P_{cr}, beyond which $\Delta G_m^* < \Delta G_s^*$. Nucleation of the metastable phase is expected at $P > P_{cr}$.

small vapor pressures such that $P_{os} < P < P_{om}$, only the stable phase will nucleate.

The main obstacle in applying Eqs. (2.144) and (2.145) is the lack of knowledge about the specific surface energies particularly at the interfaces between condensed phases. One can circumvent this obstacle in nucleation in melts by using the finding of Turnbull [1950] that the molar surface energies $\sigma_{mol} = \sigma N_A v_c^{2/3}$ of materials with the same nature of the chemical bonds are proportional to the corresponding enthalpies of melting [Jackson 1966] (see Fig. 2.15 and Table 2.1). We can assume that the proportionality constant is one and the same for the different phases and that the nuclei have one and the same shape, i.e. $b_s = b_m$. Recalling Eq. (1.12), Eq. (2.145) turns into

$$\frac{(1 - T/T_s)^2}{(1 - T/T_m)^2} < \frac{\Delta h_s}{\Delta h_m} . \tag{2.147}$$

In the case when the melting points T_m and T_s of both phases are nearly the same, Ostwald's step rule will be observed if the corresponding enthalpies of melting Δh_m and Δh_m obey the inequality

$$\Delta h_m < \Delta h_s \tag{2.148}$$

which follows from (2.144).

The problem becomes more complicated if the transient effects in nucleation have to be accounted for. The necessity of doing that arises from the fact that if the induction period of nucleation of the metastable phase is much longer than that of the stable phase, the metastable phase will not crystallize although its steady state nucleation rate could be higher. As shown above the transient effects play a considerable role in nucleation from solutions and melts, while they can be neglected in nucleation from a vapor phase. When considering the problem we will follow in general the treatment of Gutzow and Toschev [1968].

We will denote by the subscripts s and m the induction periods of the stable and metastable phases. It is seen from Eqs. (2.122) and (2.123) that τ is directly proportional to the geometric factor b and the specific surface energy σ and inversely proportional to the square of the supersaturation. Then the condition $\tau_m < \tau_s$ leads to

$$1 < \left(\frac{\Delta\mu_s}{\Delta\mu_m}\right)^2 < \frac{b_s}{b_m}\frac{\sigma_s}{\sigma_m} . \tag{2.149}$$

Then the combination of (2.145) and (2.149) gives rise in principle to the following four possibilities.

1. $J_{om} > J_{os}$ and $\tau_m < \tau_s$ (Fig. 2.32(a)):

$$1 < \left(\frac{\Delta\mu_s}{\Delta\mu_m}\right)^2 < \frac{b_s}{b_m}\frac{\sigma_s}{\sigma_m} < \frac{b_s}{b_m}\left(\frac{\sigma_s}{\sigma_m}\right)^3\left(\frac{v_s}{v_m}\right)^2 . \tag{2.150}$$

The above holds when the cube of the ratio of the specific surface energies overcompensates the square of the ratio of the molecular volumes. The metastable phase will nucleate first with a higher rate.

2. $J_{om} > J_{os}$ and $\tau_m > \tau_s$ (Fig. 2.32(b)):

$$1 < \frac{b_s}{b_m}\frac{\sigma_s}{\sigma_m} < \left(\frac{\Delta\mu_s}{\Delta\mu_m}\right)^2 < \frac{b_s}{b_m}\left(\frac{\sigma_s}{\sigma_m}\right)^3\left(\frac{v_s}{v_m}\right)^2 . \tag{2.151}$$

The metastable phase will nucleate at a later stage but with a higher rate.

3. $J_{om} < J_{os}$ and $\tau_m > \tau_s$ (Fig. 2.32(c)):

$$1 < \frac{b_s}{b_m}\frac{\sigma_s}{\sigma_m} < \frac{b_s}{b_m}\left(\frac{\sigma_s}{\sigma_m}\right)^3\left(\frac{v_s}{v_m}\right)^2 < \left(\frac{\Delta\mu_s}{\Delta\mu_m}\right)^2 . \tag{2.152}$$

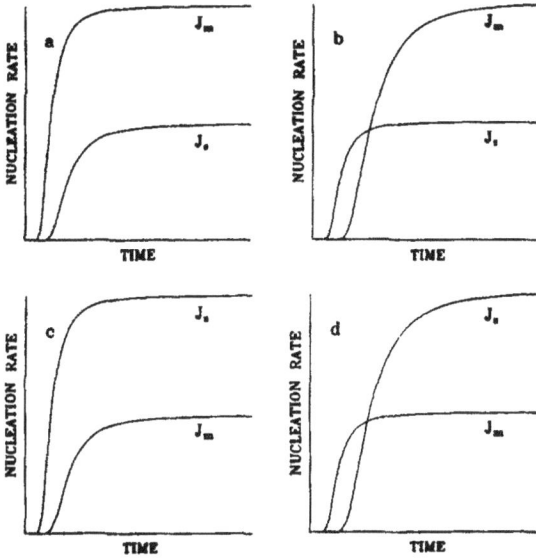

Fig. 2.32. Four possible cases of the time dependence of the nucleation rates of the stable, J_s, and the metastable, J_m, phases: (a) $J_{om} > J_{os}$, $\tau_m < \tau_s$, (b) $J_{om} > J_{os}$, $\tau_m > \tau_s$, (c) $J_{om} < J_{os}$, $\tau_m > \tau_s$, (d) $J_{om} < J_{os}$, $\tau_m < \tau_s$. τ_m and τ_s denote the incubation periods of the metastable and stable phases, respectively (after Gutzow and Toschev [1968]).

The stable phase will nucleate first and its rate of nucleation will be greater. Ostwald's step rule will not be observed.

4. $J_{om} < J_{os}$ and $\tau_m < \tau_s$ (Fig. 2.32(d)):

$$1 < \frac{b_s}{b_m}\left(\frac{\sigma_s}{\sigma_m}\right)^3\left(\frac{v_s}{v_m}\right)^2 < \left(\frac{\Delta\mu_s}{\Delta\mu_m}\right)^2 < \frac{b_s}{b_m}\frac{\sigma_s}{\sigma_m}. \qquad (2.153)$$

The metastable phase will nucleate first but its nucleation rate will be lower. The occurrence of this case requires a drastic change in chemical bonding and should be a rare event.

The theoretical analysis presented so far leads to the conclusion that in general the crystallization of the thermodynamically less stable phase is more pronounced when the square of the supersaturation ratio $\Delta\mu_m/\Delta\mu_s$ is smaller. Bearing in mind (2.146) the latter means that Ostwald's step rule will operate when crystallization takes place near the transformation temperature and/or at very high supersaturations. Going below from the transformation temperature leads to gradual transition from case 1 to case 3.

CHAPTER 3

CRYSTAL GROWTH

One can say that the building units (atoms or molecules) become a part of the crystal when their chemical potential becomes equal to the chemical potential of the crystal. As discussed in Chap. 1 the latter is equal at absolute zero to the work necessary to detach a building unit from the half-crystal or kink position taken with negative sign. In other words, when the atoms or the molecules are attached to kink positions or even stronger (positions 1 and 2 in Fig. 1.15) they become a part of the crystal. In any other position they are connected more weakly to the crystal surface than the atoms at the kink position, and their equilibrium vapor pressure and in turn their chemical potential will differ from those of the bulk crystal. In this sense the adsorption of atoms on the crystal surface or along the steps cannot be considered as crystal growth. All this is valid when the crystal is in contact with a supersaturated ambient phase, i.e. when the chemical potential of the latter is greater than that of the crystal. In equilibrium the chemical potential of the adlayer will be equal to that of the crystal and of the parent phase.

The mechanism of crystal growth is unambiguously determined by the structure of the crystal surfaces. S and K faces offer sufficient kink sites for their growth. F faces can grow without the necessity of overcoming an energy barrier beyond the roughening temperature. Below this temperature the F faces are smooth and their growth requires formation of 2D nuclei or presence of screw dislocations to ensure steps with kink sites along them. In this chapter we consider first the growth of rough faces or the so-called *normal mechanism* of growth. Then the growth of defectless crystal faces through formation and lateral spreading of 2D nuclei and the *spiral growth* of F faces containing screw dislocations are considered separately. In all cases the peculiarities of the growth from melts, solutions and vapors are accounted for.

3.1. Normal Growth of Rough Crystal Faces

We consider in this chapter the growth of rough faces without making distinction for the reason of their roughness. The latter can be due either to the crystallographic orientation of the face or to the entropy effects at sufficiently high temperatures. In any case the change of the energy of a building unit when shifting the latter across the phase boundary between the crystal and the ambient phase (vapor, melt or solution) has schematically the shape shown in Fig. 3.1. The lowest energy state at the left-hand side of the boundary represents the energy of the building unit incorporated at a kink position whereas the line at the right-hand side gives the average value of the energy of the unit in the ambient phase. The difference between both levels gives the enthalpy of the corresponding phase transition (sublimation, dissolution or melting). The barrier at the phase boundary with a height ΔU can have different nature in different media as discussed in the previous chapter. Thus in growth from vapors the barrier can be due to preceding chemical reaction, such as the pyrolysis of silane (SiH_4) in Chemical Vapor Deposition (CVD) of Si or arsine (AsH_3) in Metal-Organic Chemical Vapor Deposition (MOCVD) of GaAs. More complex molecules should overcome an energetic barrier in order to occupy the correct orientation, i.e. we have a barrier of steric character. Obviously, in the growth of simple monoatomic crystals the value of the maximum should be nearly equal to zero. In the cases of growth in solutions and melts the energy barrier ΔU can be identified with the energies of desolvation and viscous flow, respectively, as discussed in Chap. 2.

Experimental evidence concerning the roughening temperature shows that metallic crystals in contact with their vapors remain faceted up to the melting point. Heyraud and Metois [1980] observed (111) and (100) facets on the surface of rounded gold crystallites on graphite up to 1303 K ($T_m = 1337$ K). The same authors [Metois and Heyraud 1982; Heyraud and Metois 1983] found that with increasing temperature the (111) and (100) facets on the surface of Pb crystallites on graphite diminish in size but are still persistent at 300°C (T_m (Pb) = 327.5°C). Pavlovska *et al.* [1989] studied the equilibrium shape of small Pb crystals (10–20 μm) and found

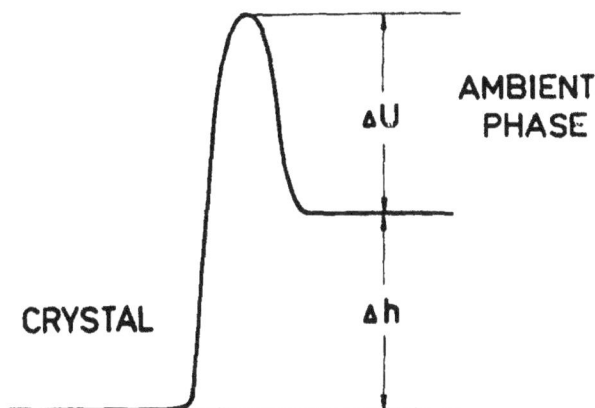

Fig. 3.1. Schematic variation of the free energy for the thermally activated transfer of building units across the interface between the ambient phase and the crystal. The lower state corresponds to a building unit in a half-crystal position. Δh is the corresponding enthalpy of the transition (sublimation, dissolution or fusion). ΔU is the kinetic barrier for the incorporation of building units into the half-crystal position connected with preceding chemical reactions, desolvation in solution growth or viscous flow in growth in melts.

that the most closely packed (111) faces were visible up to the melting point. The less closely packed (110) face which is an S face disappeared at 40 K lower than the melting temperature [Frenken and van der Veen 1985]. Tin [Zhdanov 1976], zinc [Heyer, Nietruch and Stranski 1971] and copper [Stock and Menzel 1978, 1980] in contact with their vapors did not show roughening transition up to the corresponding temperatures of melting. Well-pronounced roughening transitions below the melting point show usually organic crystals such as dyphenyl [Nenow and Dukova 1968; Pavlovska and Nenow 1971a, 1971b; Nenow, Pavlovska and Karl 1984], naphthalene [Pavlovska and Nenow 1972], carbon tetrabromide [Pavlovska and Nenow 1977] and adamantane [Pavlovska 1979]. For review see also Nenow [1984]. We can conclude that crystals of practical importance with stronger interatomic bonds in contact with their vapors should be faceted up to the melting point and should grow from vapors by the spiral or 2D nucleation mechanism, whereas organic crystals should grow by the normal mechanism at elevated temperatures and 2D nucleation or spiral mechanism at lower temperatures. We consider first the normal mechanism of growth from melts.

The rate of growth is proportional to the net flux of atoms:

$$R = a \left(\frac{a}{\delta}\right)^2 (j_+ - j_-) , \tag{3.1}$$

where δ is the average spacing of the kink sites and $(a/\delta)^2$ is the geometrical probability of a building unit arriving at the crystal surface to find a kink site. j_+ and j_- are, respectively, the fluxes of attachment and detachment of building units per site of growth to and from the growing surface given by Eqs. (2.93) and (2.94). Substituting (2.93) and (2.94) into (3.1) gives

$$R = a\nu \left(\frac{a}{\delta}\right)^2 \exp\left(-\frac{\Delta s_m}{k}\right) \exp\left(-\frac{\Delta U}{kT}\right)$$

$$\times \left\{ 1 - \exp\left[-\frac{\Delta h_m}{k} \left(\frac{1}{T} - \frac{1}{T_m}\right)\right] \right\} .$$

The term in the square brackets is precisely equal to $\Delta\mu/kT$ (see Eq. (1.12)) and

$$R = a\nu \left(\frac{a}{\delta}\right)^2 \exp\left(-\frac{\Delta s_m}{k}\right) \exp\left(-\frac{\Delta U}{kT}\right) \left[1 - \exp\left(-\frac{\Delta\mu}{kT}\right)\right] .$$

Rough faces can grow at any supersaturation higher than zero. Expanding the exponent in a Taylor series up to the linear term for small supersaturations ($\Delta\mu \ll kT$), the rate of growth becomes directly proportional to the latter:

$$R = \beta_m \Delta T , \tag{3.2}$$

where

$$\beta_m = a\nu \frac{\Delta s_m}{kT} \left(\frac{a}{\delta}\right)^2 \exp\left(-\frac{\Delta s_m}{k}\right) \exp\left(-\frac{\Delta U}{kT}\right) \tag{3.3}$$

is known as the *kinetic coefficient* for crystallization in melts [Chernov 1984]. As seen the latter depends on the entropy of the phase transition, the energy barrier ΔU and the degree of roughness accounted by the probability $(a/\delta)^2$ to find a kink site. When the average kink spacing δ tends to infinity, the kinetic coefficient and in turn the rate of growth go to zero, thus reflecting the simple fact that atomically smooth crystal faces cannot grow through the normal mechanism. Expressions similar to (3.2) and (3.3) have been derived by Wilson [1900] and Frenkel [1932].

As seen the dependence of the growth rate of rough crystal faces on the supersaturation is linear for small values of the latter. In other words, the rough crystal surface behaves as the surface of a liquid. The atomically smooth crystal faces require formation of steps to ensure kink sites along them. Then the kink spacing δ will depend on the step density and thus on the rate of formation of 2D nuclei or on the distance between the consecutive coils of the growth spirals. As will be shown below the latter is also a function of the supersaturation through the radius of the 2D nuclei which is a nonlinear function of the supersaturation. It follows that in all other cases except for the rough surfaces the growth rate will be a nonlinear function of the supersaturation.

The theory of normal growth from melts was extended to cover the case of growth of small rounded crystallites [Machlin 1953]. The derivation is exactly the same as the one given by Burton, Cabrera and Frank [1951] for lateral growth of 2D islands (see Sec. 3.2.1.1). The result (see Christian [1981]) is

$$R(r) = R\left(1 - \frac{r^*}{r}\right) ,$$

where r^* is the radius of the critical nucleus and R is given by Eq. (3.2). Obviously this equation is valid at the initial stages of the crystallization process when the radius of the growing crystal grain is comparable with the radius of the critical nucleus. It is important to note, however, that according to the above equation smaller crystallites grow more slowly than larger crystallites. Besides, this equation states that the rate of growth of a crystallite whose size is equal to that of the critical nucleus is equal to zero. In other words, such a crystallite is in equilibrium with the parent phase.

For the growth of Si from its melt with $\Delta s_m/k = 3.5$, $T = 1685$ K, $a \cong 3 \cdot 10^{-8}$ cm, $\delta = 3a$, $\nu = 1 \cdot 10^{13}$ sec^{-1} and $\Delta U/RT_m \cong 3$, the kinetic coefficient has the value $\beta \cong 0.1$ cm sec^{-1}K^{-1}. At the same time for the growth of Ag with $\Delta s_m/k = 1.2$, $\Delta U/RT_m \cong 1$ and $\beta \cong 10$ cm sec^{-1}K^{-1}.

In the case of growth from solutions the growth flux is given by

$$j_+ = \nu C v_c \exp\left(-\frac{\Delta U}{kT}\right) , \tag{3.4}$$

where C is the concentration of the solute at the crystal–solution interface in units of number of molecules in a cubic centimeter and v_c is the volume of a building unit in the crystal phase. The product $C v_c$ is thus the probability to find an atom in the vicinity of a kink site.

The reverse flux is

$$j_- = \nu(1 - C v_c) \exp\left(-\frac{\Delta h_d + \Delta U}{kT}\right) , \tag{3.5}$$

where $1 - Cv_c$ is the probability that the space around the kink site is free of solute particles and Δh_d is the enthalpy of dissolution.

In equilibrium $C = C_0$ (C_0 is the equilibrium concentration at a temperature T) both fluxes are equal and

$$\exp\left(-\frac{\Delta h_d}{kT}\right) = \frac{C_0 v_c}{1 - C_0 v_c} . \tag{3.6}$$

Making use of this relation and Eqs. (3.1), (3.4) and (3.5) gives for the growth rate

$$R = \beta_s v_c \left(C - C_0\right) , \tag{3.7}$$

where

$$\beta_s = \frac{a\nu}{C_0 v_c} \left(\frac{a}{\delta}\right)^2 \exp\left(-\frac{\Delta h_d + \Delta U}{kT}\right) \tag{3.8}$$

is the kinetic coefficient for crystallization in solutions.

Replacing Δh_d by C_0 through (3.6) gives for β_s in the case of dilute solutions ($C_0 v_c \ll 1$)

$$\beta_s \cong a\nu \left(\frac{a}{\delta}\right)^2 \exp\left(-\frac{\Delta U}{kT}\right) . \tag{3.9}$$

A classical example of normal growth in solutions is the hydrothermal growth of α-quartz (SiO_2) [Laudise 1959, 1970]. Crystals of materials like sapphire (Al_2O_3) [Laudise and Ballman 1958], ZnO and ZnS [Kolb and Laudise 1966; Laudise and Ballman 1960], yttrium–iron garnet ($Y_3Fe_5O_{12}$) [Kolb, Wood, Spencer and Laudise 1967] and many others [Demianetz, Kuznetzov and Lobachov 1984] have also been successfully grown using this method.

The growth is carried out in autoclave at high temperature under high pressure. Small pieces of the material to be grown are poured into the lower part of an autoclave in which it dissolves into the solvent. Single crystal seeds are hanged on a wire of inert material in the upper growth zone of the autoclave. Part of the latter (usually about 80%) is filled with alkaline solution of NaOH, KOH or K_2CO_3 which improves the solubility of the crystals. The autoclave is then put down vertically in a furnace which heats the lower part to a higher temperature as compared with the upper part where the growth takes place. Both dissolution and the growth zones are usually divided by a perforated metallic disk to localize the temperature gradients. Upon heating the solution fills up the whole volume of the autoclave. At temperatures 400°C and 350°C of the lower and the upper parts, respectively, the pressure usually increases up to 2000 atm. The material in the lower part dissolves into the solvent and by convection is transported into the upper part. The solution is saturated in the lower part at higher temperature and is supersaturated at lower temperature in the upper part. The supersaturation is thus determined by the difference of the solubilities C_0 and C of the material at higher and lower temperatures, respectively. Under these conditions the crystals grow at a rate of about 1 to 2 mm per day. The interested reader can find more details in Demianetz, Kuznetzov and Lobachov [1984].

In the case of growth of α-quartz Laudise [1959] found that the rate of growth is directly proportional to the difference of the temperatures, ΔT, which is in turn proportional to the difference of the concentrations $\Delta C = C - C_0$. At a temperature of growth, 347°C, and $\Delta T = 50$°C a rate of growth as high as 2.5 mm/day has been measured. The Arrhenius plot of the slopes of the straight lines $R \div \Delta T$ vs the reciprocal temperature represents a straight line whose slope can be identified by the activation energy ΔU. The value 20 kcal/mole has thus been found for the growth of the (0001) face of α-quartz. The solubilities of the α-quartz at 400°C and 347°C are found to be 2.43 and 2.28 g/100 g solvent, or 1.43×10^{20} and 1.35×10^{20} molecules/cm^3. Then for the supersaturation $\Delta C/C_0$ one obtains 0.059. Bearing in mind that the volume of a molecule is $v_c \cong 6 \times 10^{-23}$ cm^3 the approximation $C_0 v_c = 8.6 \times 10^{-3} \ll 1$ and Eq. (3.9) is justified. Then with $\delta \cong 3a$, $\nu = 1 \times 10^{13}$ sec^{-1} and $a \cong 4 \times 10^{-8}$ cm, $\beta_s \cong 5 \times 10^{-3}$ cm sec^{-1} and $R = 2.5 \times 10^{-6}$ cm/sec or 2.1 mm/day, in good agreement with the measured value.

Finally, we will derive an expression for the rate of normal growth in a vapor phase. The direct flux of atoms per kink from the vapor phase towards the growing crystal is

$$j_+ = \frac{P}{(2\pi m kT)^{1/2}} a^2 \exp\left(-\frac{\Delta U}{kT}\right) , \qquad (3.10)$$

where $P/(2\pi m kT)^{1/2}$ is the flux of atoms per unit area and a^2 is the area of a kink.

The reverse flux is given by

$$j_- = \nu \exp\left(-\frac{\Delta h_s + \Delta U}{kT}\right) . \qquad (3.11)$$

In equilibrium $(j_+ = j_-)$ $P = P_\infty$ and

$$\frac{P_\infty}{(2\pi mkT)^{1/2}}a^2 = \nu \exp\left(-\frac{\Delta h_s}{kT}\right) . \tag{3.12}$$

Then

$$j_- = \frac{P_\infty}{(2\pi mkT)^{1/2}}a^2 \exp\left(-\frac{\Delta U}{kT}\right)$$

and

$$R = \beta_v(P - P_\infty) , \tag{3.13}$$

where

$$\beta_v = \frac{a\nu}{P_\infty}\left(\frac{a}{\delta}\right)^2 \exp\left(-\frac{\Delta h_s + \Delta U}{kT}\right)$$

$$= \left(\frac{a}{\delta}\right)^2 \frac{a^3}{(2\pi mkT)^{1/2}} \exp\left(-\frac{\Delta U}{kT}\right) \tag{3.14}$$

is the kinetic coefficient in vapor. Equation (3.13) with (3.14) has been derived (without the kinetic barrier ΔU and the degree of roughness $(a/\delta)^2$) as early as the end of the last century [Hertz 1882] and the beginning of the present one [Knudsen 1909]. For dyphenyl with $\delta = 3a$, $a^3 = 2.17 \times 10^{-22}$ cm^3, $m = 2.56 \times 10^{-22}$ g, $\Delta U/kT \cong 1$ and $T = 68°$C ($T_m = 69°$C), $\beta_v = 1 \times 10^{-6}$ cm^{3}sec^{-1}dyne^{-1}. Then with P_∞ ($T = 68°$C) $\cong 1$ Torr $= 1333$ dyne/cm^2 and $P = 1343$ dyne/cm^2, $\Delta P = P - P_\infty = 10$ dyne/cm^2 and $R = 10$ μm/sec.

Comparing (3.3), (3.9) and (3.14) leads to the conclusion that in all cases the kinetic coefficient is proportional to the surface roughness in terms of the probability to find a kink, $(a/\delta)^2$, and to the exponent of the activation energy for incorporation of a building unit into the crystal lattice, ΔU. Then the latter can be determined from an Arrhenius plot of the kinetic coefficient versus the reciprocal temperature as this is done in the case of hydrothermal growth of α-quartz. Moreover, for growth from solutions and vapors the rates of growth are of the order of micrometers per second whereas from melts the growth rates are several orders of magnitude higher. A detailed analysis of the theoretical models of the normal mechanism of growth of atomically rough crystal surfaces is carried out by Rosenberger [1982] (see also Christian [1981]).

3.2. Layer Growth of Flat Faces

When the crystal face is atomically smooth its rate of growth or, in other words, the velocity of its shift parallel to itself, is determined by two independent processes: (i) formation of steps and (ii) lateral movement of these steps. One or the other of these processes can determine the overall rate of growth. In the case of a defectless crystal face the rate of growth is determined by the frequency of formation of 2D nuclei. The latter is an energetically activated process and a critical supersaturation should be overcome for the growth to take place. When screw dislocations are present they represent a nonvanishing source of steps and the process of growth is no longer limited by step formation. Then the rate of growth is determined by the rate of lateral movement of the steps, which in turn depends on their height and structure, rate of surface diffusion, interaction of the steps with each other, the encounter with crystal defects, impurity atoms, etc.

In the general case any small part of the crystal surface can be considered as a vicinal face consisting of a train of parallel steps with arbitrary height divided by smooth terraces which are parallel to the nearest singular face. When the rate of growth is determined by 2D nucleation, pyramids of growth are formed during growth by the formation of 2D nuclei one upon the other (Fig. 3.2). The side surfaces of these pyramids can be considered as vicinal faces. Hillocks with vicinal side surfaces are also formed during the growth in the presence of screw dislocations (Fig. 3.9(d)). When single crystal wafers are prepared from bulk single crystals through cutting and polishing, they can never be cut perfectly parallel to the singular faces and thus they offer in fact vicinal surfaces for further growth for geometrical reasons.

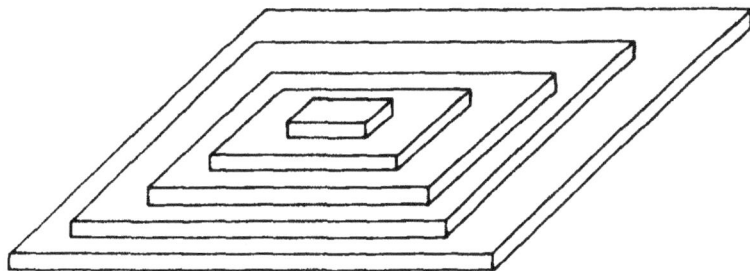

Fig. 3.2. Pyramid of growth consisting of 2D islands formed one on top of the other. The side surface of such a pyramid represents, in fact, a vicinal surface. The slope of the vicinal is determined by the rates of 2D nucleation and step propagation.

The rate of layer growth of the crystal face R in a direction normal to the singular face or to the surface of the terraces depends on the velocity of the step advance v and the density of the steps p :

$$R = pv , \tag{3.15}$$

where $p = h/\lambda = \tan\theta$ is the slope of the vicinal given by the ratio of the step height h and the step spacing λ (Fig. 3.3). The velocity of growth of the face V parallel to itself will be given by $V = R\cos\theta$, where θ is the angle the particular part of the crystal surface makes with the singular face.

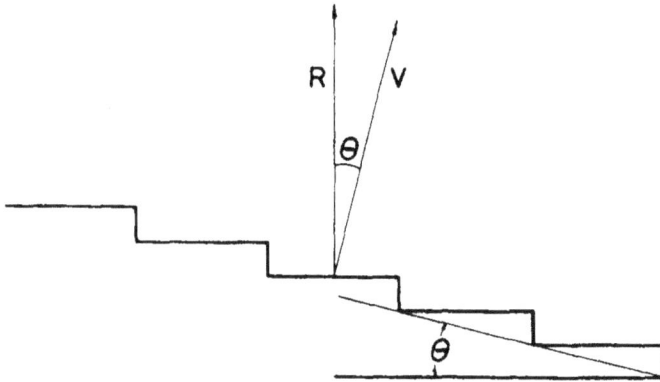

Fig. 3.3. For the determination of the rate of growth R of a vicinal surface tilted by an angle θ with respect to the nearest singular face in a direction normal to the latter. The quantity $V = R\cos\theta$ is the rate of growth of the vicinal surface parallel to itself.

Note that the angle θ and in turn the step density depends in general on the source of the steps and on the kinetics of growth, i.e. on the supersaturation. In the case of 2D nucleation growth the step distance (Fig. 3.2) depends on the rate of 2D nucleation. The higher the supersaturation (the smaller the specific edge energy) is the greater the rate of 2D nucleation will be. Then 2D nuclei are formed at an earlier moment on top of the underlying 2D islands and the step spacing is smaller. The same is true for the case of spiral growth where the step distance is directly proportional to the radius of the 2D nucleus which is inversely proportional to the supersaturation. As will be shown in the next section the velocity of step advance v is also a function of the step density p and it is our first task to find expressions for v in any particular case of growth from vapor, solution or melt.

3.2.1. Rate of advance of steps

The height of the steps on the crystal surface can vary in general from one atomic diameter (monoatomic steps) to several atomic diameters (polyatomic steps), and finally to hundreds and thousands of atomic diameters (macrosteps). In fact the latter represent ledges or even small crystal faces which are often easily visible. The formation of macrosteps from monoatomic steps can be easily explained bearing in mind that the higher the step is the lower its rate of advance will be. The latter is due to the fact that a polyatomic step requires a higher flux of atoms in order to move the same rate as monoatomic step. Two monoatomic steps can meet each other as a result of local fluctuation of the supersaturation or of the concentration of impurity atoms on the one hand, or of encountering lattice imperfections on the other. If such an event takes place irrespective of the cause a double step is formed whose rate of advance will be smaller than that of monoatomic step because it requires twice as great a flux of atoms in order to move at the same rate as a monoatomic step. Then a third monoatomic step will catch up with the double step to form a triple height step. The process continues up to the moment a macrostep is formed. Thus the initially smooth crystal surface (or the vicinal face) can under certain conditions break up into *hills* and *valleys*. These processes are usually described in terms of kinematic waves and shock waves by the kinematic theory of crystal growth [Frank 1958b; Cabrera and Vermilyea 1958; Chernov 1961]. On the other hand, the macrosteps are dissipative structures in the sense that they can turn into monoatomic steps again under certain conditions [Chernov 1961; Bennema and Gilmer 1973].

In general macrosteps are permanently present on the crystal surfaces. Their contribution to the overall rate of growth should not be great because of the smaller rate of advance. That is the reason to begin our presentation with the rate of advance of monoatomic steps. As in the case of normal growth of atomically rough faces we will consider separately the growth from different ambient phases — vapors, solutions and melts.

3.2.1.1. Growth from vapor phase

A. *Elementary processes on crystal surfaces*

Consider a vicinal crystal face (vicinal side of a growth hillock or a pyramid due to consecutive 2D nucleation, see Figs. 3.2 and 3.9) below the roughening temperature in contact with its own vapor. We assume that the steps are with monoatomic height. Now we are not interested in the origin of the steps — 2D nuclei or screw dislocations. The overall process of growth includes the following separate elementary processes (Fig. 3.4): (i) adsorption of atoms from the vapor on the terraces between the steps which gives rise to a population of adatoms, (ii) surface diffusion of the adatoms towards the steps and (iii) incorporation of the adatoms in the kinks along the steps which leads to advancement of the steps and hence to the growth of the crystal in a direction normal to its surface. The overall process of evaporation consists of the same elementary steps taken in an opposite order. We neglect the direct impingement of atoms on the steps from the vapor phase. It is easy to show, as in the case of heterogeneous nucleation from vapor, that the flux of atoms from the vapor phase going directly to the step is much smaller than the flux of atoms diffusing on the terraces to the step. (The coupled volume and surface diffusion problem has been treated by Gilmer, Ghez and Cabrera [1971]).

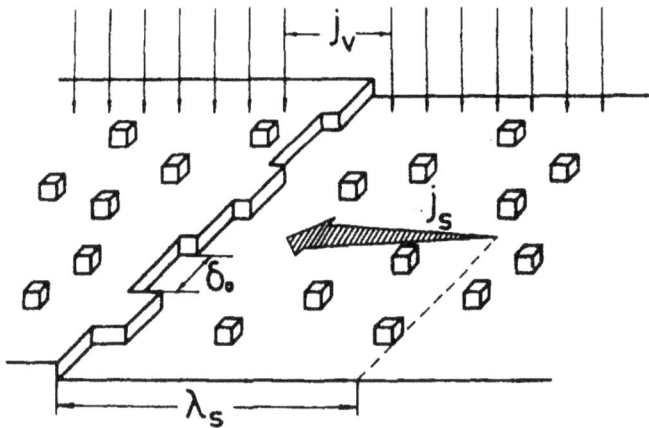

Fig. 3.4. Schematic view of an isolated single height step growing through surface diffusion. j_v is the flux of atoms from the bulk vapor phase towards the crystal surface, j_s is the flux of adatoms diffusing to the step and δ_0 is the average spacing between kinks of any sign. λ_s is the mean distance covered by the adatoms during their life time τ_s on the surface.

Under conditions of equilibrium of the crystal with its vapor phase, the fluxes of adsorption, $P_\infty/(2\pi mkT)^{1/2}$, and desorption, n_s/τ_s, of atoms are equal so that the adatom concentration n_s has the equilibrium value

$$n_{se} = \frac{P_\infty}{(2\pi mkT)^{1/2}}\tau_s \, , \qquad (3.16)$$

where P_∞ is the equilibrium vapor pressure of infinitely large crystal, m is the mass of the atoms and τ_s is the mean time of residence of the adatoms on the crystal surface before being re-evaporated and is given by

$$\tau_s = \frac{1}{\nu_\perp} \exp\left(\frac{\varphi_{\text{des}}}{kT}\right) , \tag{3.17}$$

where $\nu_\perp$ is the vibrational frequency of the adatoms normal to the surface and φ_{des} is the activation energy for desorption of an adatom from the crystal surface.

Substituting P_∞ from (1.58) and τ_s from (3.17) into (3.16) for n_{se} one obtains

$$n_{se} = N_0 \exp\left(-\frac{\varphi_{1/2} - \varphi_{\text{des}}}{kT}\right) , \tag{3.18}$$

where N_0 combines the entropy factors in (1.58), but for simple crystals is of order of the number $N_0 \cong 1/a^2$ per unit area of adsorption sites on the crystal surface ($\approx 10^{15}$ cm^{-2}), a being the mean distance between the adsorption sites. The difference $\varphi_{1/2} - \varphi_{\text{des}}$ gives the energy required to transfer an atom from a kink position on the flat surface. In other words, Eq. (3.18) expresses also the equilibrium kinks adlayer, as at equilibrium the fluxes from the adlayer to and from the kinks are equal.

The mean distance the adatoms can cover during their lifetime on the surface is

$$\lambda_s = (D_s \tau_s)^{1/2} , \tag{3.19}$$

where D_s is the surface diffusion coefficient:

$$D_s = a^2 \nu_= \exp\left(-\frac{\varphi_{\text{sd}}}{kT}\right) . \tag{3.20}$$

Here φ_{sd} is the activation energy for surface diffusion and $\nu_=$ is the vibrational frequency of the adatoms parallel to the crystal surface. Assuming $\nu_\perp = \nu_= = \nu$,

$$\lambda_s = a \exp\left(\frac{\varphi_{\text{des}} - \varphi_{\text{sd}}}{2kT}\right) . \tag{3.21}$$

The desorption energy φ_{des} is always greater than the diffusion energy barrier φ_{sd} ($\varphi_{\text{des}} = 3\psi$ for (111) face and $\varphi_{\text{des}} = 4\psi$ for (100) face of fcc crystals while $\varphi_{\text{sd}} < \psi$). Then $\lambda_s \gg a$. In order to evaluate λ_s we neglect φ_{sd} in comparison with φ_{des}. Considering the case of Ag at 1000 K ($\psi/kT = 5$) we find $\lambda_s(111) = 2 \cdot 10^3 a$ and $\lambda_s(100) = 2 \cdot 10^4 a$. We see that λ_s is much greater than the mean distance between the kinks, $\delta_0 = 7a$, under the same conditions. For (111) face of Si at the melting temperature

$\lambda_s = 2 \cdot 10^2 a \gg \delta_0 = 3.5a$. As seen, the more closely packed the surface is the smaller is λ_s and the smaller is the equilibrium adatom concentration n_{se}. Thus for Ag, $n_{se}(100) = 2 \cdot 10^{-5} N_0$, but $n_{se}(111) = 2 \cdot 10^{-7} N_0$.

B. *Kinetic coefficient of a step*

We will perform the same considerations concerning a single step as for a rough crystal face. Going back to Fig. 3.1 we identify the left-hand energetic level with the energy of an atom at a kink position as before. The upper right-hand side level is identified this time with the energy of an atom adsorbed on the smooth part of the crystal face. Then the difference between the two levels, $\Delta h = \Delta W = \varphi_{1/2} - \varphi_{des}$, gives the energy required to transfer an atom from a kink position on the flat surface. The energy barrier ΔU has the same meaning as before.

The flux of adatoms related to a kink site $(\sec^{-1})$ towards the step is

$$j_+ = \nu n_{st} a^2 \exp\left(-\frac{\Delta U}{kT}\right) , \tag{3.22}$$

where n_{st} is the adatom concentration in the step's vicinity and a^2 is the area of a kink site.

The flux of atoms leaving the kink sites to be adsorbed on the crystal face is

$$j_- = \nu \exp\left(-\frac{\Delta W + \Delta U}{kT}\right) . \tag{3.23}$$

In equilibrium $(j_+ = j_-)$ the adatom concentration attains its equilibrium value n_{se} given by Eq. (3.18).

The rate of step advance is given by

$$v_\infty = a \frac{a}{\delta_0}(j_+ - j_-) , \tag{3.24}$$

where a/δ_0 is the probability to find a kink site and δ_0 is the kink spacing defined by Eq. (1.74).

Substituting j_+ and j_- from (3.22) and (3.23) into (3.24) for v_∞ one obtains

$$v_\infty = 2a^2 \beta_{st}(n_{st} - n_{se}) , \tag{3.25}$$

where the factor 2 accounts for the arrival of atoms from both the lower and upper terraces to the step and

$$\beta_{st} = a\nu \frac{a}{\delta_0} \exp\left(-\frac{\Delta U}{kT}\right) \tag{3.26}$$

is the *kinetic coefficient of the step* or the rate of crystallization in complete analogy with the kinetic coefficient of the crystal face (3.14).

We define the rate of diffusion as the mean distance λ_s divided by the mean residence time τ_s:

$$\frac{\lambda_s}{\tau_s} = \frac{D_s}{\lambda_s} = a\nu \exp\left(-\frac{\varphi_{\text{des}} + \varphi_{\text{sd}}}{2kT}\right) . \tag{3.27}$$

Obviously, when the rate of diffusion is much lower than the kinetic coefficient of the step,

$$\frac{D_s}{\lambda_s} \ll \beta_{\text{st}} , \tag{3.28}$$

which is equivalent to $\varphi_{\text{des}} + \varphi_{\text{sd}} > 2\Delta U + 2\omega$, the velocity of step advance will be determined by the process of surface diffusion. In other words, surface diffusion is the rate controlling process. It is said that the crystal grows in a *diffusion regime.* If this is not the case, i.e. when

$$\frac{D_s}{\lambda_s} \gg \beta_{\text{st}} \tag{3.29}$$

or $\varphi_{\text{des}} + \varphi_{\text{sd}} < 2\Delta U + 2\omega$, the processes taking part when the building units are incorporated into the kink sites determine the rate of step advancement and the crystal face grows in a *kinetic regime.*

C. Rate of advance of a single step

We consider a part of the crystal face containing a single monoatomic step confined between two infinitely wide terraces (Fig. 3.4). The vapor pressure is $P > P_\infty$ and the supersaturation is given by

$$\sigma = \frac{\Delta\mu}{kT} = \ln\left(\frac{P}{P_\infty}\right) \approx \frac{P}{P_\infty} - 1 = \alpha - 1 \tag{3.30}$$

for P slightly greater than P_∞ ($\alpha = P/P_\infty$).

The adatom population in equilibrium with the vapor phase with pressure P is

$$n_s = \frac{P\tau_s}{(2\pi mkT)^{1/2}} = \alpha n_{\text{se}} . \tag{3.31}$$

The supersaturation in the adlayer is defined as

$$\sigma_s = \frac{n_s}{n_{\text{se}}} - 1 = \alpha_s - 1 , \tag{3.32}$$

where $\alpha_s = n_s/n_{\text{se}}$.

Note that σ_s is a function of the distance y normal to the step (Fig. 3.4) whereas σ is constant all over the surface.

The flux of atoms diffusing on the surface towards the step is

$$j_s = -D_s \frac{dn_s}{dy} = -D_s n_{se} \frac{d\alpha_s}{dy} \ .$$

We introduce the potential function $\Psi = \sigma - \sigma_s = \alpha - \alpha_s$. Then the surface flux reads

$$j_s = D_s n_{se} \frac{d\Psi}{dy}$$

or, in a more general form,

$$j_s = D_s n_{se} \ \mathrm{grad} \ \Psi \ . \tag{3.33}$$

The net flux of atoms arriving from the vapor phase on the crystal surface is

$$j_v = \frac{P}{(2\pi m k T)^{1/2}} - \frac{n_s}{\tau_s} = \frac{n_{se}}{\tau_s} \Psi \ . \tag{3.34}$$

Assuming the movement of the step can be neglected in diffusion problems (the justification will be given below) the equation of continuity of Ψ reads

$$\mathrm{div} \ j_s = j_v \ , \tag{3.35}$$

which in the case of diffusion in one direction is equivalent to

$$\frac{dj_s(y)}{dy} = j_v \ .$$

The latter is simply the condition for the adatom concentration at a given distance y from the step to have a time-independent (steady state) value. In other words, the difference of the surface fluxes to and from a strip parallel to the step with a width from y to $y + dy$ must be compensated by the arrival of atoms from the vapor phase.

Then

$$D_s n_{se} \ \mathrm{div} \ (\mathrm{grad} \ \Psi) = \frac{n_{se}}{\tau_s} \Psi \ .$$

The latter can be rewritten in the general form [Burton, Cabrera and Frank 1951]

$$\lambda_s^2 \Delta \Psi = \Psi \ , \tag{3.36}$$

where the symbol Δ denotes the Laplace operator

$$\Delta = \frac{d^2}{dx^2} + \frac{d^2}{dy^2} + \frac{d^2}{dz^2} \ . \tag{3.37}$$

Equation (3.36) is the governing equation which must be solved subject to various initial and boundary conditions for different symmetries and physical conditions.

Several physical possibilities should be considered when approaching the problem of the rate of advance of an isolated step:

(i) The mean path λ_s covered by the adatoms during their time of residence τ_s on the crystal face is much larger than the average kink spacing δ_0. Physically this means that the step acts as a continuous trap for the adatoms. The master equation (3.36) is then reduced to the equation of linear diffusion (see Fig. 3.4)

$$\lambda_s^2 \frac{d^2\psi}{dy^2} = \Psi \ . \tag{3.38}$$

(ii) The mean path λ_s covered by the adatoms during their time of residence τ_s is smaller than the average kink spacing δ_0. The adatoms diffuse directly to the isolated kinks. The diffusion problem is solved in polar co-ordinates as the diffusion field has a circular shape. A solution in terms of Bessel functions is obtained [Burton, Cabrera and Frank 1951].

(iii) The mean path λ_s covered by the adatoms during their lifetime is again smaller than the average kink spacing δ_0 but the adatoms diffuse on the crystal surface to join the edge of the step and then diffuse along it to be incorporated into the kinks [Burton, Cabrera and Frank 1951].

As has been shown in Chap. 1 the steps are rough long before the critical temperature is reached and the condition $\lambda_s \gg \delta_0$ is practically always fulfilled. It follows that case (i) is the most probable one. As for the remaining cases the reader is referred to the original paper [Burton, Cabrera and Frank 1951] as well as to the review paper of Bennema and Gilmer [1973] for more details.

We solve first the particular case (3.38) of linear diffusion to a single isolated straight step. In order to find a solution of the master equation (3.36) we have to specify the boundary conditions. At a distance large enough from the step the adlayer is unaffected by the presence of the step and $n_s = \alpha n_{se}$ (Eq. 3.31). Then $\sigma = \sigma_s$ and $\Psi = 0$. In the near vicinity of the step ($y \to 0$) the adatom concentration is determined by the processes of attachment and detachment of adatoms to and from the kink sites. If the activation energy ΔU is negligible the kinetic coefficient will be large enough and (3.28) will be fulfilled. Then the exchange of atoms between the kinks and the adlayer will be rapid enough and the concentration of adatoms in the near vicinity of the step will be equal to the equilibrium concentration

n_{se}. Then $\sigma_s = 0$ and $\Psi = \sigma$. In the opposite case of considerable value of ΔU, the kinetic coefficient of the step will be very small and the condition (3.29) will hold.

In the general case at $y = 0$,

$$\sigma_s = \sigma_{st} = \frac{n_{st} - n_{se}}{n_{se}} . \tag{3.39}$$

Then $\Psi = \sigma - \sigma_{st} = \chi\sigma$ where [Bennema and Gilmer 1973]

$$\chi = \frac{\sigma - \sigma_{st}}{\sigma} , \tag{3.40}$$

whence

$$\sigma_{st} = \sigma(1 - \chi) .$$

Equation (3.25) becomes

$$v_{\infty} = 2a^2 \beta_{st} n_{se} \sigma (1 - \chi) . \tag{3.41}$$

The solution of Eq. (3.38) subject to the boundary conditions

$$y = 0, \qquad \Psi = \chi\sigma ,$$
$$y \to \pm\infty, \qquad \Psi = 0$$

reads

$$\Psi = \chi\sigma \exp\left(\pm\frac{y}{\lambda_s}\right) , \tag{3.42}$$

where the $+$ and $-$ signs refer to $y < 0$ and $y > 0$, respectively.

Then the rate of advance of the single step,

$$v_{\infty} = \frac{j_s\ (y = 0)}{N_0} ,$$

is

$$v_{\infty} = 2\chi\sigma a^2 n_{se}\frac{D_s}{\lambda_s} = 2\chi\sigma\lambda_s\nu \exp\left(-\frac{\varphi_{1/2}}{kT}\right) . \tag{3.43}$$

Equation (3.41) has been derived under the assumption that n_{st} is the adatom concentration in the near vicinity of the step. We did not specify what was the reason for the deviation of the concentration from its equilibrium value. Then we can determine the parameter χ by equating the expressions (3.41) and (3.43) to obtain

$$\chi = \left(1 + \frac{D_s}{\lambda_s\beta_{st}}\right)^{-1} . \tag{3.44}$$

As seen the unknown parameter χ depends only on the ratio of the rate of diffusion D_s/λ_s and the rate of crystallization β_{st}.

Finally for the rate of advance of a single step one obtains

$$v_\infty = 2\sigma\lambda_s\nu\exp\left(-\frac{\varphi_{1/2}}{kT}\right)\left(1+\frac{D_s}{\lambda_s\beta_{st}}\right)^{-1}. \tag{3.45}$$

Applying the condition (3.28) leads to the expression

$$v_\infty = 2\sigma\lambda_s\nu\exp\left(-\frac{\varphi_{1/2}}{kT}\right) \tag{3.46}$$

valid for the advance of the step in a purely diffusive regime. The condition (3.29) leads to Eq. (3.41) with $\chi \to 0$:

$$v_\infty = 2a^2\beta_{st}n_{se}\sigma , \tag{3.41'}$$

which describes the behavior of the step in a kinetic regime and where the adatom concentration around the step is determined solely by the processes taking part at the step edge. No diffusion gradient exists in this case and the adatom concentration preserves the value n_s determined by the adsorption–desorption balance all over the crystal surface except for a narrow strip around the step. It is seen, however, that in both cases the rate of advance of an isolated step is a linear function of the supersaturation σ.

The movement of the step, when solving the diffusion problem, can obviously be neglected when the mean velocity of the motion of an adatom on the surface, $v_{diff} = \lambda_s/\tau_s$, is much greater than the rate of advance of the step, $v_\infty = 2\sigma D_s n_{se}a^2/\lambda_s$. The ratio $v_\infty/v_{diff} = 2\sigma n_{se}/N_0$ is obviously smaller than unity as the supersaturation $\sigma < 1$ and the equilibrium adatom concentration n_{se} is usually a small part of the density of the adsorption sites N_0 [Bennema and Gilmer 1973].

D. Rate of advance of a train of parallel steps

We consider a train of parallel equidistant steps as shown in Fig. 3.5 where y_0 is the step separation. We again assume that $\lambda_s \gg \delta_0$. Obviously, the adatom concentration has its maximum value at the midpoint between the steps so that $(dn_s/dy)_{y=0} = 0$ (the distance y being measured from the midpoint between the steps) or $(d\Psi/dy)_{y=0} = 0$. The concentration in the near vicinity of the steps is again equal to n_{st} and $\Psi(y \to \pm y_0/2) = \chi\sigma$. Then the solution of Eq. (3.38) reads

$$\Psi = \chi\sigma\frac{\cosh\left(\dfrac{y}{\lambda_s}\right)}{\cosh\left(\dfrac{y_0}{2\lambda_s}\right)} \tag{3.47}$$

and for v_∞ one obtains

$$v_\infty = 2\chi\sigma a^2 n_{se}\frac{D_s}{\lambda_s}\tanh\left(\frac{y_0}{2\lambda_s}\right) = 2\chi\sigma\lambda_s\nu\exp\left(-\frac{\varphi_{1/2}}{kT}\right)\tanh\left(\frac{y_0}{2\lambda_s}\right) \tag{3.48}$$

which reduces to (3.43) when $y_0 \to \infty$.

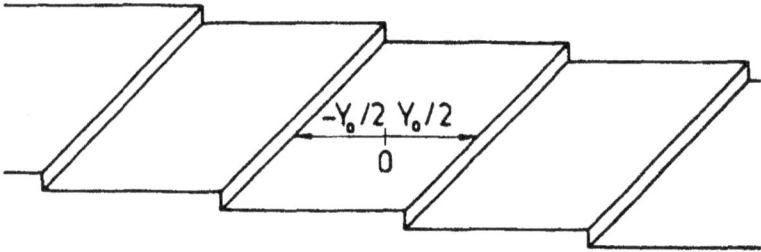

Fig. 3.5. Train of parallel equidistant steps spaced at a distance y_0 from each other. To solve the diffusion problem it is convenient to consider distances from the midpoint between the steps.

Bearing in mind that in a kinetic regime the adatom concentration on the terraces between the steps is not affected by the presence of the latter and the steps do not interact with each other through the diffusion fields, we can perform the same operation as the above to find an expression for χ. Equating (3.41) and (3.48) gives

$$\chi = \left[1 + \frac{D_s}{\lambda_s\beta_{st}}\tanh\left(\frac{y_0}{2\lambda_s}\right)\right]^{-1}$$

and

$$v_\infty = 2\sigma\nu\lambda_s\exp\left(-\frac{\varphi_{1/2}}{kT}\right)\frac{\tanh\left(\dfrac{y_0}{2\lambda_s}\right)}{1 + \dfrac{D_s}{\lambda_s\beta_{st}}\tanh\left(\dfrac{y_0}{2\lambda_s}\right)}\;. \tag{3.49}$$

It is immediately seen that the condition (3.28) leads to an expression valid for the purely diffusion regime of growth:

$$v_\infty = 2\sigma\lambda_s\nu\exp\left(-\frac{\varphi_{1/2}}{kT}\right)\tanh\left(\frac{y_0}{2\lambda_s}\right)\;, \tag{3.50}$$

whereas the condition (3.29) leads again to Eq. (3.41′) for the kinetic regime of step advance.

The hyperbolic tangent $\tanh(x)$ initially increases linearly with its argument x, and at large enough values of the latter $(x > 2)$ it goes asymptotically to unity. So if the step distance y_0 is sufficiently larger than the mean free path λ_s of the adatoms on the crystal surface $\tanh(y_0/2\lambda_s) \to 1$ and the adatom concentration in the middle parts of the terraces far from the steps will be unaffected by the presence of the latter, i.e. it will be equal to $n_s = P(2\pi m k T)^{-1/2}\tau_s$. The diffusion fields will not overlap, the steps will not interact with each other and the rate of step advance will be equal to that of the isolated steps. Equation (3.49) reduces to (3.45) and Eq. (3.50) reduces to (3.46). At the other extreme $y_0/2\lambda_s \to 0$ (it is enough if $y_0/2\lambda_s < 0.1$), the hyperbolic tangent can be approximated by its argument and Eq. (3.50) turns into $v_\infty = \nu\sigma y_0 \exp(-\varphi_{1/2}/kT)$. As will be shown in the next chapter, the step separation y_0 is inversely proportional to the supersaturation and $v_\infty = \text{const}$. Physically this means that the overlapping of the diffusion fields is so strong that the adatom concentration on the terraces between the steps is practically equal to the equilibrium adatom concentration n_{se} and the diffusion gradient becomes equal to zero. The steps move under conditions which are very near to equilibrium and their rate of advance ceases to depend on the supersaturation.

E. *Rate of advance of curved steps*

We consider the rate of lateral growth of a circular 2D cluster with radius ρ. Its shape is determined by the differences in the velocities for different orientations. If the velocity is orientationally independent the shape will be circular.

The flux of atoms towards the curved step will be given now by

$$j_+ = \frac{2\pi\rho}{\delta_0}\nu n_{st} a^2 \exp\left(-\frac{\Delta U}{kT}\right) , \qquad (3.51)$$

where $2\pi\rho/\delta_0$ is the number of the kinks at the island's periphery.

In order to find the reverse flux we recall that the equilibrium of small 2D islands with the parent phase is determined not by the work of separation from the half-crystal position, but by the mean separation work $\bar{\varphi}_2$ (Eq. (1.65)). Then the work to transfer an atom from a kink position along the edge of the 2D island to the adlayer on the terrace will be given by

$$\Delta W(\rho) = \bar{\varphi}_2 - \varphi_{\text{des}} = \varphi_{1/2} - \frac{\varkappa a^2}{\rho} - \varphi_{\text{des}} = \Delta W - \frac{\varkappa a^2}{\rho} , \qquad (3.52)$$

where $\varkappa$ is the specific edge energy of the step (Eq. (1.67)). In fact Eq. (3.52) reflects the enhanced chemical potential of a cluster of finite size, or in other words, the Thomson–Gibbs effect (see Eq. (1.66′)).

The reverse flux reads

$$j_- = \frac{2\pi\rho}{\delta_0} \nu \exp\left(-\frac{\Delta W(\rho) + \Delta U}{kT}\right) . \qquad (3.53)$$

At equilibrium both fluxes are equal, $n_{\text{st}} = n_{\text{se}}(\rho)$, and the equation of Thomson–Gibbs for the two-dimensional case

$$n_{\text{se}}(\rho) = n_0 \exp\left(-\frac{\Delta W(\rho)}{kT}\right) = n_{\text{se}} \exp\left(\frac{\varkappa a^2}{\rho kT}\right) \qquad (3.54)$$

results where n_{se} is given by (3.18).

Assuming the 2D island is large enough (low supersaturation) the net fluxes from the upper surface of the island and the surrounding crystal face to the circular step should be equal. Then the radial velocity of the step advance will be

$$v(\rho) = 2\frac{j_+ - j_-}{2\pi\rho N_0}$$

and

$$v(\rho) = 2a^2\beta_{\text{st}}[n_{\text{st}} - n_{\text{se}}(\rho)] , \qquad (3.55)$$

where the kinetic coefficient β_{st} is again given by (3.26). As seen, (3.55) reduces to (3.25) when $\rho \to \infty$. Bearing in mind that $n_{\text{se}}(\rho) > n_{\text{se}}$ it follows that the velocity of advance of a curved step is smaller than the rate of advance of a straight step under the same conditions.

The difference $n_{\text{st}} - n_{\text{se}}(\rho)$ can be rearranged as follows:

$$n_{\text{st}} - n_{\text{se}}(\rho) = n_{\text{se}}\left[\left(\frac{n_{\text{st}}}{n_{\text{se}}} - 1\right) - \left(\frac{n_{\text{se}}(\rho)}{n_{\text{se}}} - 1\right)\right] = n_{\text{se}}[\sigma_{\text{st}} - \sigma(\rho)]$$

$$= n_{\text{se}}\{[\sigma - \sigma(\rho)] - [\sigma - \sigma_{\text{st}}]\} = n_{\text{se}}\sigma\left(1 - \frac{\sigma(\rho)}{\sigma} - \chi\right) .$$

The radius ρ_c of the critical radius is defined by the Thomson–Gibbs equation (1.66′)

$$\sigma = \frac{\varkappa a^2}{kT\rho_c} . \qquad (3.56)$$

From (3.54) and (3.56) it follows that

$$\frac{\sigma(\rho)}{\sigma} = \frac{\rho_c}{\rho}$$

and finally

$$v(\rho) = 2a^2 \beta_{st} n_{se} \sigma \left(1 - \frac{\rho_c}{\rho} - \chi\right) . \tag{3.57}$$

The diffusion equation (3.36) in polar coordinates reads

$$\frac{d^2 \Psi(r)}{dr^2} + \frac{1}{r}\frac{d\Psi(r)}{dr} = \frac{\Psi(r)}{\lambda_s^2} , \tag{3.58}$$

where $\Psi(r) = \sigma - \sigma_s(r)$, and is subject to the boundary conditions $\Psi(r \to \infty) = 0$, $[d\Psi(r)/dr]_{r=0} = 0$ and $\Psi(r = \rho) = \Psi(\rho) = \sigma - \sigma_{st}(\rho) = \chi\sigma$ $(\sigma_{st}(\rho) = n_{st}/n_{se} - 1)$. The solution of (3.58) reads

$$\Psi(r) = \Psi(\rho)\frac{I_0\left(\dfrac{r}{\lambda_s}\right)}{I_0\left(\dfrac{\rho}{\lambda_s}\right)} \qquad \text{for } r < \rho , \tag{3.59'}$$

$$\Psi(r) = \Psi(\rho)\frac{K_0\left(\dfrac{r}{\lambda_s}\right)}{K_0\left(\dfrac{\rho}{\lambda_s}\right)} \qquad \text{for } r > \rho , \tag{3.59''}$$

where $I_0(x)$ and $K_0(x)$ are the Bessel functions of the first and the second kind with imaginary argument.

The flux of atoms towards the edge of the cluster is

$$j_s(\rho) = 2\pi\rho D_s n_{se}\left(\frac{d\Psi}{dr}\right)_{r=\rho} = 4\pi\rho n_{se}\frac{D_s}{\lambda_s}\Psi(\rho) , \tag{3.60}$$

where the formulae $I_0'(x) = I_1(x)$, $K_0'(x) = -K_1(x)$, $I_1(x)K_0(x) + I_0(x)K_1(x) = 1/x$ and the approximation $I_0(x)K_0(x) = 1/2x$ valid for $x > 1$ have been used.

The radial velocity of advance of a curved step is then

$$v(\rho) = \frac{j_s(\rho)}{2\pi\rho N_0} = 2a^2 n_{se}\chi\sigma\frac{D_s}{\lambda_s} . \tag{3.61}$$

From (3.57) and (3.61) one obtains

$$\chi = \left(1 - \frac{\rho_c}{\rho}\right)\left(1 + \frac{D_s}{\lambda_s \beta_{st}}\right)^{-1}$$

and

$$v(\rho) = 2\sigma\lambda_s\nu \exp\left(-\frac{\varphi_{1/2}}{kT}\right)\left(1 - \frac{\rho_c}{\rho}\right)\left(1 + \frac{D_s}{\lambda_s \beta_{st}}\right)^{-1}$$

or

$$v(\rho) = v_\infty \left(1 - \frac{\rho_c}{\rho}\right) \, , \tag{3.62}$$

where v_∞ is the rate of advance of a straight step given by (3.45).

Finally, for the interesting case of growth of concentric circular clusters with edges spaced y_0 from each other the rate of advance will be given by (see Eq. (3.49))

$$v(\rho) = 2\sigma\lambda_s\nu \exp\left(-\frac{\varphi_{1/2}}{kT}\right) \frac{\tanh\left(\dfrac{y_0}{2\lambda_s}\right)}{1 + \dfrac{D_s}{\lambda_s \beta_{st}}\tanh\left(\dfrac{y_0}{2\lambda_s}\right)} \left(1 - \frac{\rho_c}{\rho}\right) \, . \tag{3.63}$$

This is a general expression for the rate of advance of monoatomic steps. All limiting cases of curved and straight steps or train of steps in both diffusion and kinetic regimes of growth can be easily derived from it.

3.2.1.2. *Growth from solutions*

In the case of growth from solutions the supply of growth units takes place predominantly through diffusion in the bulk of the solution [Burton, Cabrera and Frank 1951; Chernov 1961] although there is evidence that the growth units reach the growth sites at least partly by surface diffusion as well [Bennema 1974; Vekilov *et al.* 1992; Zhang and Nancollas 1990]. The problem of growth by surface diffusion resembles very much that from vapors, and in the following presentation we will take into consideration the bulk diffusion only. The problem of propagation of steps simultaneously by surface and bulk diffusion has been treated theoretically by Van der Eerden [1982, 1983].

The solutions are usually stirred. If the solution is still, the exhaustion of the solution near the growing crystal will give rise to convection flows. Thus in all cases the solution moves with respect to the growing crystal. When a fluid moves tangentially to a plane surface the velocity of the fluid decreases towards the surface and in the near vicinity of the latter an immobile boundary layer is formed as shown in Fig. 3.6 [Schlichting

1968]. The latter is frequently called a *stagnant layer*. The thickness of the stagnant layer depends on the velocity of the fluid, ϑ, its viscosity η and density ρ, and the distance x from the leading edge of the crystal surface according to the approximate formula

$$d \cong 5 \left(\frac{x\eta}{\vartheta\rho} \right)^{1/2} .$$

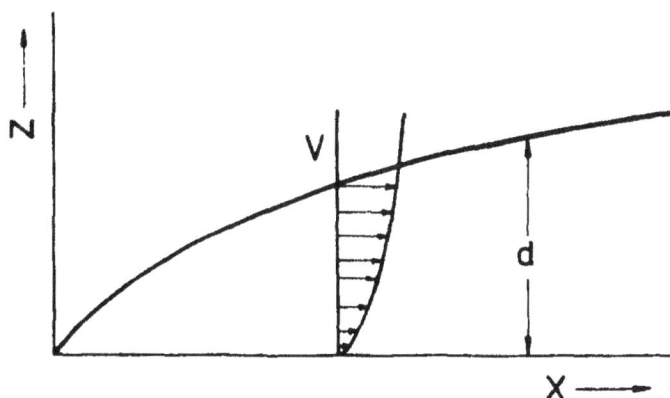

Fig. 3.6. Schematic representation of the stagnant boundary layer above the surface of a crystal in a tangentially moving fluid. The arrows give an impression of the decrease of the velocity of the fluid when approaching the crystal surface. The thickness of the stagnant layer d depends on the distance x from the leading edge of the crystal face.

For values of the parameters involved, typical for aqueous solutions at room temperature, $\eta = 1 \times 10^{-2}$ g cm^{-1}sec^{-1}, $\rho = 1$ g cm^{-3}, $\vartheta = 40$ cm sec^{-1} and $x = 0.1$ cm, $d \cong 0.25$ mm. The abovementioned formula gives only a qualitative indication as the real situation in stirred solution or around rotating crystal can be quite different. The concept of stagnant boundary layer is also widely used for the description of processes taking part in reactors for Chemical Vapor Deposition (CVD) [Carra 1988].

It is usually assumed that within a stagnant layer the transport of the growth species to the surface of the crystal occurs by diffusion, while at the upper boundary of the layer the concentration of the solute is maintained constant and equal to the bulk concentration C_∞. Assuming again that the rate of the movement of the steps is sufficiently smaller than the rate of diffusion the concentration of the solute in the boundary layer is described by the equation of Laplace $\Delta C = 0$, where Δ is the Laplace operator. When considering the movement of a single step as a result of incorporation of growth units into kink sites along the step, the mean kink spacing is obviously much smaller than the thickness of the boundary layer. Then the step acts as a linear sink for the growth units and the diffusion field has the form of a semicylinder oriented with its axis along the step (Fig. 3.7). It is thus convenient to express the Laplace equation in cylindrical coordinates:

$$\frac{d^2C}{dr^2} + \frac{1}{r}\frac{dC}{dr} = 0 \ ,$$

where r is the radius vector. When we consider the growth of a vicinal crystal face with equidistant steps we have to take into account the overlapping of the diffusion fields as is shown in Fig. 3.8 [Chernov 1961]. We will consider these two cases separately as was done above.

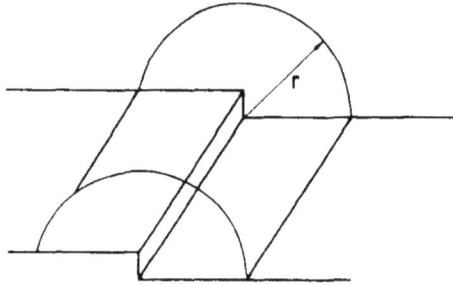

Fig. 3.7. Cylindrical symmetry of the volume diffusion field around an isolated step. The distance from the step is characterized by the radius vector r.

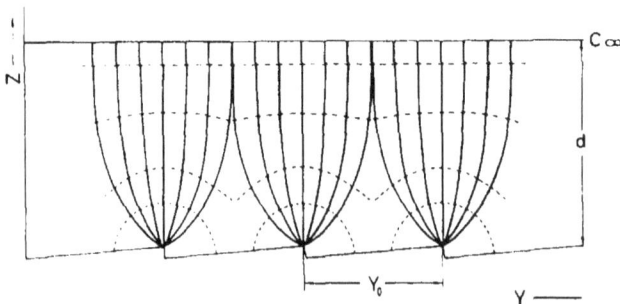

Fig. 3.8. Schematic view of a train of parallel steps along the y direction spaced at an average distance y_0. The transport of building units through bulk diffusion takes place along the solid lines. The dashed lines represent surfaces with equal solute concentration. Far from the steps at the upper boundary d of the stagnant layer the solute concentration C_∞ is equal to that of the bulk of the solution (after Chernov [1961]).

A. *Rate of advance of a single step*

We consider this simpler case for illustrating the approach. A solution of the Laplace equation is the function [Carslaw and Jaeger 1960]

$$C(r) = A \ln(r) + B , \qquad (3.64)$$

which can be verified by inspection and where A and B are constants. The latter can be found from the boundary conditions

$$r = \frac{a}{\pi}, \qquad C = C_{st} , \qquad (3.65')$$

$$r = d, \qquad C = C_\infty , \qquad (3.65'')$$

where C_{st} and C_∞ are the concentrations at the step vicinity and the upper boundary of the stagnant layer or in the bulk of the solution, respectively (Fig. 3.8).

The condition (3.65′) means that we approximate the step with height a by a semicylindrical surface of radius $r = a/\pi$. The condition (3.65″) as it is written means that the concentration has the value C_∞ at the semicylindrical surface with radius $r = d$. As there are no other steps nearby and therefore no other sinks for the solute species, this condition is a direct consequence of the assumption that the transport occurs by volume diffusion towards the sites of growth. In other words, there is no flux of atoms to other parts of the surface, and the concentration far from the step is constant and equal to C_∞ in all directions.

Then for the concentration profile around the step one obtains

$$C(r) = C_\infty - (C_\infty - C_{st}) \frac{\ln\left(\dfrac{r}{d}\right)}{\ln\left(\dfrac{a}{\pi d}\right)} . \qquad (3.66)$$

The rate of the step advance is

$$v_\infty = v_c D \left(\frac{dC}{dr}\right)_{r=\frac{a}{\pi}} = \frac{\pi v_c D C_0}{a \ln\left(\pi \dfrac{d}{a}\right)} \chi \sigma , \qquad (3.67)$$

where v_c is the volume of a growth unit in the crystal, D is the bulk diffusion coefficient and C_0 is the equilibrium concentration of the solute at the given temperature. $\sigma = C_\infty/C_0 - 1$ is the supersaturation, $\sigma_{st} = C_{st}/C_0 - 1$, $\chi = (\sigma - \sigma_{st})/\sigma$ and $C_\infty - C_{st} = \chi C_0 \sigma$ in complete analogy with the previous case.

On the other hand,

$$v_\infty = \beta_{st} v_c (C_{st} - C_0) = \beta_{st} C_0 v_c \sigma_{st} = \beta_{st} C_0 v_c \sigma (1 - \chi) , \qquad (3.68)$$

where

$$\beta_{st} = a\nu \frac{a}{\delta_0} \exp\left(-\frac{\Delta U}{kT} \right)$$

is the kinetic coefficient of the step, in complete analogy with that of the crystal face (Eq. 3.9). As seen, the only difference is in the dimensionality of the probability to find a kink site.

Equating (3.67) with (3.68) gives

$$\chi = \frac{a\beta_{st} \ln \left(\pi \frac{d}{a} \right)}{\pi D + a\beta_{st} \ln \left(\pi \frac{d}{a} \right)}$$

and

$$v_\infty = \frac{\beta_{st} C_0 v_c \sigma}{1 + \dfrac{a\beta_{st}}{\pi D} \ln \left(\pi \dfrac{d}{a} \right)} . \qquad (3.69)$$

As before, when the rate of diffusion, $\pi D/a$, is sufficiently greater than the rate of crystallization, β_{st}, the latter controls the rate of the step advance. The latter is given in the kinetic regime by

$$v_\infty = \beta_{st} C_0 v_c \sigma . \qquad (3.70)$$

At the other extreme of the diffusion regime ($\pi D/a \ll \beta_{st}$),

$$v_\infty = \frac{\pi D C_0 v_c}{a \ln \left(\pi \dfrac{d}{a} \right)} \sigma \qquad (3.71)$$

and v_∞ is a linear function of the supersaturation as in the case of vapor growth.

B. *Rate of advance of a step in a train of steps*

This problem was solved for the first time by Chernov [1961]. The solution of the Laplace equation $\Delta C = 0$ (see Fig. 3.8 for the orientation of the coordinate system) subject to the boundary conditions

$$y = 0 \qquad \text{and} \qquad z = d, \quad C = C_\infty , \qquad (3.72')$$

$$y = 0 \qquad \text{and} \qquad z = \frac{a}{\pi}, \quad C = C_{\text{st}} \qquad (3.72'')$$

reads

$$C = A \ln \left[\sin^2 \left(\frac{\pi}{y_0} y \right) + \sinh^2 \left(\frac{\pi}{y_0} z \right) \right]^{1/2} + B , \qquad (3.73)$$

where A and B are constants:

$$A = \frac{C_\infty - C_{\text{st}}}{\ln \left[\sinh \left(\frac{\pi d}{y_0} \right) \right] - \ln \left[\sinh \left(\frac{a}{y_0} \right) \right]} \cong \frac{C_\infty - C_{\text{st}}}{\ln \left[\frac{y_0}{a} \sinh \left(\frac{\pi d}{y_0} \right) \right]} , \qquad (3.74)$$

$$B = C_\infty + A \ln \left[\sinh \left(\frac{\pi d}{y_0} \right) \right] , \qquad (3.75)$$

where the approximation $\sinh(a/y_0) \cong a/y_0$ ($a/y_0 \ll 1$) has been used.

The function $\sin(\pi y/y_0)$ in (3.73) reflects the periodicity in C due to the sequence of equidistant steps. The hyperbolic sine $\sinh(\pi z/y_0)$ accounts for the dependence of C in a direction normal to the growing surface. It is immediately seen that (3.73) reduces to (3.64) at $y = 0$ and large distances between the steps so that $\sinh(\pi z/y_0) \cong \pi z/y_0$. The boundary condition (3.72'), $C = C_\infty$, is, strictly speaking, valid for $z = d$ at any y. As discussed by Chernov [1961], when $\pi d \gg y_0$ the concentration does not depend anymore on y for large values of z and the condition $C(0, d) = C_\infty$ becomes equivalent to $C(d) = C_\infty$. In the opposite case where $\pi d \ll y_0$ we have in fact steps far apart and the solution for single step is valid.

In order to calculate the rate of step advance we have to find the concentration gradient dC/dr. The latter is given by $dC/dr = (dC/dz)(dz/dr) + (dC/dy)(dy/dr)$, where $dz/dr = r/z$, $dy/dr = r/y$ and $r = (y^2 + z^2)^{1/2}$. Making use of (3.67) and the above relations, we find that $(dC/dr)_{r=a/\pi} = \pi A/a$ and

$$v_\infty = \frac{\pi v_c D C_0 \chi \sigma}{a \ln \left[\frac{y_0}{a} \sinh \left(\frac{\pi d}{y_0} \right) \right]} .$$

Equating (3.68) and the above expression gives

$$\chi = \frac{a \beta_{\text{st}} \ln \left[\frac{y_0}{a} \sinh \left(\frac{\pi d}{y_0} \right) \right]}{\pi D + a \beta_{\text{st}} \ln \left[\frac{y_0}{a} \sinh \left(\frac{\pi d}{y_0} \right) \right]}$$

and finally

$$v_\infty = \frac{v_c \beta_{st} C_0 \sigma}{1 + \dfrac{a\beta_{st}}{\pi D} \ln \left[\dfrac{y_0}{a} \sinh \left(\dfrac{\pi d}{y_0} \right) \right]} . \qquad (3.76)$$

It is immediately seen that the condition $\pi d/y_0 \ll 1$ reduces the equation (3.76) to (3.69), valid for single isolated steps.

The corresponding limit cases for diffusion and kinetic regime are easy to obtain. In the first case $(\pi D/a \ll \beta_{st})$,

$$v_\infty = \frac{\pi D v_c C_0 \sigma}{a \ln \left[\dfrac{y_0}{a} \sinh \left(\dfrac{\pi d}{y_0} \right) \right]} , \qquad (3.77)$$

whereas in the second case $(\pi D/a \gg \beta_{st})$, Eq. (3.70) results.

The reciprocal of the function in the denominator in (3.77) has qualitatively the same behavior as the hyperbolic tangent. The velocity of advance of a step in a train is a linear function of the supersaturation in the diffusion regime only when the step spacing y_0 is sufficiently larger than the thickness of the boundary layer, which means in practice isolated steps. At the other extreme $\pi d \gg y$, the hyperbolic sine $\sinh(x)$ can be approximated by $\exp(x)/2$ and $v_\infty \cong D v_c C_0 \sigma y_0 / ad = \text{const}$ $(y_0 \approx 1/\sigma)$ as in the case of growth from vapors.

Let us consider as an example the growth of the prismatic face of $NH_4H_2PO_4$ (ADP) crystals from an aqueous solution at room temperature [Chernov 1989]. With $a \cong 4 \times 10^{-8}$ cm, $\nu \cong 1 \times 10^{13}$ sec^{-1}, $\delta_0 \cong 4a$, $D \cong 1 \times 10^{-6}$ cm^{2}sec^{-1} and $\Delta U \cong 10$ kcal/mole, $\beta_{st} \cong 4 \times 10^{-3}$ cm sec^{-1}, $a\beta_{st}/\pi D \cong 5 \times 10^{-5} \ll 1$ (the logarithm can contribute no more than an order of magnitude) and the growth proceeds in the kinetic regime. The saturation concentration $C_0 \cong 3.5$ mole l^{-1}, $C_0 v_c \cong 0.2$ and with $\sigma = 0.03$, $v_\infty \cong 2.4 \times 10^{-5}$ cm sec^{-1}, in good agreement with the experimentally measured value 3×10^{-5} cm sec^{-1}. The reader can find more details in the excellent review paper of Chernov [1989] and the original papers quoted therein [Chernov *et al.* 1986; Kuznetsov, Chernov and Zakharov 1986; Smol'sky, Malkin and Chernov 1986].

Another interesting example is the growth of (111) faces of $Ba(NO_3)_2$ crystals [Maiwa, Tsukamoto and Sunagawa 1990]. The experimentally measured rate of step advance depends linearly on the supersaturation at high rates of flow of the solution (40 cm sec^{-1}) and nonlinearly at low flow rates (5 cm sec^{-1}). Assuming $\pi d/y_0 \ll 1$ at high flow rates, the rate of the step advance will be given by Eq. (3.69), which is a linear function of the supersaturation. At low flow rates, $\pi d/y_0 \gg 1$ and the hyperbolic sine can be approximated by an exponent, thus giving rise to a linear term of the supersaturation in the denominator of Eq. (3.76). The latter leads to a clear nonlinear dependence of v_∞ on σ at small supersaturations as observed in the experiment whose slope at $\sigma = 0$ is approximately equal to $\beta_{st}C_0 v_c$.

3.2.1.3. Growth from melts

As mentioned in Chap. 1 the entropies of melting, $\Delta s_m/k$ (in terms of Bolzmann's constant), of most of the metals have an average value of about 1.2 and their surfaces near to the melting point are expected to be rough according to the simplest criterion of Jackson. The $\Delta s_m/k$ values for semiconductors are usually greater — from 3.6 and 3.7 for Si and Ge to 5.7 for InP, 7.4 for InSb, 7.6 for InAs and 8.5 for GaAs, etc. So the surfaces of the binaries mentioned above are expected to be smooth and to grow by the layer mechanism. As for elemental semiconductors it is difficult to predict the structure of their surfaces with sufficient accuracy.

In general the growth from simple one-component melts is similar to the growth from solutions [Chernov 1961, 1984]. When a building unit is incorporated into a site of growth, heat of crystallization is released and the local temperature becomes higher. Assuming as before that the mean kink spacing δ_0 is small enough the step will act as a linear source of heat. The undercooling around the step will decrease just like the supersaturation in the solution in the near vicinity of the steps. Then a temperature gradient in a semicylindrical space around the steps arises. The mathematical equations which govern the conduction of heat in condensed phases are exactly the same as the diffusion equations [Carslaw and Jaeger 1960] and we have to solve precisely the same mathematical problem as for solution growth [Chernov 1961, 1984]. The heat of crystallization can be taken away through the melt or, more typically, through the crystal as the thermal conductivity of the crystals is usually higher than that of the liquids. For instance, the thermal conductivities of solid and liquid aluminum at the melting point are 0.51 and 0.21 cal/cm sec K respectively. In the first case when the heat is taken away through a stirred melt we have precisely the same solution of the master equation as for growth in stirred solution and the thickness of the boundary layer has the same physical meaning as before. In the second case, the thickness of the single crystal wafer can be taken instead. Obviously, the same expressions are obtained as in the previous section in which the supersaturation $\sigma = (C_\infty - C_0)/C_0$ should be replaced by the undercooling $\sigma = (T_m - T_\infty)/T_m$ and the diffusion coefficient D by the coefficient of the temperature conductivity, $\varkappa_T = k_T/C_p\rho$ (cm^{2}sec^{-1}) (k_T is the coefficient of the thermal conductivity, C_p is the specific heat capacity (cal/g K) and ρ (g/cm^3) is the density). Instead of the kinetic coefficient of the crystal face, β_{st}, for solutions one has to take the kinetic coefficient of the step for melts,

$$\beta_{st}^T = a\nu \frac{\Delta s_m}{kT} \frac{a}{\delta_0} \exp\left(-\frac{\Delta s_m}{k}\right) \exp\left(-\frac{\Delta U}{kT}\right) , \qquad (3.78)$$

multiplied by a characteristic temperature $T_q = \Delta H_m/C_p^{liq}$ which is given by the ratio of the heat of crystallization (or melting) to the molar heat capacity of the liquid. As C_p^{liq} is the quantity of heat required to increase the temperature of one mole of the melt by one degree and ΔH_m is the quantity of heat per mole that is introduced into the melt as a result of the crystallization process, T_q is the temperature up to which the melt will be heated up if the heat of crystallization is not taken away from the system. For metal melts it has the value of several hundreds of degrees (445 K for Ag) but is considerably higher for semiconductors (1860 K for Si and 1490 K for GaAs).

Then the expression for the rate of advance of the step reads

$$v_\infty = \frac{\beta_{st}^T T_m \sigma}{1 + \dfrac{a\beta_{st}^T T_q}{\pi \varkappa_T} \ln\left[\dfrac{y_0}{a} \sinh\left(\dfrac{\pi d}{y_0}\right)\right]} . \qquad (3.79)$$

This expression is valid for both cases of removing the heat of crystallization, through the melt and through the crystal. One has to bear in mind that in the first case the coefficient of the temperature conductivity, $\varkappa_T$, has the value for the liquid, and vice versa. In the case of growth of Si at $\Delta T = T_m - T = 1$ K, $k_T = 0.356$ cal/cm sec K, $C_p^{sol} = 5.455$ cal/g-atom K $= 0.194$ cal/g K, $C_p^{liq} = 6.5$ cal/g-atom K, $\rho = 2.328$ g/cm^3, $\Delta H_m = 12082$ cal/mole and $\varkappa_T = 0.787$ cm^2/sec, $T_q = 1860$ K. With $\Delta U \cong 5000$ cal/mole, $\delta_0 = 3.5a$ and $\Delta s_m/k = 3.6$, $\beta_{st}^T = 3.7$ cm/sec K and $a\beta_{st}^T T_q/\pi \varkappa_T \cong 1 \times 10^{-4} \ll 1$, i.e. the growth proceeds in a kinetic regime. For the rate of advance of the step one obtains $v_\infty = \beta_{st}^T \Delta T = 3.7$ cm/sec. The value for GaAs is two orders of magnitude smaller ($v_\infty \cong 0.1$ cm/sec, $\varkappa_T = 0.267$ cm^2/sec, $T_q = 1490$ K, $\beta_{st}^T \cong 0.1$ cm/sec K) due to the higher entropy of melting. Comparing the above values with the one valid for growth in solutions one can see that they are about five orders of magnitude higher. The latter is due to the better supply of growth units in melts as compared with (dilute) solutions on the one hand, and to the smaller barrier for crystallization, $\Delta U/kT$, in the melts on the other.

3.2.2. *Spiral growth of F faces*

As discussed in Chap. 1 screw dislocations offer nonvanishing steps on crystal surface. A growth hillock is formed (Fig. 3.9) and in order to calculate the rate R of growth of the crystal face we have to find an expression for the step density p of the side face of the hillock, or in other words, the distance y_0 between the successive turns of the spiral.

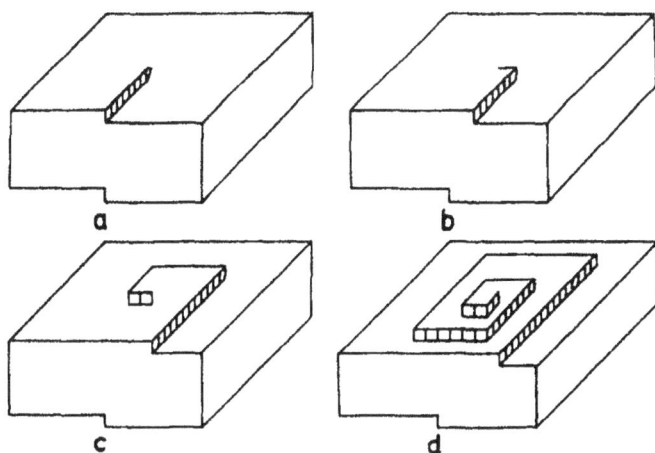

Fig. 3.9. Consecutive stages from (a) to (d) of the formation of a growth pyramid around the emergency point of a single screw dislocation. As in Fig. 3.2 the side faces of such pyramids represent in fact the vicinal surfaces. Their slopes are proportional to the supersaturation.

3.2.2.1. *Shape of the growth spiral*

Let us consider first for simplicity the formation of a growth hillock around a single screw dislocation (Fig. 3.10). We assume that the spiral is polygonized with a square shape and that the rate of advance v_∞ of the steps in every direction is one and the same. In the initial moment (Fig. 3.10(a)) the dislocation offers a single step with a lower terrace to the right of it. Atoms diffuse towards the step (on the crystal surface or in the bulk of the solution) and join kink sites along it. As a result, the step moves to the

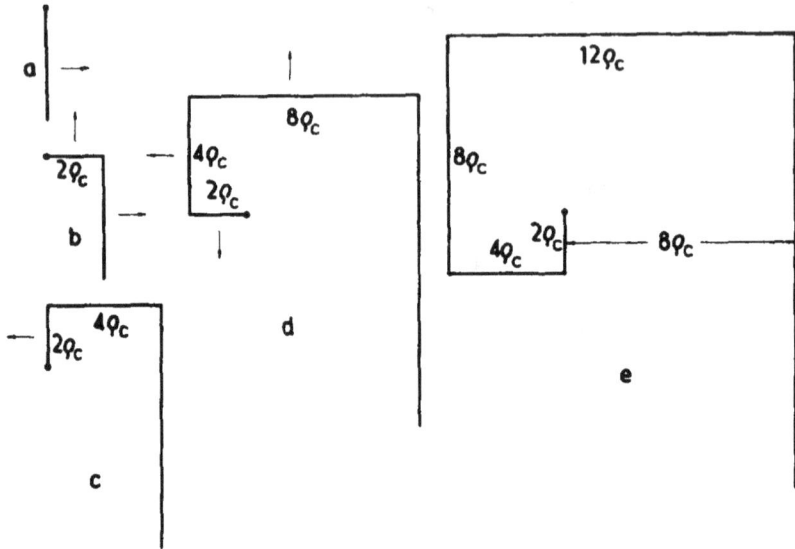

Fig. 3.10. Consecutive stages from (a) to (e) of the formation of a growth spiral. (a) shows a single step originating from the emergency point of the dislocation denoted by the black point. The arrow shows the direction of step advance. During the growth of the initial step a new step is formed as shown in (b). When the length of this step exceeds $2\rho_c$, the second step begins to grow and a new step is formed as seen in (c). As seen in (e), the distance between the consecutive turns of the growth spiral is proportional to the radius of the critical 2D nucleus. The radius of curvature of the step which comes out from the emergency point is always equal to the radius of the critical 2D nucleus. These are in fact the properties of the Archimedean spiral.

right and a new step normal to the first one appears (Fig. 3.10(b)). As long as the second step is shorter than the edge of the critical 2D nucleus, $l_2 = 2\rho_c$, at the given supersaturation it will not move because a 2D cluster smaller than the critical nucleus is thermodynamically unfavored and has a greater tendency to decay than to grow (the step "does not know" whether it belongs to a 2D nucleus or to a growth spiral). Once the size $2\rho_c$ is reached the second step begins to grow with a velocity v_∞ and a third step appears which is parallel to the first one (Fig. 3.10(c)). This third step will begin to grow when its length becomes greater than $2\rho_c$. At that moment the length of the second step will be equal to $4\rho_c$. Then a fourth step will appear, and so on. Following this procedure further we arrive at the picture in Fig. 3.10(e) and see that the distance between two successive turns of the growth spiral is equal to $8\rho_c$. Obviously, this consideration is oversimplified and gives only an indication of the real processes which take place during growth. It reflects correctly, however, two very important facts: (i) the spacing of the steps originating from a single dislocation is proportional to the size (radius) of the critical 2D nucleus at the given supersaturation and (ii) the length of the step which comes out from the emerging point of the screw dislocation is always equal to $2\rho_c$. If we assume that the growth is isotropic, the spiral will be rounded. Then the conclusion (ii) above means that the radius of curvature of the step at the emerging point of the screw dislocation is always equal to the radius ρ_c of the critical nucleus.

In order to find a more accurate expression for the shape of the spiral and, in turn, for the interstep distance we will follow the approach of Burton, Cabrera and Frank [1951].

The radius of the curvature, ρ, in polar coordinates (the polar angle φ and radius vector r) is given by

$$\rho = \frac{(r^2 + r'^2)^{3/2}}{r^2 + 2r'^2 - rr''} , \tag{3.80}$$

where $r' = dr/d\varphi$ and $r'' = d^2r/d\varphi^2$.

We consider further a spiral with center at the point O as shown in Fig. 3.11. The rate of advance $v(r)$ in the direction of the radius vector r is

$$v(r) = \frac{dr}{dt} = \frac{dr}{d\varphi}\frac{d\varphi}{dt} = \omega r' , \tag{3.81}$$

where $\omega = d\varphi/dt$ is the angular velocity of spiral winding.

The rate of advance in a direction of the radius of curvature, $v(r)$, is related to the rate of advance $v(\rho)$ in a direction normal to the step by

$$v(\rho) = v(r)\cos\gamma = \frac{v(r)r}{(r^2 + r'^2)^{1/2}} = \frac{rr'\omega}{(r^2 + r'^2)^{1/2}} . \tag{3.82}$$

Substituting (3.80) and (3.82) into (3.62) gives the equation

$$v_\infty \left(1 - \rho_c \frac{r^2 + 2r'^2 - rr''}{(r^2 + r'^2)^{3/2}}\right) = \frac{\omega rr'}{(r^2 + r'^2)^{1/2}} , \tag{3.83}$$

whose solution $r = r(\varphi)$ will give us the shape of the spiral.

In order to simplify this expression we consider the case of small r, i.e. around the center of the spiral. We neglect all terms containing r^2 and r and integrate the remaining differential equation $r' = 2\rho_c$. As a result the

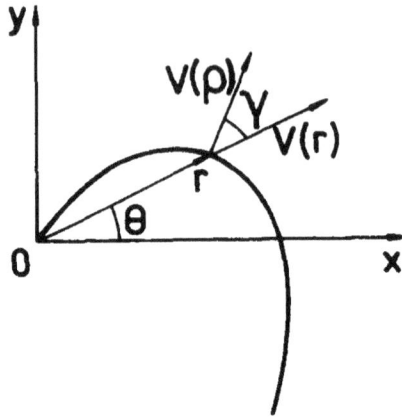

Fig. 3.11. Schematic view of the growth spiral around the emergency point of a screw dislocation. $v(r)$ and $v(\rho)$ are the rates of step advance in directions along the radius vector and normal to the step, respectively. γ denotes the angle between them.

equation for the simplest spiral, which is well known as an Archimedean spiral, is obtained:

$$r = 2\rho_c \varphi \ . \tag{3.84}$$

Let us consider this equation more carefully. Substituting $r' = 2\rho_c$ and $r'' = 0$ into (3.80) we find that the radius of the curvature of the Archimedean spiral is given by

$$\rho = \rho_c \frac{(1 + x^2)^{3/2}}{1 + \frac{1}{2}x^2} \ ,$$

where $x = r/2\rho_c$. It is immediately seen that the radius of the curvature is equal to the radius of the 2D nucleus in the center of the spiral ($x = 0$) and goes linearly to infinity far from the spiral center ($x \to \infty$). For comparison the radius of curvature $\rho = r(1 + m^2)^{1/2}$ of the logarithmic spiral ($r = a\exp(m\varphi)$) tends to infinity for $r \to \infty$ but is equal to zero at the spiral center.

On the other hand, the multiplier

$$1 - \frac{\rho_c}{\rho} = 1 - \frac{1 + \frac{1}{2}x^2}{(1 + x^2)^{3/2}}$$

in the equation for the rate of the step advance, (3.62), is equal to zero at the spiral center and tends asymptotically to unity far from the spiral center. It follows that the rate of the step advance will vary from zero at the spiral center to the rate of advance of straight steps far from the spiral center. Thus the first turns of the spiral will move more slowly (slow kinetics) than the more distant ones (fast kinetics).

The distance Λ between two successive turns of the single spiral or two successive steps is

$$\Lambda = r(\varphi + 2\pi) - r(\varphi) = 2\rho_c[(\varphi + 2\pi) - \varphi] = 4\pi\rho_c \ . \tag{3.85}$$

A more detailed analysis performed by Cabrera and Levine [1956] based on a better approximation for the spiral shape $r = r(\varphi)$ resulted in the expression

$$\Lambda = 19\rho_c = \frac{19\varkappa a^2}{kT\sigma} \tag{3.86}$$

which we will use further. It follows that the interstep distance is inversely proportional to the supersaturation and the increase of the latter makes the slope $p = a/\Lambda$ of the cone (or pyramid) of growth steeper and vice versa. Note that $p = \tan\theta$ where θ is the tilt angle with respect to the corresponding singular face and hence p is equal to the step density.

The existence of only one screw dislocation on a given crystal face is usually less probable than the existence of many dislocations and even groups of dislocations. Two neighboring dislocations can have in general equal or opposite signs. This means that they can either both turn clockwise (or counter-clockwise) or one of them clockwise and the other counter-clockwise. If they have opposite signs and are spaced farther than $2\rho_c$ they will make loops as shown in Fig. 3.12(a) [Frank 1949a]. The problem when the dislocations have like signs is more complicated. Two cases are distinguished. In the first one the dislocation spacing l is smaller than $2\rho_c$, in the second one l is greater than $2\rho_c$. Imagine now a group of n dislocations ordered in a straight line as along a grain boundary. Let L denote its length so that the dislocation spacing $l = L/n$. Figures 3.12(b) and (c) show parts of the dislocation group (the grain boundary) consisting, for clarity, only of two dislocations spaced at a distance AB. The condition $l \gg 2\rho_c$ is equivalent to $L \gg \Lambda$, where Λ is the interstep distance determined by a single dislocation and is given by Eq. (3.86). The second condition $l \ll 2\rho_c$ is equivalent to $L \ll \Lambda$ (Fig. 3.12(c)). As seen in the figures the real interstep spacing y_0 is equal either to $l = L/n$ (Fig. 3.12(b)) or to Λ/n (Fig. 3.12(c)), when $L \gg \Lambda$ or $L \ll \Lambda$, respectively. In the general case [Burton, Cabrera and Frank 1951; Bennema and Gilmer 1973; Chernov 1989]

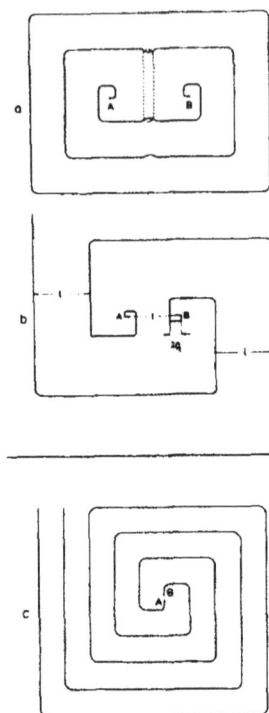

Fig. 3.12. Shape of growth spirals due to pairs of dislocations. (a) Closed loops due to dislocations with opposite signs when the distance AB between the emergency points is greater than the radius of the critical 2D nucleus. (b) Shape of the spiral due to a pair of dislocations of like sign, separated by a distance $l = AB \gg 2\rho_c$. The step separation is equal to the distance l and does not depend on the supersaturation. (c) Shape of the spiral due to a pair of dislocations of like sign, separated by a distance $l = AB \ll 2\rho_c$. The step separation is two times smaller than the distance Λ originating from a single dislocation and is inversely proportional to the supersaturation (after Burton, Cabrera and Frank [1951]).

$$y_0 = \frac{\Lambda}{\mathfrak{n}} , \qquad (3.87)$$

where

$$\mathfrak{n} = n \left(1 + \frac{L}{\Lambda}\right)^{-1}$$

is the so-called "strength" of the dislocation source. It follows from above that when $L \gg \Lambda$ the interstep spacing is simply equal to the interdislocation distance $l = L/n$ and does not depend on the supersaturation. In the second limiting case the interstep spacing depends on the supersaturation but is n times smaller than $19\rho_c$. In the case when the dislocations in the source are not ordered in a straight line but are grouped in a spatial region L denotes the perimeter of the region [Chernov 1989].

3.2.2.2. Growth from a vapor phase

The rate of growth normal to the surface is given by Eq. (3.15), $R = p v_\infty = a v_\infty / y_0$, where v_∞ / y_0 is the flux of steps with thickness a passing in a direction parallel to the singular crystal face over any point of the latter. We assume first that the crystal face is entirely covered by a growth pyramid formed by a single dislocation (or by a group of dislocations such that $L \ll \Lambda$ and $y_0 = \Lambda/n$). Substituting (3.50), (3.86) and (3.87) into (3.15) (diffusion regime of growth and far from the spiral center) gives, for R,

$$R = C \frac{\sigma^2}{\sigma_c} \tanh \left(\frac{\sigma_c}{\sigma} \right) , \qquad (3.88)$$

where

$$\sigma_c = \frac{19 \varkappa a^2}{2 n k T \lambda_s} \qquad (3.89)$$

is a characteristic supersaturation and

$$C = a \nu \exp \left(- \frac{\varphi_{1/2}}{kT} \right) \qquad (3.90)$$

is a rate constant.

Let us study Eq. (3.88) more closely. For typical values of the parameters included in (3.89): $\varkappa \cong 3 \times 10^{-5}$ erg/cm, $a \cong 3 \times 10^{-8}$ cm, $T = 1000$ K, $n = 1$ and $\lambda_s = 2 \times 10^3 a$, $\sigma_c = 3 \times 10^{-2}$. However, growth of crystals is observed at supersaturations σ as low as 1×10^{-4}. Obviously, two limiting cases can be distinguished. At small supersaturations such that $\sigma_c/\sigma \gg 1$, $\tanh(\sigma_c/\sigma) \to 1$ and R obeys the famous parabolic law of Burton, Cabrera and Frank [1951]:

$$R = C \frac{\sigma^2}{\sigma_c} . \qquad (3.91)$$

At supersaturations sufficiently higher than σ_c, $\tanh(x \to 0) = x$ and R obeys a linear law

$$R = C \sigma . \qquad (3.92)$$

The dependence of R on σ is shown in Fig. 3.13. As seen the parabolic law holds up to the characteristic supersaturation σ_c. Beyond it a linear relationship is gradually established. The condition for the latter is $\sigma \gg \sigma_c$ or $\lambda_s \gg y_0/2$. Physically this means that the density of rough steps and hence of kinks on the surface is so high that every adatom is incorporated into a growth site before succeeding to re-evaporate. Hence, R is directly proportional to σ as in the case of normal growth of rough F faces. In the other extreme ($\lambda_s \ll y_0/2$) the diffusion fields of the neighboring steps do not overlap and a large fraction of adatoms re-evaporate before joining the growth sites. The proportionality of v_∞ and the step density $1/y_0$ with respect to σ result in the parabolic law (3.91).

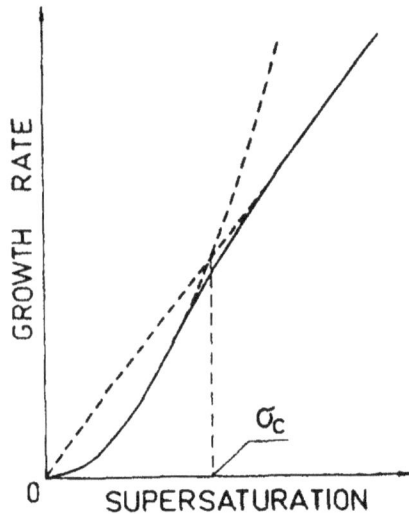

Fig. 3.13. Plot of the rate of spiral growth versus the supersaturation. For supersaturations smaller than the characteristic one, σ_c, the growth obeys the parabolic law of Burton, Cabrera and Frank [1951]. Beyond σ_c the growth rate is a linear function of the supersaturation (after Burton, Cabrera and Frank [1951]).

Let us consider now the case when $L \gg \Lambda$ ($l \gg 2\rho_c$) and $y_0 = l = L/n$ (Fig. 3.12(b)) does not depend on the supersaturation. A linear law for the rate of growth R results instead of the parabolic one:

$$R = C'\sigma ,\tag{3.93}$$

where

$$C' = C\frac{2\lambda_s}{l}\tanh\left(\frac{l}{2\lambda_s}\right) .\tag{3.94}$$

This is the so-called *second linear law* [Bennema and Gilmer 1973]. It results from the specific interrelation of the spacing of the dislocations in the source and the radius of the critical 2D nucleus. The interstep distance and in turn the slope of the pyramids no longer depend on the supersaturation. Two limiting cases are distinguished: (i) $l \ll 2\lambda_s$, $\tanh(x) \cong x$ and $C' = C$, and (ii) $l \gg 2\lambda_s$, $\tanh(l/2\lambda_s) = 1$ and $C' = 2\lambda_s C/l \ll C$. Thus linear dependence of the growth rate versus the supersaturation should be observed which is characteristic for normal growth of rough faces.

On the other hand, a parabolic dependence of the growth rate versus the supersaturation (*the second parabolic law*) [Bennema and Gilmer 1973] can be observed under conditions of a kinetic regime of growth. Then the velocity of the step advance is given by Eq. (3.41') $v_\infty = 2a^2\beta_{st}n_{se}\sigma$, which combined with (3.15), (3.86) and (3.87) gives

$$R = C''\sigma^2 , \tag{3.95}$$

where

$$C'' = \frac{2nkTa}{19\varkappa}n_{se}\beta_{st} . \tag{3.96}$$

It is interesting to compare the constants C and C'' in order to distinguish the diffusion and kinetic mechanisms. The ratio $C''/(C/\sigma_c)$ is given by

$$\frac{C''}{C/\sigma_c} = \frac{a}{\lambda_s}\frac{a}{\delta_0}\frac{n_{se}}{N_0}\exp\left(\frac{\varphi_{1/2} - \Delta U}{kT}\right) .$$

Taking into account Eqs. (3.21), (3.18) and (1.74) the above equation turns into

$$\frac{C''}{C/\sigma_c} = \exp\left(\frac{\varphi_{des} + \varphi_{sd} - 2\Delta U - 2\omega}{2kT}\right) ,$$

where ω is the work required to produce a kink on the step edge. As seen the value of the ratio $C''/(C/\sigma_c)$ depends on the interrelation of the activation energies for desorption and crystallization. Obviously, when $\varphi_{des} + \varphi_{sd} > 2\Delta U + 2\omega$, $C'' \gg C/\sigma_c$ and vice versa. However, in order to derive Eq. (3.95) we assumed a kinetic regime of growth, i.e. the condition (3.29) which is equivalent to $\varphi_{des} + \varphi_{sd} < 2\Delta U + 2\omega$. Then $C'' \ll C/\sigma_c$ and the second parabolic law dependence will cross the straight line of the linear BCF law at a characteristic supersaturation

$$\sigma_c'' \cong \sigma_c \exp\left(-\frac{\varphi_{des} + \varphi_{sd} - 2\Delta U - 2\omega}{2kT}\right)$$

which is much greater than σ_c. It may happen that the second parabola $R = C''\sigma^2$ would not cross the BCF linear dependence in the experimental interval of the supersaturation.

A. *The back stress effect*

It follows from Eq. (3.86) that the higher the supersaturation is the smaller the interstep distance Λ will be. The latter in turn leads to the linear law of growth (3.92). Cabrera and Coleman [1963] discussed this question and found that the analysis of Cabrera and Levine [1956] underestimates the interstep distance, particularly at the center of the growth spiral. The center of the spiral "will see" a supersaturation smaller than σ because of the diffusion field which is due to the first turn of the spiral. The higher the supersaturation is, the smaller should be the radius of the first turn and the stronger should be its influence on the adatom concentration at the center which in turn leads to an increase of the radius of the first turn of the spiral. Thus we should observe a feedback effect which is known in the literature as a "back stress" effect [Cabrera and Coleman 1963]. In principle we should observe the back stress effect not only at the center of the spiral as each step is under the influence of the diffusion fields of the neighboring steps at higher supersaturations.

In order to estimate it we approximate the first turn of the spiral by a circular step with a radius Λ_0 (Fig. 3.14). The supersaturation at the center can be found easily from Eq. (3.59'). Under the condition $r = 0$, the Bessel function $I_0(0) = 1$ and the supersaturation in the center, σ_{so}, reads

$$\sigma_{so} = \sigma \left[1 - I_0^{-1} \left(\frac{\Lambda_0}{\lambda_s} \right) \right] . \tag{3.97}$$

$I_0(x)$ is always greater than unity and the supersaturation at the spiral center, σ_{so}, will always be smaller than σ. We multiply both sides of Eq. (3.97) with Λ_0/λ_s and bearing in mind that $\sigma_{so}\Lambda_0/\lambda_s = \sigma_c$ (Eq. (3.89)) with $n = 1$ (elementary steps), we rewrite (3.97) in the form

$$\frac{\sigma}{\sigma_c} = \frac{1}{\dfrac{\Lambda_0}{2\lambda_s} \left[1 - I_0^{-1} \left(\dfrac{\Lambda_0}{\lambda_s} \right) \right]} . \tag{3.98}$$

We then tabulate the left-hand side of (3.98) and construct a plot of Λ_0/λ_s vs σ/σ_c (curve 1 in Fig. 3.15). As seen, the interstep distance at the spiral center is always larger than the one predicted by Eq. (3.86) (curve 2, $\Lambda = 19\rho_c$ or $\Lambda/\lambda_s = 2/(\sigma/\sigma_c)$).

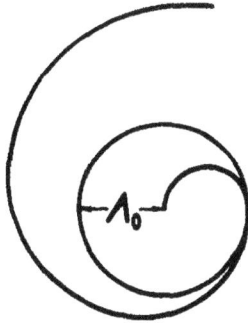

Fig. 3.14. For the evaluation of the "back stress effect." The spiral center is approximated by a circle of radius Λ_0 (after Cabrera and Coleman [1963]).

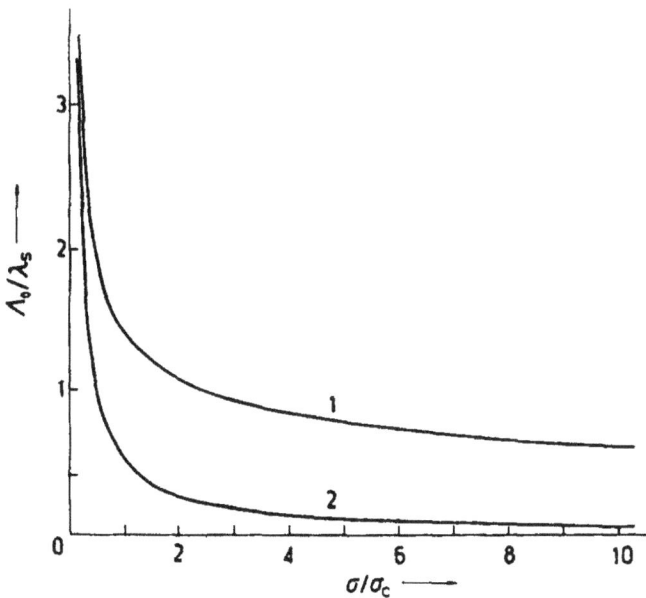

Fig. 3.15. Dependence of step spacing in units of λ_s on the supersaturation σ in units of σ_c when the back stress effect is accounted for (curve 1). The dependence given by Eq. (3.86) is also shown for comparison (curve 2).

In the more interested region of high supersaturations (small values of Λ_0/λ_s ($\Lambda_0/\lambda_s \to 0$)) the reciprocal of the Bessel function, $I_0^{-1}(x)$, can be approximated by the parabola

$$I_0^{-1}(x) \cong 1 - \frac{1}{4}x^2 \; ,$$

which results in [Cabrera and Coleman 1963]

$$\frac{\Lambda_0}{\lambda_s} \cong 2 \left(\frac{\sigma_c}{\sigma}\right)^{1/3} \; . \tag{3.99}$$

One can conclude that with increasing supersaturation the interstep distance decreases much more slowly than required by the simple hyperbolic law $\Lambda \sim 1/\sigma$. The step spacing practically no longer depends on the supersaturation, but on the temperature through the mean free path λ_s. What is more important, however, is that the slopes p of the growth pyramids can never become too steep and the growing surface will remain macroscopically more or less smooth. As for the $R(\sigma)$ dependence the back stress effect leads to a more gradual transition from the parabolic to the linear growth law than that required by Eq. (3.88).

One can conclude that when the source of the steps on the crystal surface is due to the presence of screw dislocations one can observe a parabolic as well as a linear dependence of the rate of growth on the supersaturation. In the diffusion regime one should observe a parabolic dependence at small supersaturations which gradually becomes a linear one at high supersaturations. The latter is due to the strong overlapping of the diffusion fields around each step. The back stress effect makes the transition from a parabolic to a linear dependence more gradual. A linear dependence should be observed also from the beginning when the length or the perimeter of the step source is much smaller than the interstep spacing due to a single dislocation. In this case the constant of the proportionality should be smaller than that due to the diffusion fields overlapping provided the interdislocation distance is greater than the mean free path of the adatoms and equal to it in the opposite case. In the case of kinetic regime of growth a parabolic dependence of the growth rate on the supersaturation should be observed with a rate constant much smaller than that in the diffusion regime.

3.2.2.3. *Growth in solutions*

Combining (3.15), (3.86) and (3.76) gives an expression for the rate of growth in a diffusion regime:

$$R = C \frac{\sigma^2}{\sigma_c} \frac{1}{\ln\left[\dfrac{d}{\pi a} \dfrac{\sigma_c}{\sigma} \sinh\left(\dfrac{\sigma}{\sigma_c}\right)\right]} , \qquad (3.100)$$

where

$$\sigma_c = \frac{19 \varkappa a^2}{\pi n k T d} \qquad (3.101)$$

is the characteristic supersaturation and

$$C = \frac{D v_c C_0}{d} \qquad (3.102)$$

is the rate constant.

The condition $\sigma \ll \sigma_c$ ($\sinh(\sigma/\sigma_c) \cong \sigma/\sigma_c$) results in the parabolic dependence

$$R = C \frac{\sigma^2}{\sigma_c} \frac{1}{\ln\left(\dfrac{d}{\pi a}\right)} . \qquad (3.103)$$

At $\sigma \gg \sigma_c$ the hyperbolic sine transforms into $\exp(\sigma/\sigma_c)/2$ and, neglecting $\ln(d\sigma_c/2\pi a\sigma)$ with respect to σ/σ_c, one obtains the linear dependence

$$R = C\sigma \qquad (3.104)$$

as in the case of growth from a vapor phase.

In the kinetic regime of growth ($a\beta_{st}/\pi D \ll 1$) one obtains the second parabolic law

$$R = C''\sigma^2 , \qquad (3.105)$$

where

$$C'' = \frac{nkT}{19\varkappa a} \beta_{st} v_c C_0 . \qquad (3.106)$$

Bearing in mind (3.101), (3.102) and (3.103) we find that $C''/[C/\sigma_c \ln(d/\pi a)] = (a\beta_{st}/\pi D)\ln(d/\pi a) \ll 1$ ($\ln(d/\pi a) \cong 10$), i.e. the rate constant in the second parabolic law in a kinetic regime of growth is again smaller than that in the diffusion regime.

Finally, when $L \gg \Lambda$, $y_0 = L/n = l$ and $p = an/L$ we obtain the second linear law of growth

$$R = C'\sigma , \qquad (3.107)$$

where

$$C' = \frac{a}{l} \frac{\beta_{st} C_0 v_c}{1 + \dfrac{a\beta_{st}}{\pi D} \ln\left[\dfrac{l}{a} \sinh\left(\dfrac{\pi d}{l}\right)\right]} . \qquad (3.108)$$

We consider again the example of growth of ADP crystals in aqueous solutions at room temperature [Chernov 1989]. We can estimate the radius ρ_c of the critical nucleus and in turn the interstep spacing from independent measurements of the rate of growth of a perfect crystal face without screw dislocations. As will be shown in the next chapter, such measurements allow the evaluation of the work of nucleus formation and of all parameters connected with it, as the rate of growth is limited by 2D nucleation. Thus the interpretation of the experimental results gave for the specific edge energy the value $\varkappa \cong 5.5 \times 10^{-7}$ erg cm^{-1}. Then at $\sigma = 0.03$, $\rho_c \cong 0.95 \times 10^{-6}$ cm, $y_0 = \Lambda = 18 \times 10^{-6}$ cm $(n = 1)$ and $p \cong 2.6 \times 10^{-3}$. With the value $v_\infty = 2.4 \times 10^{-5}$ cm sec^{-1} calculated in the previous chapter for $\sigma = 0.03$; one obtains for R the value 6.24×10^{-8} cm sec^{-1}. The latter is in good agreement with the experimentally measured value 5.8×10^{-8} cm sec^{-1}.

3.2.2.4. *Growth in melts*

The rate of growth in melts obeys the same equations as in solutions. We assume that the interstep distance is again given by (3.86). In Sec. 3.2.1.3 we evaluated the rates of step advance of Si and GaAs to be 3.7 cm sec^{-1} and 0.1 cm sec^{-1}, respectively, at undercooling $\Delta T = 1$ K. In order to evaluate the slopes p we need data for the specific edge energies. We can estimate the latter from data of the specific surface energies. However, such data concerning the crystal–melt boundary for semiconductor substances are scarce in the literature. It is believed that they are around some hundreds of ergs per square centimeter (181 erg cm^{-2} [Turnbull 1950] and 251 erg cm^{-2} [Skripov, Koverda and Butorin 1975] for Ge). Adopting the value 200 erg cm^{-2} for both Si and GaAs we find y_0 (Si) $= 1.52 \times 10^{-4}$ cm, y_0 (GaAs) $= 1.44 \times 10^{-4}$ cm and p (Si) $= 2 \times 10^{-4}$, p (GaAs) $= 2.8 \times 10^{-4}$ at $\Delta T = 1$ K. Then for the rate of growth, $R = p v_\infty$, the values 7.4×10^{-4} cm sec^{-1} and 2.8×10^{-5} cm sec^{-1} can be obtained for Si and GaAs, respectively. As seen, they are 3 to 5 orders of magnitude higher than the respective values for growth in solutions.

3.2.3. *Growth by 2D nucleation*

The growth of the defectless crystals of Si, Ge, GaAs, CdTe, etc. to meet the demand of microelectronics stresses the necessity of developing in more detail the theory of growth through formation and lateral propagation of 2D nuclei. Historically it was the first theory of crystal growth whose foundations were laid down more than a hundred years ago by Gibbs [1928]. He pointed out in his famous footnote (see also Frank 1958a) that the continuous growth of a given crystal face is impossible if new molecular layers could not be built. The building of a new layer is particularly *difficult at the beginning or immediately after the beginning of the new layer formation.* The change of the Gibbs free energy necessary for the growth of the crystal face to take place is not, however, one and the same on different crystal faces. It is possible that it is greater for surfaces with smaller surface energies. Thus without even mentioning the term "nucleus" Gibbs gave the concept of the growth of perfect crystals through the formation and lateral spreading of two-dimensional nuclei. He even gave a hint concerning the effect of the surface structure of the crystal faces with respect to the rate of 2D nucleation.

We will consider first the *layer-by-layer growth* or, in other words, the growth when the next atomic plane is nucleated after the completion of the previous one. Then we will consider the case when the nucleation of the next atomic plane takes place before the completion of the previous one. This is the so-called *multilayer growth* when two or more monolayers grow simultaneously. We will consider first the simpler case of constant nucleation rate and constant rate of propagation of the 2D islands (constant rate of step advance) and then we will allow the nucleation rate and the rate of step advance to depend on time through the size of the underlying 2D islands. In fact the latter takes place during the growth from a vapor phase or MBE growth. In doing all that we will follow one and the same approach which will be outlined when considering the simplest case of layer-by-layer growth with constant rates of 2D nucleation and step advance.

3.2.3.1. *Constant rates of nucleation and step advance*

A. *Layer-by-layer growth*

Consider a face of a perfect defectless crystal with size L (Fig. 3.16) [Chernov 1984]. At the given supersaturation 2D nuclei are formed with a rate $J_0 = $ const $(cm^{-2}sec^{-1})$. We define first the frequency (sec^{-1})

$$\tilde{J}_0 = J_0 L^2 \qquad (3.109)$$

of 2D nucleation on the crystal face with an area L^2.

If the rate of lateral growth or the rate of step advance is $v = $ const $(cm\ sec^{-1})$ then the time for complete coverage of the face by one monolayer will be

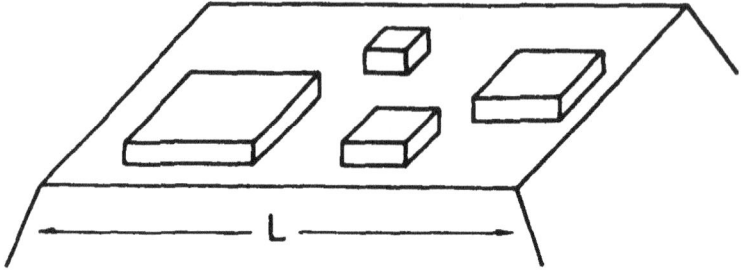

Fig. 3.16. Crystal face of linear size L with growing 2D islands on top of it.

$$T = L/v .$$

The number of nuclei formed during this time interval is a product of the nucleation frequency and the time for complete coverage of the face, or

$$N = \tilde{J}_0 T = J_0 L^2 \frac{L}{v} = J_0 \frac{L^3}{v} . \tag{3.110}$$

One can distinguish two cases. In the first one, $N < 1$ which is equivalent to $L < (v/J_0)^{1/3}$. This means that every succeeding nucleus will be formed after complete coverage of the crystal face by the monolayer initiated by the preceding nucleus. As a result we will observe layer-by-layer growth. Hence the growth of the crystal face will be a periodic process of successive formation of 2D nuclei and their lateral propagation (see Fig. 3.21). The rate of growth of the crystal face will be determined by the rate of 2D nucleation and will be given by

$$R = \tilde{J}_0 a = J_0 L^2 a , \tag{3.111}$$

where a is the height of the step originated by the nucleus and J_0 is given by Eq. (2.86), (2.89) or (2.96) in the particular cases of growth from vapors, solutions or melts, respectively.

Bearing in mind that v depends linearly on the supersaturation while the nucleation rate increases exponentially with it, one can conclude from (3.110) that layer-by-layer growth should be observed at low enough supersaturations. Besides, the smaller the crystal face is the more pronounced the layer-by-layer growth will be. Bearing in mind the estimates of J_0 and v made in the previous chapters (Eqs. (2.89) and (3.76)) we find that the linear size L of the face of a crystal growing in solution should be smaller than 2×10^{-4} cm at $\sigma = 0.01$, or 2×10^{-6} cm at $\sigma = 0.02$, or 4×10^{-7} cm at $\sigma = 0.03$, etc., in order to grow layer after layer.

The growth rate R vs supersaturation according to Eq. (3.111) is illustrated in Fig. 3.17 (curve 1). As seen a critical supersaturation

$$\sigma_c = \frac{\pi \varkappa^2 a^2}{(kT)^2 \ln(K_1 L^2 a / R_c)} \qquad (3.112)$$

should be overcome for visible growth to take place. In the above equation K_1 is the pre-exponential term in Eq. (2.86) (or (2.89)) and R_c is a more or less arbitrary chosen critical value of the growth rate. For example, in the growth of ADP crystals in aqueous solution at room temperature [Chernov 1989] with $\varkappa \cong 11.8$ erg cm^{-2} $\times 8 \times 10^{-8}$ cm $\cong 1 \times 10^{-6}$ erg cm^{-1}, $a \cong 8 \times 10^{-8}$ cm, $K_1 \cong 1 \times 10^{19}$ cm^{-2}sec^{-1}, $R_c \cong 1 \times 10^{-9}$ cm sec^{-1} and $L = 1 \times 10^{-4}$ cm, $\sigma_c = \Delta\mu_c / kT \cong 0.38$ or $C/C_0 \cong 1.5$. Thus a critical supersaturation as high as 50% is required for the layer-by-layer growth of the prismatic face of ADP crystal to take place.

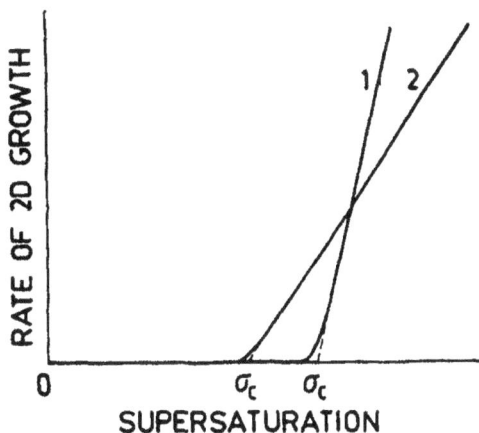

Fig. 3.17. Supersaturation dependence of the rate of 2D growth in the case of layer-by-layer growth (curve 1) and multilayer growth (curve 2).

B. *Multilayer growth*

At the other extreme $N > 1$ (or $L > (v/J_0)^{1/3}$), new nuclei will form on top of the growing monolayer before the latter is completed and the situation given in Fig. 3.18 results. Several monolayers grow simultaneously. The theoretical analysis of this case is much more complicated [Chernov and Lyubov 1963, Nielsen 1964, Hillig 1966]. An approximate treatment will be given here following Chernov [1984].

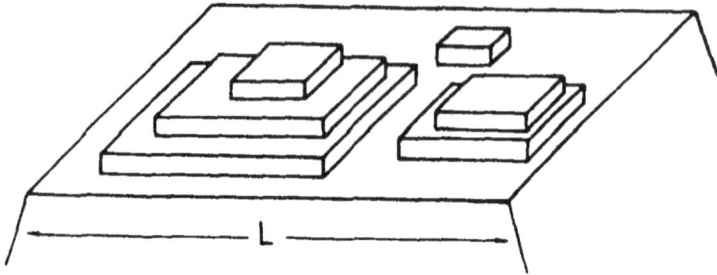

Fig. 3.18. Illustration of the multilayer growth of a crystal face with a size L by 2D nucleation.

Consider a crystal face with a monolayer island growing onto it at a rate v. At a given moment when the size of the island reaches a lateral size l a 2D nucleus appears on top of it. The frequency of nucleation on top of the island is now $\tilde{J}_0 = J_0 l^2$ and the average time elapsed from the moment of nucleation of the first island to the moment of nucleation of the second island is $l/v \cong 1/\tilde{J}_0 = 1/J_0 l^2$. We then find that the mean size of the lower 2D island when a new 2D nucleus is formed on top is $l = (v/J_0)^{1/3}$. The rate of growth of the crystal face is proportional to $J_0 l^2 a$, or

$$R \cong J_0 l^2 a = a(J_0 v^2)^{1/3} . \tag{3.113}$$

Recollecting that the pre-exponential factor in the equations (2.86), (2.89) and (2.96) for the rate of 2D nucleation is proportional to the square root of the supersaturation and that the rate of step advance is a linear function of the latter, we find

$$R = \text{const} \, (\Delta\mu)^{5/6} \exp\left(-\frac{\Delta G^*}{3kT}\right) , \tag{3.114}$$

where the constant is equal to $[K_1'(\beta_{st}C_0 v_c)^2]^{1/3}$ ($K_1' = K_1/\sqrt{\sigma}$) (in the particular case of growth from solutions). The rate of growth does not depend any more on the size of the crystal, but still a critical supersaturation should be overcome in order for the growth to take place. Obviously this critical supersaturation should be smaller than the one calculated on the basis of the layer-by-layer growth (Fig. 3.17, curve 2). Taking into account the values of $v = 2.4 \times 10^{-5}$ cm sec^{-1}, $K_1 = 1 \times 10^{19}$ cm^{-2}sec^{-1} and $R_c = 1 \times 10^{-9}$ cm/sec calculated before, we find a critical supersaturation of about 8.6%, instead of 50%. Comparison with the experiment [Chernov 1989] shows that this value is still an overestimation, but it gives the right tendency.

One concludes that initially the perfect crystals grow by the layer-by-layer mechanism up to a critical size determined by the condition $N = 1$ or $L = (v/J_0)^{1/3}$. Beyond this size, a gradual transition from layer-by-layer to multilayer mechanism takes place. Large enough crystal faces grow by the multilayer mechanism and the growth front consists of several monomolecular layers which grow simultaneously. This question is closely connected with the RHEED intensity oscillations of the specular beam during MBE growth and will be considered here in some more detail.

We denote the surface coverage of the first monolayer by Θ_1. The growth rate is $R_1 = ad\Theta_1/dt$ and from Eq. (2.141) we find that the completion with time of the first monolayer follows the time law

$$R_1 = \pi a J_0 v^2 t^2 \exp\left(-\frac{\pi}{3} J_0 v^2 t^3\right) , \qquad (3.115)$$

which is illustrated in Fig. 3.19 (curve 1). As seen the rate of deposition of the first monolayer increases parabolically at short times, then displays a maximum at $t_m = (2/\pi J_0 v^2)^{1/3}$, and after that decreases exponentially at large times up to the completion of the layer.

It is useful to express the growth rate (3.115) in terms of the surface coverage Θ_1. Making use of (2.141) gives

$$R_1 = a(9\pi J_0 v^2)^{1/3}(1 - \Theta_1)[-\ln(1 - \Theta_1)]^{2/3} . \qquad (3.115')$$

The growth rate displays a maximum at $\Theta_m = 1 - \exp(-2/3) = 0.4866$. As seen the maximum is slightly shifted to the left from $\Theta_1 = 1/2$ when constant growth v and nucleation J_0 rates are assumed.

In fact Eq. (3.115) describes the behavior in time of the layer-by-layer growth when each layer is initiated by several nuclei growing simultaneously (multinuclei layer-by-layer growth, see Fig. 3.23). Then the maximum value $R_m = 1.19a(J_0 v^2)^{1/3}$ gives the amplitude of the growth oscillations. The theoretical treatment of the multinuclei multilayer growth is much more complicated and has not been solved analytically up to now. That is why the time evolution of the separate monolayers and of the growth front is usually studied numerically by the method of Monte Carlo [Gilmer 1980a, 1980b].

The analysis is based again on the approach of Kolmogorov [1937] and Avrami [1939, 1940, 1941]. Recalling (2.137), the surface coverage Θ_1 of the first layer now reads

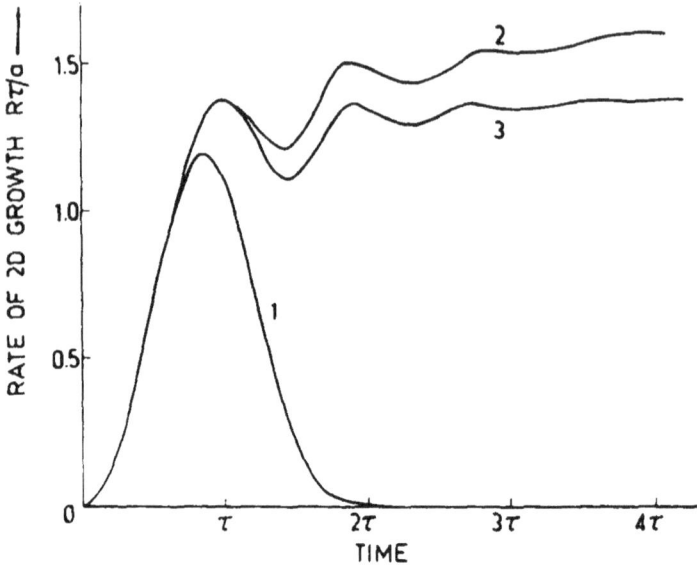

Fig. 3.19. Plot of the rate of 2D growth as a function of time in terms of the characteristic time $\tau = (J_0 v^2)^{-1/3}$. Curve 1 gives the layer-by-layer growth. Curve 2 represents the multilayer growth calculated with the mean field approximation of Borovinski and Tzindergosen [1968]. Curve 3 is a result of Monte Carlo simulation by Gilmer [1980] (after Gilmer [1980]).

$$\Theta_1 = 1 - \exp\left[-\pi J_0 c^2 \int\limits_0^t \left(\int\limits_{t'}^t k(\tau - t')d\tau\right)^2 dt'\right].$$

Assuming $k(t) = 1$ results in (2.141) and (3.115). For every succeeding monolayer the above expression is not valid any more as the formation of 2D nuclei of each new layer depends on the surface coverage of the preceding one, or, more precisely, on the probability that there is a crystallized part of the underlying layer just under the 2D nuclei. Instead, one writes the expression

$$\Theta_n = 1 - \exp\left[-\pi J_0 c^2 \int\limits_0^t p_{n-1}(t') \left(\int\limits_{t'}^t k(\tau - t')d\tau\right)^2 dt'\right], \qquad (3.116)$$

where $p_{n-1}(t')$ is the probability at the moment t' of formation of a 2D nucleus of the nth layer, the latter to find a crystallized part under itself in the preceding $n-1$ layer. Obviously $p_{n-1}(t') = 1$ for the first monolayer $n = 1$.

In the more general case when the nucleation rate is time dependent the latter enters into the integral and Eq. (3.116) turns into

$$\Theta_n = 1 - \exp\left[-\pi c^2 \int_0^t J(t')p_{n-1}(t') \left(\int_{t'}^t k(\tau - t')d\tau\right)^2 dt'\right] . \qquad (3.116')$$

Differentiation of (3.116') gives an expression which is often used in calculations of the growth rate:

$$\frac{d\Theta_n}{dt} = [1 - \Theta_n(t)] \int_0^t J(t')p_{n-1}(t')2\pi\rho_n(t')v_n(t')dt' , \qquad (3.116'')$$

where $v_n(t',t) = d\rho_n(t',t)/dt$ is the rate of growth of the 2D island of the nth layer.

In order to find a solution of (3.116) one has to determine the probability $p_{n-1}(t')$ and here is the main problem. To solve it Borovinski and Tzindergosen [1968] used the mean field approximation assuming that the probability $p_{n-1}(t')$ is equal to the surface coverage of the preceding monolayer, i.e. $p_{n-1}(t') = \Theta_{n-1}(t')$, and they computed the set of recurrent equations for Θ_n. First the expression for Θ_1 is substituted into Eq. (3.116'') for Θ_2 and the latter is solved numerically, then the result is substituted into the equation for Θ_3, etc. A set of S-shaped curves for Θ_n going from zero to unity and a set of bell-shaped curves $R_n/a = d\Theta_n/dt$ are obtained. The integration is carried out up to the moment at which a steady state is reached such that the shift of each curve with respect to the preceding one remains constant (Fig. 3.20), i.e.

$$\Theta_{n+1}(t) = \Theta_n(t-T) , \qquad (3.117)$$

where T is the period or the time of advance of the front of crystallization by one monolayer. Numerical calculations have shown that $T = 0.63(J_0 v^2)^{-1/3}$. The steady state rate of growth is then given by

$$R = \frac{a}{T} = 1.59a(J_0 v^2)^{1/3} , \qquad (3.118)$$

which coincides up to a constant with Eq. (3.113).

A very important question is the number of monolayers which grow simultaneously or, in other words, the thickness of the front of crystallization.

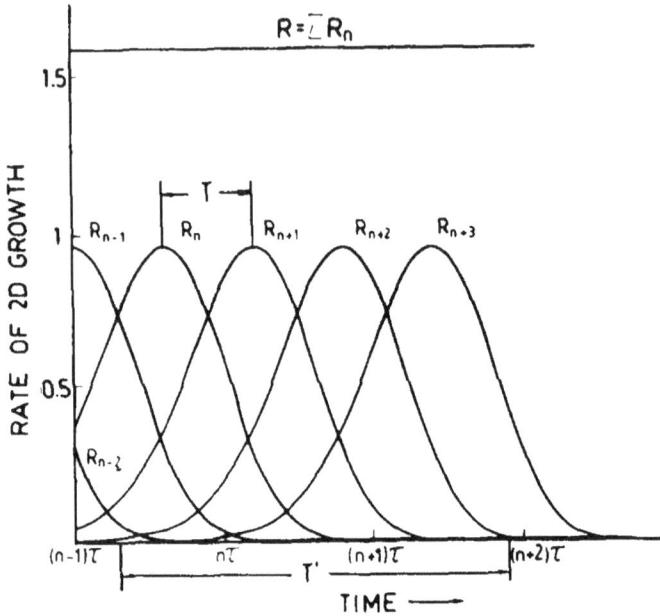

Fig. 3.20. Variation with time of the rates of growth of the separate monolayers, $R_n = ad\Theta_n/dt$, in the case of steady state multilayer growth. The curves are shifted by a period T. One monolayer is completed in a time interval T'. The overall rate of growth, $R = \Sigma R_n$, is also shown by the straight line (after Borovinski and Tzindergosen [1968]).

Numerical calculations have shown that an S-shaped curve giving the time dependence of Θ_n in the steady state varies from 0.001 to 0.999 in a time interval $T' = 2.6(J_0 v^2)^{-1/3}$ (or so does the width of the bell-shaped curves in Fig. 3.20). The number of the simultaneously growing monolayers is then given by $T'/T \cong 4$. Note that the growth front thickness does not depend on the rates of nucleation and step advance in the model under study.

Figure 3.20 shows also the sum of the bell-shaped curves

$$R = \sum_{n=1}^{\infty} R_n = a \sum_{n=1}^{\infty} \frac{d\Theta_n}{dt} , \qquad (3.119)$$

which represents the overall rate of growth of the crystal face.

Figure 3.20 gives the steady state, i.e. after enough time is elapsed from the beginning of the growth. At the beginning of the process the growth rate displays several oscillations (or a series of maxima and minima) which gradually attenuate (Fig. 3.19, curve 2) due to the thickening of the growth front. The value of $R(t)$ at long enough times goes to a time-independent value given by Eq. (3.119).

Numerical simulations performed using the method of Monte Carlo [Gilmer 1980a,b] (Fig. 3.19, curve 3) have shown that the mean field approximation used by Borovinski and Tzindergosen (Fig. 3.19, curve 2) overestimates the rate of growth. This is the reason why different approximations for the probability $p_{n-1}(t)$ have been used [Armstrong and Harrison 1969], which give the same qualitative behavior but different asymptotic values for the steady state growth rate [Gilmer 1980].

3.2.3.2. Time-dependent rates of nucleation and step advance

Molecular Beam Epitaxy (MBE) [Chang and Ludeke 1975; Ploog 1986] is a powerful method for the investigation of the elementary processes of crystal growth in detail which are inaccessible by other methods. Besides, *in situ* measurements with surface analytical methods such as RHEED and LEED are easily performed during growth and detailed information concerning the mechanism of growth is easily gathered. That is why we will consider the MBE growth in more detail. It is worth noting that the MBE growth represents simply a crystal growth via 2D nucleation. This is the reason why we will consider it in this chapter rather than later in Chap. 4.

As was shown above the basic postulates J_0 = const and v = const lead to several, in fact not more than 4 or 5, oscillations of the overall growth rate (Fig. 3.19). The latter is in agreement with experimental observation for electrolytic growth of Ag from aqueous solutions where the above postulates are fulfilled. This is not, however, the case of MBE growth. Strong oscillations of the RHEED intensity have been reported during the growth of a series of materials such as Si, Ge and GaAs [Wood 1981; Harris, Joyce and Dobson 1981; Neave *et al.* 1983; Van Hove *et al.* 1983]. 700 oscillations of the specular beam during the growth of $Al_x Ga_{1-x} As(100)$ face (x = 0.41) in the [100] azimuth [Sakamoto *et al.* 1985b, 1990] and 2200 oscillations of the specular beam during the growth of Si(100) in the [110] azimuth (Fig. 3.21) [Sakamoto *et al.* 1986a, 1987] have been reported.

The decoding of the true nature of the oscillations leads to accumulation of new knowledge about crystal growth processes. It was proved that one period of the oscillations corresponds exactly to the time T of growth of one complete monolayer. We thus measure exactly the rate of growth. On the other hand, we could use the oscillations in order to tailor more precisely the epitaxial layers or superlattices. Obviously, if the deposition of one material is interrupted in order to deposit another material when

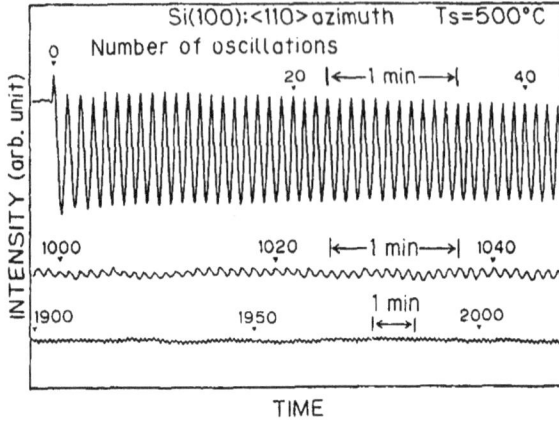

Fig. 3.21. RHEED intensity oscillations on Si(100) taken from the [110] azimuth at 500°C. (T. Sakamoto, N. J. Kawai, T. Nakagawa, K. Ohta, T. Kojima and G. Hashiguchi, *Surf. Sci.* **174**, 651 (1986). By permission of Elsevier Science Publishers B.V. and courtesy of T. Sakamoto.)

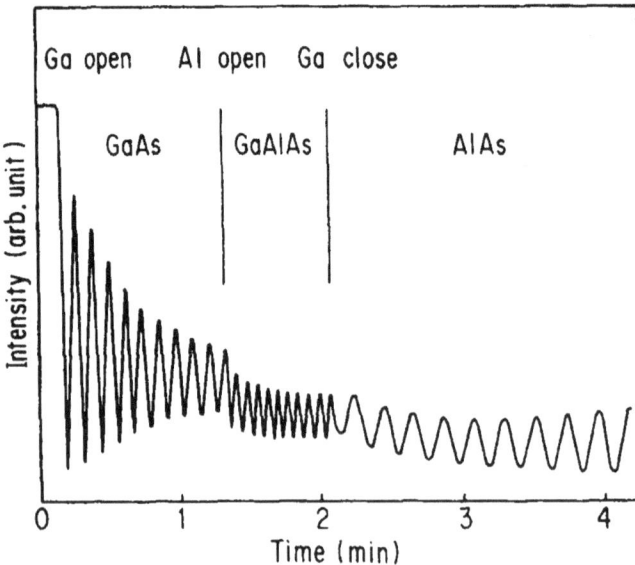

Fig. 3.22. RHEED intensity oscillations of the specular beam observed in the (100) azimuth of (001) GaAs substrate during the continuous growth of GaAs, $Al_x Ga_{1-x}As$ and AlAs. (T. Sakamoto, H. Funabashi, K. Ohta, T. Nakagawa, N. G. Kawai, T. Kojima and Y. Bando, *Superlatt. Microstruct.* **1**, 347 (1985). By permission of Academic Press Ltd. and courtesy of T. Sakamoto.)

the RHEED intensity goes through a maximum the possibility to produce a sharp interface is much greater. This is usually known as "phase-locked epitaxy" [Sakamoto *et al.* 1985b].

Another example of the use of the RHEED intensity oscillations is the measurement of the composition of, say, a ternary alloy and in turn the natural misfit between it and the underlying binary [Sakamoto *et al.* 1985b; Chang *et al.* 1991]. If for example a superlattice $Al_xGa_{1-x}As/GaAs(100)$ is grown [Sakamoto *et al.* 1985b] the opening and closing of the shutter of the Al source lead to increase and decrease of the growth rate and, in turn, of the oscillations frequency (Fig. 3.22). Measuring the oscillations frequencies $\mathfrak{f}(GaAs)$ and $\mathfrak{f}(Al_xGa_{1-x}As)$ one can calculate the mole fraction x through the relation [Sakamoto *et al.* 1985b]

$$x = \frac{\mathfrak{f}(Al_xGa_{1-x}As) - \mathfrak{f}(GaAs)}{\mathfrak{f}(Al_xGa_{1-x}As)} \; .$$

Then the natural misfit f can be easily estimated through Vegard's law [1921]. According to the latter the relative change of the lattice parameter (increase or decrease) of the host crystal is directly proportional to the concentration of the solute atoms within a certain interval. In other words,

$$f = x \frac{a_0(AlAs) - a_0(GaAs)}{a_0(GaAs)} \; .$$

In this section we will consider the growth of a defectless crystal face via 2D nucleation in the more realistic case when J_0 and v depend on the size of the underlying 2D islands through the adatom concentration on top of it. First we will consider the multinuclei layer-by-layer growth (Fig. 3.23) and the simultaneous growth of two monolayers (Fig. 3.24). Then we will generalize the model for an arbitrary number of simultaneously growing monolayers and study the dependence of the total step density on the thickness of the growth front. After that, the transition from layer-by-layer to bilayer growth will be considered as a first step to the thickening of the growth front and damping of the RHEED intensity oscillations. At the end of this section the influence of the anisotropy of the growing surface on the mode of growth will be considered using the example of the growth of the (001) face of Si.

A. *Multinuclear layer-by-layer growth*

We make the following assumptions. First, we consider the case of complete condensation or absence of re-evaporation. This means that all material

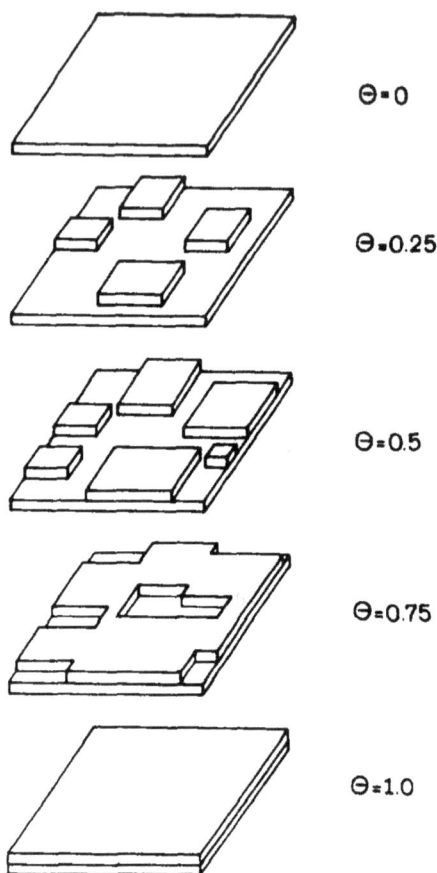

Fig. 3.23. Different stages of growth of a crystal face by multinuclear layer-by-layer growth. After deposition of a monolayer the face restores its initial state.

deposited joins the growing 2D islands. In fact this is usually the case when semiconductor films are grown due to the high binding energies. Second, we assume that the nucleation of the first monolayer takes place in a short (in fact negligible) period of time in the beginning of deposition (this is the so-called *instantaneous nucleation*). In other words, N_s nuclei are formed in the initial moment $t = 0$ of growth of each monolayer [Toschev, Stoyanov and Milchev 1972]. After complete coverage of the crystal face N_s nuclei are again formed in a short interval of time, and so on. In considering this problem we will follow the treatment given by Stoyanov [1988].

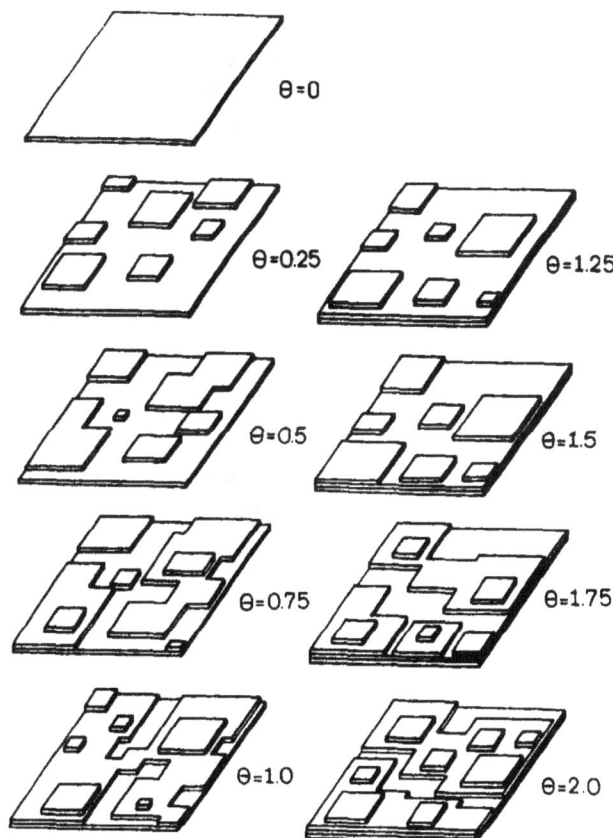

Fig. 3.24. Different stages of growth of a crystal face by bilayer growth. In this case the crystal face never restores its initial state. Instead, states with higher and lower total step densities alternate.

Before going into details a certain point should be clarified. As discussed above, any deviation from the layer-by-layer growth (Fig. 3.23), i.e. transition to simultaneous growth of two, three, etc. monolayers, leads to a decrease of the amplitude of the RHEED intensity oscillations with time. Two parameters vary periodically with time during growth: first the rate of growth of each monolayer, $R_n = ad\Theta_n/dt$ (Fig. 3.20), and second the total step density. The overall growth rate is a sum of the growth rates of the separate monolayers (see Eq. (3.119) and Fig. 3.20) and should vary periodically with time also. It will be shown in this section that the rate of growth of each monolayer is directly proportional to the step density and the proportionality constant is simply the rate of advance v of the separate 2D islands. In the model considered above the rate of advance v was assumed constant and hence both the step density and the rate of growth of each monolayer R_n oscillate in a phase. When more than 4 or 5 layers grow simultaneously the growth rate ceases to oscillate visibly. A more elaborate model of growth should account for the surface diffusion of adatoms between the steps originated from 2D islands on different levels and the rate of advance of the steps depends on the island size and hence on time. Then the total step density and the rates of growth of the separate monolayers cease to oscillate in a phase. The second important assumption made is that the re-evaporation of adatoms is negligible. This means that at a constant rate of deposition the overall rate of growth should be constant, i.e. it will not oscillate irrespective of the oscillations of the growth rates of the separate monolayers. However, the total step density oscillates and the amplitude of the oscillations is a decreasing function of the growth front thickness. It follows that the oscillations of the RHEED intensity during MBE growth are due solely to the oscillations of the total step density but not to the oscillations of the surface coverages or the rate of growth.

With the assumption for instantaneous nucleation the surface coverage of the first monolayer Θ_1 is given by (see Chap. 2)

$$\Theta_1 = 1 - \exp\left[-\pi N_s \left(\int_0^t v(t')dt'\right)^2\right] . \tag{3.120}$$

The growth rate $R_1 = ad\Theta_1/dt$ of the first monolayer then reads

$$R_1 = 2av(t)\sqrt{\pi N_s}(1 - \Theta_1)\sqrt{-\ln(1 - \Theta_1)} . \tag{3.121}$$

The total step density is given by (see Fig. 3.25)

$$L = \frac{1}{v(t)}\frac{d\Theta_1}{dt} = \frac{R_1}{av(t)} . \tag{3.122}$$

Equation (3.122) gives the proportionality between the step density and the growth rate discussed above. As seen if the rate of step advance v is time-independent both L and R_1 should have the same time behavior.

Substituting (3.121) into (3.122) gives

$$L = 2\sqrt{\pi N_s}(1 - \Theta_1)\sqrt{-\ln(1 - \Theta_1)} . \tag{3.122'}$$

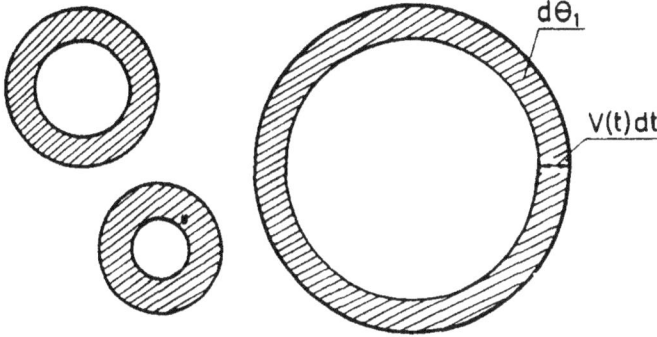

Fig. 3.25. For the determination of the step density $L(t)$ through the surface coverage $\Theta(t)$. The change of $\Theta(t)$ with time is represented by the shadowed areas.

The step density displays a maximum $L_{\max} = (2\pi N_s/e)^{1/2}$ at a certain value of the surface coverage $\Theta_{\max} = 1 - \exp(-0.5) = 0.393$. The maximum is shifted much more to the left of $\Theta_1 = 0.5$ than in the case of constant rate of nucleation and step propagation. As seen the variation of the step density is reduced to that of the number of nuclei N_s.

We consider a system of N_s regularly spaced 2D islands and at a moment t the surface coverage is

$$\Theta_1 = \pi\rho_1^2 N_s = \frac{\mathcal{R}t}{N_0} , \qquad (3.123)$$

where N_0 (cm^{-2}) is the density of a monolayer and $\mathcal{R}$ $(\text{cm}^{-2}\text{sec}^{-1})$ is the atom arrival rate. The above equation means that all atoms arriving from the vapor phase join the 2D islands, or in other words, all material deposited until the moment t, $\mathcal{R}t/N_0$, is equally distributed among the growing 2D islands.

Further, we will follow the same approach as developed by Chernov [1984] and Borovinski and Tzindergosen [1968]. Analogously to the frequency of nucleation on a crystal face with a lateral size L, $J_0 L^2$ (see Eq. 3.109), the frequency of nucleation on the surface of a growing island with a radius $\rho_1(t)$ is

$$\bar{J}_0(\rho_1) = \int_0^{\rho_1} J_0(r) 2\pi r dr . \qquad (3.124)$$

In the above equation $J_0(r) = J_0[n_s(r)]$ is the nucleation rate which is now a function of time through the size dependence of the adatom concentration $n_s(r)$ on the surface of the latter.

The adatom concentration on the surface of the growing 2D island can be found by solving the diffusion problem in polar coordinates (Eq. 3.58), which in the case of complete condensation reads

$$\frac{d^2 n_s}{dr^2} + \frac{1}{r}\frac{dn_s}{dr} + \frac{\mathcal{R}}{D_s} = 0 \ . \tag{3.125}$$

This equation differs from (3.58) only by the absence of the desorption flux n_s/τ_s. It must be solved subject to the boundary conditions

$$n_s(r = \rho_1) = n_{se} \ , \tag{3.126'}$$

$$\left(\frac{dn_s}{dr}\right)_{r=0} = 0 \ . \tag{3.126''}$$

The condition (3.126') means that the exchange of atoms between the island edges and the adlayer is fast enough and near the island edge the adatom concentration has its equilibrium value n_{se}. In other words the growth proceeds in a diffusion regime.

The solution of the problem reads

$$n_s(r) = n_{se} + \frac{\mathcal{R}}{4D_s}(\rho_1^2 - r^2) \ . \tag{3.127}$$

It is immediately seen that the adatom concentration has its highest value $n_s(r = 0) = n_{s,max} = n_{se} + \mathcal{R}\rho_1^2/4D_s$ at the island center and that the larger the island is the higher this adatom concentration will be. Obviously, the nucleation on the island center is most probable. Besides, the flux of atoms towards the edge of the island, $j_s = -2\pi\rho_1 D_s(dn_s/dr)_{r=\rho_1} = \pi\mathcal{R}\rho_1^2$, also increases with the island size, and so does the rate of growth (c.f. Eq. (3.62)). It follows that the nucleation rate and the rate of step advance become greater with increasing island size.

At high temperatures the equilibrium adatom concentration n_{se} increases whereas the second term in (3.127) decreases (see Eq. (3.20)). Bearing in mind that $\mathcal{R} \cong 1 \times 10^{13}\text{--}1 \times 10^{14} \text{ cm}^{-2}\text{sec}^{-1}$, $\nu \cong 3 \times 10^{13} \text{ sec}^{-1}$ and $\rho_1 \cong 1 \times 10^{-6}\text{--}1 \times 10^{-5}$ cm, we find that at high enough temperatures n_{se} becomes much greater than $\mathcal{R}\rho_1^2/4D_s$ and $n_{s,max} \rightarrow n_{se}$. The latter means that the supersaturation tends to zero at high temperatures. At low temperatures we can neglect n_{se} in comparison with $\mathcal{R}\rho_1^2/4D_s$ and $n_s \cong (\mathcal{R}/4D_s)(\rho_1^2 - r^2)$. Bearing in mind Eq. (3.18) it follows that for Si $(\varphi_{1/2} - \varphi_{des} \cong 1.8$ eV$)$ n_{se} can be neglected in the whole temperature interval of interest (300–800°C). This is valid also for many other materials.

The nucleation rate as a function of the adatom concentration reads (see Eq. 2.109)

$$J_0 = \alpha^* D_s \frac{n_s^{n^*+1}}{N_0^{n^*-1}} \exp\left(\frac{U^*}{kT}\right) .$$ (3.128)

Substituting (3.127) into (3.128) (neglecting n_{se}) and the latter into (3.124) gives upon integration for the nucleation frequency $\tilde{J}_0$

$$\tilde{J}_0(\rho_1) = A\rho_1^{2(n^*+2)} ,$$ (3.129)

where A combines all the quantities which are independent of the island size:

$$A = \frac{\pi \alpha^*}{n^* + 2} D_s N_0^2 \left(\frac{\mathcal{R}}{4D_s N_0}\right)^{n^*+1} \exp\left(\frac{U^*}{kT}\right) .$$ (3.129')

The condition for layer-by-layer growth in complete analogy with condition (3.110) is

$$N = \int_0^T \tilde{J}_0(\rho_1) dt = 1 ,$$ (3.130)

where the integral gives the number of nuclei which can be formed on top of the growing island for the time of deposition of one monolayer, $T = N_0/R$. In other words, this condition states that a nucleus of the second monolayer will be formed exactly at the moment of completion of the first one.

We then substitute ρ_1^2 from (3.123) into (3.129) and (3.130) and after integration and rearrangement of the results we obtain

$$N = \frac{1}{4\pi} \left[\left(\frac{4\pi\alpha^*}{(n^*+2)(n^*+3)}\right) \left(\frac{\mathcal{R}}{D_s N_0}\right)^{n^*} N_0^2\right]^{1/(n^*+2)} \exp\left(\frac{U^*}{(n^*+2)kT}\right)$$

$$= \frac{1}{4\pi} \left[\left(\frac{4\pi\alpha^*}{(n^*+2)(n^*+3)}\right) N_0^2 \left(\frac{\mathcal{R}}{\nu}\right)^{n^*}\right]^{1/(n^*+2)} \exp\left(\frac{U^* + n^*\varphi_{sd}}{(n^*+2)kT}\right) .$$ (3.131)

We have just obtained an expression for the density of nuclei which give rise to 2D islands belonging to the first monolayer. This was possible thanks to the conditions of complete condensation (3.123) and layer-by-layer growth (3.130) of a perfect crystal face. Let us evaluate it. In the case of growth of Si(001) at $T = 600$ K (high supersaturation), $n^* = 1$, $U^* = 0$,

$\Delta H_e = 4.33$ eV, $\psi/kT \cong 31$, $\varphi_{sd} \cong 0.67$ eV, $\mathcal{R} = 1 \times 10^{13}$ cm^{-2}sec^{-1}, $\nu = 1 \times 10^{13}$ sec^{-1}, $\alpha^* = 4$, $N_0 = 0.68 \times 10^{15}$ cm^{-2}, $N_s \cong 7.4 \times 10^{10}$ cm^{-2}. In the case of a higher temperature $T = 1200$ K (lower supersaturation), $n^* = 3$. The nucleus represents a cluster consisting of three atoms situated at the apexes of a rectangular triangle. The cluster consisting of four atoms in the shape of a quadrate is a stable cluster with a greater probability to grow than to decay. It should be noted that the bonds between the atoms are second nearest bonds and we can assume that $\psi_2 \cong \psi/10$. Then with $U^* = 2\psi_2 = 0.2\psi$, $\psi/kT = 25$, $\mathcal{R} = 1 \times 10^{13}$ cm^{-2}sec^{-1} and $\alpha^* = 8$, $N_s = 1 \times 10^7$ cm^{-2}. Thus a decrease of the supersaturation leads to an increase of the size of the critical nucleus, which in turn leads to a sharp decrease of the saturation nucleus density.

The maximum step density reads (the minimum step density is equal to zero)

$$L_{\max} = \frac{1}{\sqrt{2e}} \left[\left(\frac{4\pi\alpha^*}{(n^*+2)(n^*+3)} \right) N_0^2 \left(\frac{\mathcal{R}}{\nu} \right)^{n^*} \right]^{\frac{1}{2(n^*+2)}}$$

$$\times \exp\left(\frac{U^* + n^* E_{sd}}{2(n^*+2)kT} \right) . \tag{3.132}$$

Values of 4×10^5 and 5×10^3 cm^{-1} for the maximum step density on Si(001) are obtained at high $(n^* = 1)$ and low $(n^* = 3)$ supersaturations, respectively. It follows that the amplitude of the oscillations of the step density during the periodic process of consecutive 2D nucleation and lateral propagation will decrease with decreasing supersaturation (increasing temperature). Obviously at too high a temperature (too low a supersaturation) the RHEED intensity oscillations should disappear because of the inhibition of the 2D nucleation process. The same should be expected at too low a temperature or at high supersaturations, but for another reason. The step density becomes very high and the step spacing too small. In other words, the crystal surface will behave as a rough one although the temperature is lower than the critical temperature for thermodynamic roughness (see Chap. 1). As discussed at the end of Chap. 1 this phenomenon is called kinetic roughness. What follows is that the oscillations of the step density and in turn the RHEED intensity oscillations will disappear as the steps are practically no more detectable.

B. Simultaneous growth of two monolayers

In considering this case [Stoyanov and Michailov 1988] we follow the same procedure as described above. We consider again the case of complete condensation and instantaneous nucleation. The latter now means that before the complete coverage of the crystal face by 2D islands of the nth monolayer 2D nuclei of the $(n + 1)$th monolayer are formed in a short period of time (as if by a pulse) on top of the islands of the nth monolayer. Then the growth of the crystal face is realized by simultaneous growth of pyramids with two monolayers thickness. We assume further that the number of 2D nuclei giving rise to the islands of the nth monolayer is equal to that of the $(n + 1)$th monolayer, or in other words, the growth proceeds by simultaneous growth of N_s bilayer pyramids (Fig. 3.26). This pattern is preserved indefinitely and every deviation from it leads to a further increase of the growth front thickness and, in turn, to further damping of the oscillations of the step density.

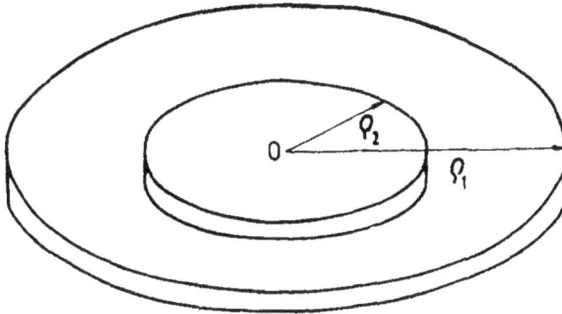

Fig. 3.26. Bilayer pyramid of growth. ρ_1 and ρ_2 denote the radii of the lower and upper 2D islands, respectively.

Making use of (3.122) gives the step density for each monolayer

$$L_n(t) = \frac{1}{v_n(t)}\frac{d\Theta_n}{dt} = 2\sqrt{\pi N_s}(1 - \Theta_n)\sqrt{-\ln(1 - \Theta_n)} . \qquad (3.133)$$

The total step density then is

$$L(t) = 2\sqrt{\pi N_s}\sum_n [1 - \Theta_n(t)]\sqrt{-\ln[1 - \Theta_n(t)]} . \qquad (3.134)$$

Note that in this case the rates of step advance $v_n(t)$ and the surface coverages $\Theta_n(t)$ $(n = 1, 2)$ are not independent but are interconnected through the diffusion fields. Then our next task is to solve again the diffusion problem and to find the density of the bilayer pyramids.

The solution for the adatom concentration on top of the upper island ($r < \rho_2$) (see Eq. (3.127)) is

$$n_s(r) = n_{se} + \frac{\mathcal{R}}{4D_s}(\rho_2^2 - r^2) \ . \tag{3.135}$$

The adatom concentration on the terrace ($\rho_2 < r < \rho_1$) is given by

$$n_s(r) = A - \frac{\mathcal{R}}{4D_s}r^2 + B\ln r \ , \tag{3.136}$$

where A and B are constants which can be determined from the boundary conditions $n_s(\rho_1) = n_s(\rho_2) = n_{se}$. The solution reads

$$n_s(r) = n_{se} + \frac{\mathcal{R}}{4D_s}(\rho_1^2 - r^2) - \frac{\mathcal{R}}{4D_s}(\rho_1^2 - \rho_2^2)\frac{\ln\left(\dfrac{r}{\rho_1}\right)}{\ln\left(\dfrac{\rho_2}{\rho_1}\right)} \ . \tag{3.136'}$$

The flux of adatoms diffusing to the edges of the lower islands on the surface of the complete $(n-1)$th layer is

$$j_1 = \frac{\mathcal{R}}{N_s}\left(1 - \pi\rho_1^2 N_s\right) \ . \tag{3.137}$$

In the above equation it is assumed that the atoms arriving from the vapor phase on the area $1 - \pi\rho_1^2 N_s$ uncovered by growth pyramids are equally distributed among the latter.

Correspondingly, the flux of adatoms diffusing to the same edges but on the terraces is

$$j_2 = -2\pi\rho_1 D_s\left(\frac{dn_s(r)}{dr}\right)_{r=\rho_1} \ .$$

The rate of growth v_1 is then given by

$$v_1 = \frac{d\rho_1}{dt} = \frac{1}{2\pi\rho_1 N_0}(j_1 + j_2)$$

or

$$\frac{d\rho_1}{dt} = \frac{\mathcal{R}}{2\pi\rho_1 N_0 N_s}\left(1 - \pi N_s\frac{\rho_1^2 - \rho_2^2}{2\ln(\rho_1/\rho_2)}\right) \ .$$

In the same way one obtains for the upper island

$$\frac{d\rho_2}{dt} = \frac{\mathcal{R}}{2\rho_2 N_0} \frac{\rho_1^2 - \rho_2^2}{2\ln(\rho_1/\rho_2)} \, .$$

We have to solve now a set of two nonlinear differential equations for the rates of advance of the steps. The latter can be written in terms of surface coverages

$$\Theta_n = \pi\rho_n^2 N_s \quad (n = n, n+1) \tag{3.138}$$

as a function of a dimensionless time $\theta = Rt/n_0$, which is in fact the number of monolayers deposited, in the form

$$\frac{d\Theta_n}{d\theta} = 1 - \frac{\Theta_n - \Theta_{n+1}}{\ln(\Theta_n/\Theta_{n+1})} \, ,$$

$$\frac{d\Theta_{n+1}}{d\theta} = \frac{\Theta_n - \Theta_{n+1}}{\ln(\Theta_n/\Theta_{n+1})} \, . \tag{3.139}$$

As seen the overall rate of growth, $R = \Sigma R_n = a\Sigma d\Theta_n/dt = \mathcal{R}a/n_0$, does not vary with time. This is a direct consequence of the lack or re-evaporation or complete condensation.

An approximate analytical solution of (3.139) can be obtained if the equations are linearized assuming $\ln(\Theta_n/\Theta_{n+1}) = \text{const} = 2$. Then the solution of (3.139) subject to the boundary conditions $\Theta_{n+1}(\theta = 0) = 0$ and $\Theta_n(\theta = 1) = 1$ reads [Kamke 1959]

$$\Theta_n = 1 + \frac{1}{2}\theta - \frac{e}{2(e+1)}(1 + e^{-\theta}) \, , \tag{3.140'}$$

$$\Theta_{n+1} = \frac{1}{2}\theta - \frac{e}{2(e+1)}(1 - e^{-\theta}) \, , \tag{3.140''}$$

where $e = 2.71828$ is the base of the Naperean logarithms.

As seen, at $\theta = 0$, $\Theta_n = 1/(e+1) = 0.27$ and $\Theta_{n+1} = 0$, while at $\theta = 1$, $\Theta_n = 1$ and $\Theta_{n+1} = 1/(e+1) = 0.27$. In other words, the solution reflects the periodicity inherent in the front of growth consisting of two monolayers.

Substituting (3.140) into (3.134) results in a periodic curve with amplitude $L_{\max} = 0.23 \, (4\pi N_s)^{1/2}$ [Stoyanov and Michailov 1988]. So our next task is to find the number N_s of the growth pyramids, and we follow exactly the procedure used in the previous section. The frequency of nucleation on top of the upper islands is obtained by substituting the solution (3.135) (neglecting n_{se}) into the expression (3.128) for the nucleation rate and the latter into (3.124). The integration gives an expression completely equivalent to (3.129) with the only exception being that ρ_1 is replaced by ρ_2:

$$\tilde{J} = A\rho_2^{2(n^*+2)} , \tag{3.141}$$

where A is given by (3.129').

From (3.138) and (3.140'') we have

$$\rho_2^2 = \frac{1}{2\pi N_s} \left(\theta - \frac{e}{e+1}(1 - e^{-\theta}) \right) . \tag{3.142}$$

Substituting (3.142) into (3.141) and the latter into (3.130) and carrying out the integration gives after rearrangement of the result

$$N_s = \frac{1}{8\pi} \left[\left(\frac{4\pi\alpha^* \mathfrak{J}^*}{n^* + 2} \right) \left(\frac{\mathcal{R}}{\nu} \right)^{n^*} N_0^2 \right]^{\frac{1}{(n^*+2)}} \exp \left(\frac{U^* + n^*\varphi_{sd}}{(n^* + 2)kT} \right) , \tag{3.143}$$

where the definite integral

$$\mathfrak{J}^* = \int\limits_0^1 \left(\theta - \frac{e}{e+1}(1 - e^{-\theta}) \right)^{n^*+2} d\theta$$

is a function only of the number of atoms in the critical nucleus, n^*, and has values 0.03, 0.0125, 0.0056 and 0.00056 for $n^* = 1, 2, 3$ and 6, respectively.

We can estimate now the decrease of the amplitude of the step density oscillations when the growth front increases from one to two monolayers. The latter depends on the height and the number density of bilayer pyramids as given by (3.143) and monolayer 2D islands as given by (3.131). The ratio of the square root of the islands densities is not very sensitive to the number of atoms in the critical nuclei, n^*, and is approximately equal to 0.5. In addition, the decrease of the amplitude due to the fact that the surface never reaches a state without steps when two monolayers grow simultaneously is also approximately equal to 0.5, so the overall decrease of the amplitude of the total step density is about 0.2–0.25. Obviously further decrease of the amplitude should be expected when further increase of the growth front thickness takes place.

C. *Simultaneous growth of an arbitrary number of monolayers*

We consider now the problem of simultaneous growth of $N > 2$ monolayers where N is an integer. This means that the instantaneous nucleation of 2D islands of the Nth monolayer coincides exactly with the completion of the zeroth monolayer. We assume further that the nucleation of the $(N+1)$th monolayer takes place before a significant coalescence of the 2D islands of the first monolayer occurs. Then N_s pyramids of growth, each consisting of N 2D islands one on top of the other, will be formed as shown in Fig. 3.2. This model is a variation of the "birth–death model" proposed by Cohen et al. [1989] (see also Kariotis and Lagally [1989]).

We solve the same diffusion problem as above. The adatom concentration on top of the islands of the Nth monolayer is given by (3.135) in which ρ_2 is replaced by ρ_N. The adatom concentration on the terraces is given by (3.136') where ρ_1 and ρ_2 are replaced by ρ_n and ρ_{n+1}, respectively. Then, instead of (3.139), one obtains

$$\frac{d\Theta_1}{d\theta} = 1 - \frac{\Theta_1 - \Theta_2}{\ln(\Theta_1/\Theta_2)} \; ,$$

$$\frac{d\Theta_n}{d\theta} = \frac{\Theta_{n-1} - \Theta_n}{\ln(\Theta_{n-1}/\Theta_n)} - \frac{\Theta_n - \Theta_{n+1}}{\ln(\Theta_n/\Theta_{n+1})} \; , \qquad (3.144)$$

$$\frac{d\Theta_N}{d\theta} = \frac{\Theta_{N-1} - \Theta_N}{\ln(\Theta_{N-1}/\Theta_N)} \; .$$

As above, the overall rate of growth $R = \Sigma R_n = \mathcal{R}a/n_0 = $ const.

A remarkable property of the system (3.144) is that the dependence of the surface coverages on the number of the monolayers does not involve any other parameter. This means that the solution of the system subject to the steady state boundary conditions $\Theta_n(\theta = 0) = \Theta_{n+1}(\theta = 1)$ is unique. The solutions of (3.144) for $N = 2$ and $N = 3$ in a reduced form are shown in Fig. 3.27. The solutions for $N = 2$ and $N = 4$ are shown in Fig. 3.28 in an unfolded form. As seen they represent S-shaped curves with an inflection point between the surface coverages of the second and first monolayers. It follows that with increasing growth front thickness N the rates of growth of the separate monolayers, $R_n = ad\Theta_n/d\theta$, will become more and more asymmetric with a maximum which shifts more and more to the right of $\theta = 0.5$. In other words, the bell-shaped curves obtained by differentiation of the curves shown in Fig. 3.28 will have a downward branch after the maximum steeper than their upward branch before the maximum.

The total step density is given by (3.134) and is shown in Fig. 3.29 in terms of $2\sqrt{\pi N_s}$ for N varying from 1 to 6. The curves are obviously asymmetric with a maximum shifted to the left of 0.5.

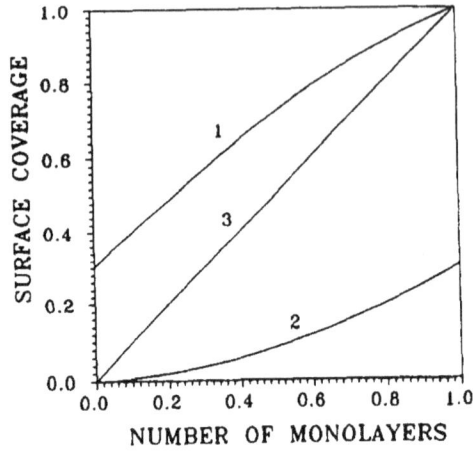

(a)

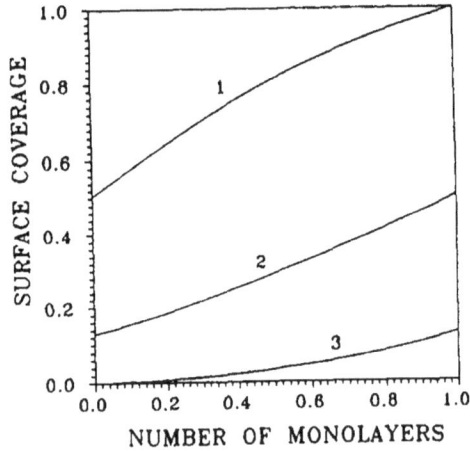

(b)

Fig. 3.27. Steady state variation with time in a folded form of the surface coverage of the separate monolayers in (a) bilayer and (b) trilayer growth. In (a), curves 1 and 2 correspond to the first and second monolayers, respectively, and the straight line 3 gives the layer-by-layer growth. In (b) the surface coverages of the first, second and third monolayers are shown by curves 1, 2 and 3, respectively. The curves are obtained by numerical solution of the system of equations (3.144).

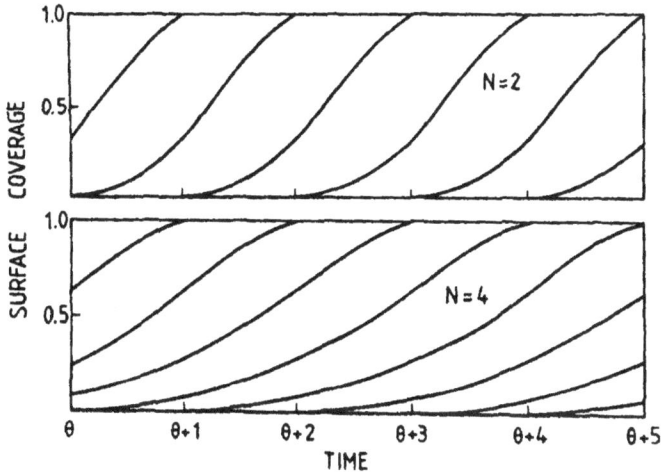

Fig. 3.28. Unfolded view of the steady state variation with time of the surface coverages of the separate monolayers in bilayer ($N = 2$) and tetralayer ($N = 4$) growth. The curves are obtained by numerical solution of the system of equations (3.144).

Making use of the same approach as above (Eqs. (3.124) and (3.130)) we can calculate the number N_s of the growth pyramids at different values of the growth front thickness N and then the amplitudes of the time variations of the total step density. The latter are shown in Fig. 3.30 relative to that of $N = 1$ for N varying from 1 to 6. It is seen that the total step density decreases by an order of magnitude when the number of the simultaneously growing monolayers, N, increases from 1 to 4, and about 20 times when $N = 6$.

The most important conclusions we can draw are the following:

(i) The shape of the oscillations of the total step density depends only on the surface coverage. In that sense it is unique.

(ii) The amplitude of the total step density oscillations is a function of the height and density of the growth pyramids. In the extreme case of layer-by-layer growth, stepped and completely smooth surfaces alternate (Fig. 3.23) and the amplitude has its highest value. The increase of the number of the simultaneously growing crystal planes leads to alternation of states with higher and lower step density (see Fig. 3.24), thus decreasing the overall amplitude. In addition, the density of the growth pyramids is a decreasing function of the thickness of the growth front (the density of the higher pyramids must be obviously smaller than that of the lower ones), this leading to additional decrease of the amplitude of the step density.

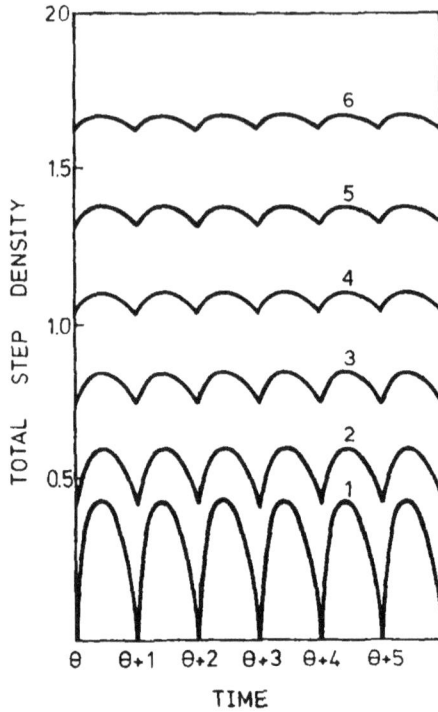

Fig. 3.29. Variation with time of the total step density in units of $(4\pi N_s)^{1/2}$ at different thicknesses of the growth front denoted by the figure at each curve. As seen the curves are visibly asymmetric. The amplitude decreases with increasing thickness of the growth front. In the case of layer-by-layer growth $(N = 1)$, the step density varies from zero to the maximum density, thus reflecting the fact that the crystal face restores its initial (smooth) state after deposition of one monolayer. The variation of the step density in the case of multilayer growth never reaches zero. This means that the crystal face never becomes smooth again.

(iii) The decrease of the amplitude due to a decrease of the density of the growth pyramids is a function only of the size n^* of the critical nucleus at the particular temperature.

(iv) The amplitude of the step density oscillations is not very sensitive to the atom arrival rate R as the exponent $n^*/2(n^* + 2)$ is smaller than unity.

(v) A decrease of temperature leads to a sharp increase of the number of the growth pyramids, N_s, and to a kinetic roughness of the crystal face. The latter leads in turn to disappearance of the oscillations.

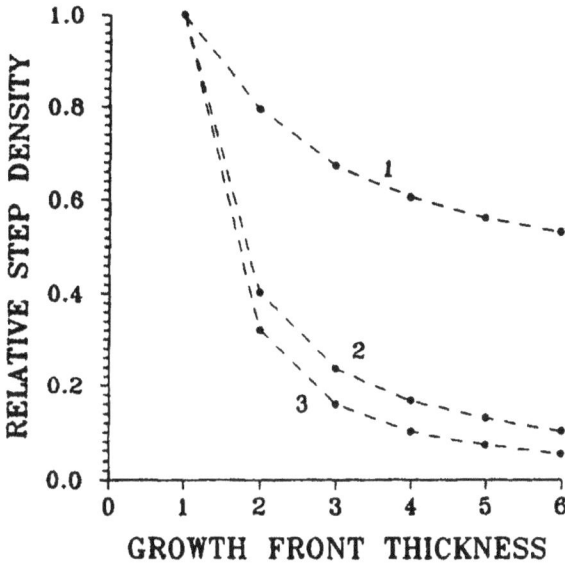

Fig. 3.30. Dependence on the thickness of the growth front of the amplitude of the total step density L_{max} relative to the step density in layer-by-layer growth (curve 3), the amplitude of the total step density L_{max} relative to the step density in layer-by-layer growth in units of $(4\pi N_s/e)^{1/2}$ (curve 2), and the square root of the density of the pyramids of growth $(4\pi N_s/e)^{1/2}$ relative to the layer-by-layer growth (curve 1).

(vi) An increase of temperature leads to a decrease of the supersaturation. The nucleation is suppressed and the density of growth pyramids and in turn the step density can become smaller than the resolution capabilities of the surface analytical tools. The latter leads again to disappearance of the oscillations. It follows that the oscillations of the step density can be observed in a limited interval of the temperature, which is in qualitative agreement with the experimental observations [Neave et al. 1985].

Note that the size of the critical nucleus depends on temperature but remains constant in a comparatively large interval of the latter (see Sec. 2.2.9). The lower the temperature is the broader the intervals of constant nucleus size will be. This is in agreement with the statement of Neave et al. [1983] that the damping of the oscillations is not sensitive to temperature. It is, however, worth noting that at high enough temperatures the condensation can become incomplete. In other words, significant part of the material deposited can re-evaporate before being incorporated into the growth sites.

D. *Damping of oscillations*

As discussed above the damping of the step density oscillations is due to the gradual transition from layer-by-layer growth to multilayer growth. In this section we will consider only the transition from layer-by-layer growth to the simultaneous growth of two monolayers [Stoyanov and Michailov 1988].

The problem is simplified if we take into account the periodicity connected with the multilayer growth. The latter was demonstrated when the case of the multilayer growth with $J_0 = $ const and $v = $ const was considered. As shown in Fig. 3.20 (Eq. 3.117) the surface coverage is repeated after a period of time T but each monolayer is fully completed after a period of time T' so that the number of the simultaneously growing monolayers is equal to $N = T'/T$. Obviously when $N = 1, 2, 3$, etc., $T' = T, 2T, 3T$ and so on. The damping of the step density oscillations is evidently characterized by the derivative dN/dt and it is the aim of this chapter to estimate it.

The transition from layer-by-layer growth to bilayer growth means that T' increases gradually from T to $2T$. As shown in Fig. 3.24, after a deposition of one monolayer, $\Theta_n(T) = 1 - \Theta_{n+1}(T)$. In other words, in the transition from layer-by-layer to bilayer growth, T' becomes longer than T, or $T' = T + \Delta t$, where Δt is just the time necessary to deposit the material of the upper monolayer. Then one can write $\Theta_{n+1}(T) \cong \mathcal{R}\Delta t/N_0$, or $\Delta t \cong N_0 \Theta_{n+1}(T)/\mathcal{R} = T\Theta_{n+1}(T)$, and

$$N = \frac{T'}{T} = 1 + \frac{\Delta t}{T} = 1 + \Theta_{n+1}(T) \ . \qquad (3.145)$$

One can also write $N = 1 + \Delta N$, $\Delta N = \Theta_{n+1}(T)$. The increment of N takes place in a time interval of T, and hence $dN/dt \cong \Delta N/T = \Theta_{n+1}(T)/T$, or $TdN/dt = dN/d\theta = \Theta_{n+1}(T)$.

In order to calculate $\Theta_{n+1}(T)$ we can use Eq. (3.116') or, better, its equivalent in the differential form (3.116''). We will use the mean field approximation $p_n(t') = \Theta_n(t')$ as the latter is quite correct at the beginning of the formation of the second monolayer (compare curves 2 and 3 in Fig. 3.19). Besides, we should calculate the rate of nucleation on the islands of the lower monolayer, $J(t')$, and the rate of growth of an island of the upper monolayer, $v_n = ck(t - t')$, where t' is the moment of its nucleation.

The nucleation rate is given by the nucleation frequency $\bar{J}_0(\rho_1)$, as given by Eq. (3.129) divided by the area of the underlying island $\pi\rho_1^2$. The growth rate v_n can be found by solving Eq. (3.139'') in its linearized form

$$\frac{d\Theta_{n+1}}{d\theta} = \frac{1}{2}(\Theta_n - \Theta_{n+1})$$

subject to the boundary condition $\theta = \theta' = \mathcal{R}t'/N_0$, $\Theta_{n+1} = 0$ and under the simplifying condition $\Theta_n \cong \Theta_n(t') \cong \mathcal{R}t'/N_0$. The latter means that in the beginning of the formation and growth of the upper monolayer the material deposited in the time interval $[0, t']$ has been consumed practically completely by the lower monolayer. Then for ρ_{n+1} and v_{n+1} one obtains

$$\rho_{n+1} = \frac{\mathcal{R}}{N_0} \frac{1}{(2\pi N_s)^{1/2}} t'^{1/2}(t - t')^{1/2} \,, \qquad (3.146)$$

$$v_{n+1} = \frac{1}{2} \frac{\mathcal{R}}{N_0} \frac{1}{(2\pi N_s)^{1/2}} \frac{t'^{1/2}}{(t - t')^{1/2}} \,. \qquad (3.147)$$

As seen, the size of the island and its growth rate are proportional to the square root of the time of nucleation, t', of the upper island. This dependence reflects the size of the underlying island, $\rho_1(t')$. As discussed above the larger ρ_1 is the higher the adatom concentration on top of it and the greater the flux of adatoms to the edge of the upper island will be. It follows that when the upper 2D island is formed at a later moment t' the greater will be its rate of growth and the larger will be its radius at a moment t.

Substituting (3.129), (3.146) and (3.147) into (3.116″) and carrying out the integration (with upper limit $t = T = N_0/\mathcal{R}$) give

$$\frac{dN}{d\theta} = \Theta_{n+1}(T) = 1 - \exp\left(-\frac{1}{2} \frac{(n^* + 3)}{(n^* + 4)(n^* + 5)}\right)$$

$$\cong \frac{1}{2} \frac{(n^* + 3)}{(n^* + 4)(n^* + 5)} \,. \qquad (3.148)$$

Then the transition from the layer-by-layer to the bilayer growth will take place after the deposition of θ_{tr} monolayers, where

$$\theta_{tr} = \frac{1}{dN/d\theta} = 2\frac{(n^* + 4)(n^* + 5)}{(n^* + 3)} \,. \qquad (3.149)$$

It follows that the damping of the oscillations of the step density depends only on the size of the critical nucleus, and hence is not much sensitive to the temperature (recollecting that the number of atoms in the nucleus remains one and the same in a broad interval of temperatures). Thus in the case of MBE growth of Si(111) a fourfold decrease of the amplitude of the step density oscillations is expected after 15 oscillations at low enough temperatures when $n^* = 1$, after 17 oscillations at some intermediate temperatures when $n^* = 2$ and after 24 oscillations at high enough temperatures when $n^* = 6$.

3.2.4. *Influence of surface anisotropy—Growth of*
Si(001) *vicinal surface*

A simplification of the model outlined above is obtained by assuming isotropy of the growing crystal face, which leads to the circular shape of the islands at high enough temperatures (Fig. 3.26) and the corresponding shape of the diffusion fields around them. This model describes well the case of growth of materials with central forces like metals. This is not the real case, however, when, particularly, materials with diamond lattice, such as GaAs and Si, are grown. The (001) surfaces of such crystals show significant anisotropy, which in turn affects strongly the parameters controling the growth process, namely, the height and roughness of the consecutive steps and the surface diffusion. The problems discussed in the present section are still under intensive study and thus illustrate the difficulties the theory encounters in treating real experimental observations.

We consider as an example the vicinal face of Si which is obtained by a slight tilt with respect to the (100) direction towards the [$\bar{1}$10] azimuth. Then the monoatomic steps which accommodate the macroscopic inclination are directed along the [110] azimuth and have a height $a_0/4$, where $a_0 = 5.4307$ Å is the lattice parameter of silicon (Fig. 3.31(a)). The projections of the dangling bonds on the unreconstructed surface on the upper terrace are directed along the same [$\bar{1}$10] direction, and those on the lower terrace, along the perpendicular [110] direction. In other words, assuming the crystal surface preserves its bulk structure, the projections of the dangling bonds will rotate by 90° on every next terrace. This means that the consecutive steps will be parallel or normal to the projections of the dangling bonds on the upper terraces.

This is the case when the (100) surface is unreconstructed, i.e. it preserves its bulk structure. The latter is characterized by two dangling bonds per atom on the surface and the surface energy is very high. In order to reduce the surface free energy, the dangling bonds of two neighboring atomic rows interact with each other thus forming π bonds [Levine 1973]. As a result only one bond per atom remains unsaturated on the surface. The π-bonded atoms move closer than required by lattice geometry (the bulk 1×1 spacing is $a = a_0/\sqrt{2} = 3.84$ Å) thus forming "dimers" (Fig. 3.32) which in turn are spaced broader than the normal interatomic spacing a. The dimers form rows which rotate by 90° on every next terrace (Fig. 3.31(b)). The latter leads to the appearance of strong elastic deformations which spread deep under the crystal surface. The so-called 2×1 and 1×2 reconstructed surfaces result.

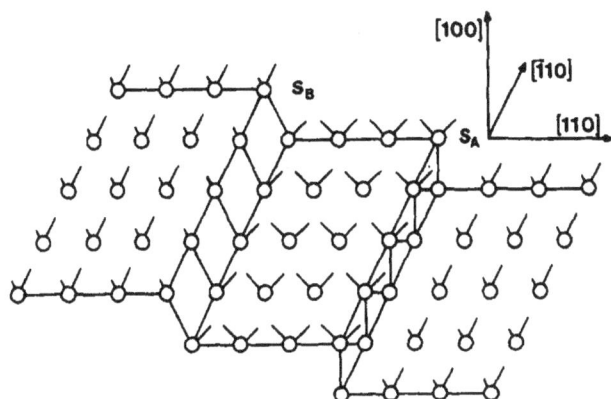

(a)

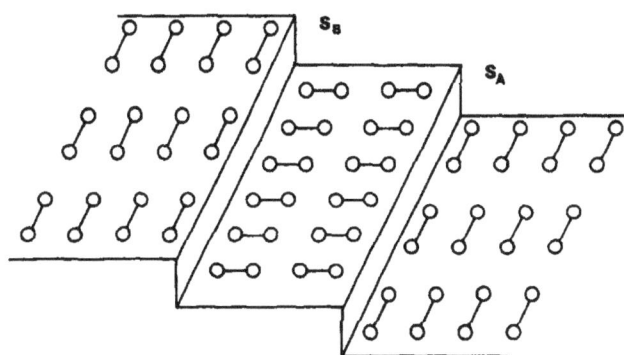

(b)

Fig. 3.31. (a) Bulk and (b) reconstructed view of a Si(001) vicinal surface tilted towards the [$\bar{1}$10] direction. A rotation by 90° of the projections of the dangling bonds is clearly seen in (a). The dimers also rotate by 90° on every next terrace. The terraces are separated by single layer steps denoted by S_A and S_B according to the notation of Chadi [1987]. The structure of the steps also alternate due to the rotation of the chemical bonds. The step height is $a_0/4 = 1.36$ Å, where $a_0 = 5.4307$ Å is the bulk lattice constant of Si.

They alternate on every second terrace and it is said that 1×2 and 2×1 domains alternate. Such a surface is often called nonprimitive. It is worth noting that a π-bonded chain model has also been proposed for the Si(111) surface [Pandey 1981].

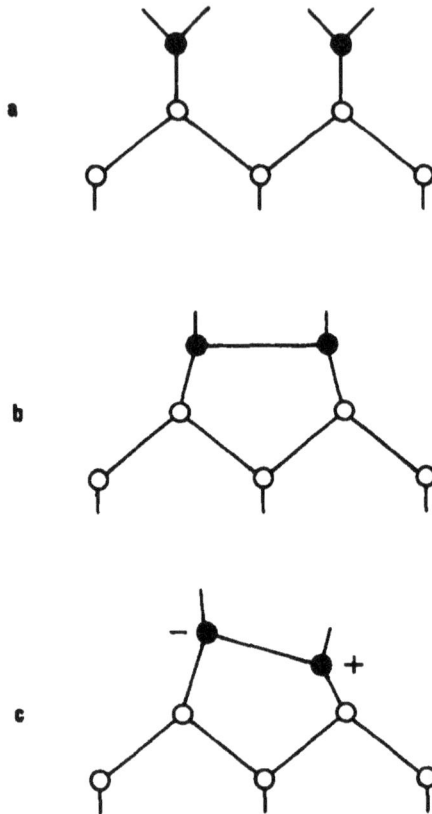

Fig. 3.32. Side view of a dimer. (a) shows the bulk structure. (b) shows a symmetric (nonbuckled) dimer. (c) shows an asymmetric (buckled) dimer. The buckling causes a partial charge transfer from the "down" to the "up" atom and the dimer's bond is partially ionic.

3.2.4.1. *Dimer's structure*

Chadi [1979] concluded that when the atoms which constitute the dimer are situated in one plane which is parallel to the surface (the so-called symmetric dimer, Fig. 3.33(b)) the dimer is unstable. The lowest energy state is reached when the dimer's atoms are displaced in a direction normal to the surface in addition to the in-plane displacements towards one another. Thus one of the atoms is displaced upwards and the other, downwards (Fig. 3.33(c)). Such "buckled" dimer is called asymmetric. Whereas the bonding between the atoms of symmetric dimers is covalent the bonding in asymmetric dimers is partially covalent and partially ionic. The formation of an asymmetric dimer results in a charge transfer from the "down" to the "up" atom of the dimer [Chadi 1979]. The same conclusion concerning the dimer's geometry has been drawn by Pauling and Herman [1983] (see also Lin, Miller and Chiang [1991]).

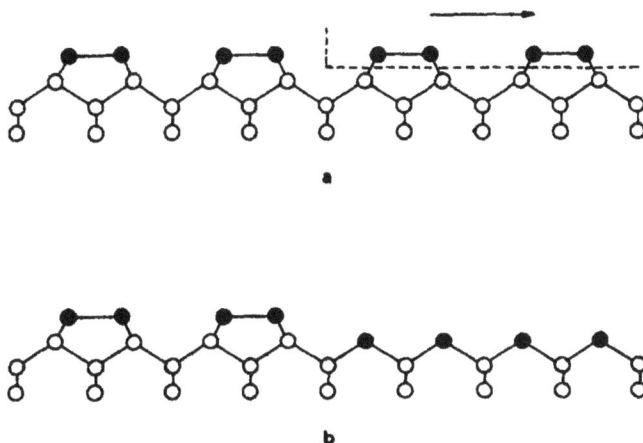

Fig. 3.33. Illustration of an imaginary process of formation of a S_A step. First, we cleave the uppermost atomic plane between two dimer rows and shift (to infinity) the right-hand half-plane to the right as shown in (a). Two S_A steps are formed (the second not shown) as shown in (b). Dimers on the lower B type terrace are formed and are directed perpendicular to the plane of the figure. The atoms which form dimers are shown by solid circles.

Both symmetric and asymmetric dimers have been observed to exist simultaneously on Si(100) by Tromp, Hamers and Demuth [1985] (see also Hamers, Tromp and Demuth [1986]) with the help of scanning tunneling microscopy (STM). They found that the (100) surface of Si has many defects, particularly vacancies or missing dimers, which in turn give rise to additional elastic strains. Far from the defects, only symmetric (nonbuckled) dimers were observed, while near the defects, asymmetric (buckled) dimers were observed as a rule. It was concluded that the vacancy type defects stabilize the dimer asymmetry, and often zigzag patterns were observed near large defect sites. These zigzag structures were explained as rows of asymmetric dimers in which the direction of buckling alternates from dimer to dimer along the row. Rows in which the dimers are buckled in one direction only have never been detected. Moreover, the degree of buckling (or asymmetry) is not always the same as that predicted by theory. A gradual transition from symmetric to asymmetric dimers has been observed when going from defectless area to an area consisting of large defects. It is interesting to note that the dimers which belong to a row at the edge of a step are always strongly buckled. For more details the reader is referred to an excellent review paper of Griffith and Kochanski [1990].

3.2.4.2. *Structure and energy of steps*

Consecutive steps on a vicinal (100) surface will be either parallel or normal to the dimer rows. Adopting the notation suggested by Chadi [1987], monoatomic steps which are parallel to the dimers rows on the upper terrace (or normal to the dimers bonds) are labelled S_A (single A) steps and those perpendicular to the dimers rows (parallel to the dimers bonds) on the upper terrace are labelled S_B (single B) steps (Fig. 3.31(b)). The corresponding upper terraces with 2×1 and 1×2 reconstructed surfaces are labelled type A and type B terraces, respectively (Fig. 3.31(b)). We will consider the two types of steps separately.

Two S_A steps can be produced by imaginary cleaving of the uppermost lattice plane between two dimers rows (parallel to the dimers rows) and shifting of one of the half-planes far enough from the other (Fig. 3.33). It is very important to note that strong first neighbor bonds are not broken during this process which means that extra dangling bonds are not created. It follows that the edge energy of such steps should be very small. In order to produce two S_B steps we cleave the uppermost lattice plane in a direction normal to the dimers rows between two neighboring dimers and shift apart the two half-planes. To do this we break a first neighbor σ bond per atom between the atoms of the uppermost and the underlying layers (Fig. 3.34(a)). Then an extra dangling bond per atom of the underlying layer is created and the specific edge energy of the S_B step should be much greater than that of the S_A steps (Fig. 3.34(b)). Such a step is called a nonbonded S_B step. The dangling bonds at the step edges can interact with the dangling bonds belonging to the atoms of the neighboring parallel row of the lower terrace. As a result an additional π bond per atom is created to reduce the step energy. The so-called rebonded S_B step is formed as shown in Fig. 3.34(c) [Chadi 1987]. The calculations of Chadi [1987] gave the values $\varkappa_{SA} = 0.01$ eV$/a = 4.16 \times 10^{-7}$ erg cm^{-1} and $\varkappa_{SB} = 0.15$ eV$/a = 6.24 \times 10^{-6}$ erg cm^{-1} for the specific edge energies of

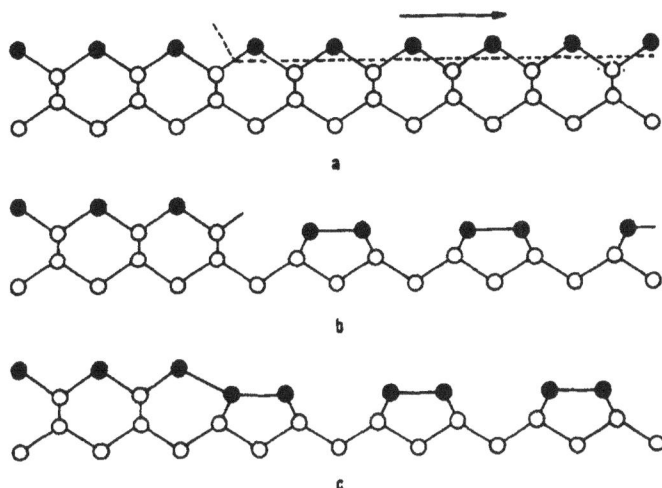

Fig. 3.34. Illustration of an imaginary process of formation of a S_B step. First, we cleave the uppermost atomic plane perpendicular to the dimer rows as shown in (a) and shift (to infinity) the right-hand half-plane to the right. Two S_B steps are formed (the second not shown) as shown in (b) and (c). The atoms on the exposed surface then form dimers. Two configurations of the S_B steps are possible. In (b) the bonds belonging to the atoms at the step edge do not take part in dimer formation. Nonbonded S_B step results. In (c) the bonds belonging to the atoms at the step edge take part in dimer formation and rebonded S_B step results.

the S_A and the rebonded S_B steps, respectively, in qualitative agreement with the above considerations. Obviously, the edge energy of a nonbonded S_B step (Fig. 3.34(b)) should be greater than that of a rebonded one. Note that no rebonding of the S_A steps can take place as no extra dangling bonds are formed.

Monoatomic steps have been studied by Hamers, Tromp and Demuth [1986] with the help of STM. As mentioned above the dimer row which forms the upper S_A step edge is strongly buckled. An interesting picture is observed in the vicinity of a kink site along the S_A step. Before the kink the dimers which constitute the row on the upper edge are strongly asymmetric. After the kink the same row is spaced at a distance $2a = 7.68$ Å from the edge and the dimers are no longer buckled. Besides, simultaneous existence of rebonded and nonbonded S_B steps has been established although minimum energy considerations [Chadi 1987] showed that the nonbonded steps are energetically unfavored. The experimental observations of Hamers, Tromp and Demuth [1986] are in good qualitative agreement with theoretical conclusions which follow from the calculations of the electronic states of the Si(100) stepped surface [Yamaguchi and Fujima 1991].

An immediate consequence of the calculations of the specific edge free energies of the monoatomic steps is that a 2D island with a monolayer height will be surrounded by two S_A and two S_B steps. As $\varkappa_{SA} < \varkappa_{SB}$ the equilibrium shape of the island will be elongated along the dimer rows according to the Gibbs–Curie–Wulff's theorem (see Chap. 1).

It was found by LEED measurements that in the case of highly mis-oriented Si(100) surfaces ($6° \leq \theta \leq 10°$) the macroscopic inclinations were accommodated in all cases by steps with double height $a_0/2$ [Henzler and Clabes 1974; Kaplan 1980]. It is worth noting that the cleaning procedure included annealing for 2 min at 1100°C and for 30 min at 950°C [Kaplan 1980]. Double steps have been observed in the case of smaller inclinations ($2° \leq \theta \leq 4°$) [Sakamoto *et al.* 1985a] after annealing for 85 min at 1000°C. Even well-oriented surfaces ($\theta < 0.5°$) [Sakamoto and Hashiguchi 1986] showed double steps after sufficiently prolonged annealing at high temperatures. Double height steps D_A and D_B in the notation of Chadi [1987] are shown in Fig. 3.35. As seen, only one type of terraces, either of type A or type B, exists in these cases. The double steps can also be rebonded. We can imagine the D_A step as having been formed by a S_A step which has caught up with a S_B step. Then the π bonds are formed between the atoms in the lower edge and the neighboring atoms belonging to the lower terrace. In the reverse case, when a S_A step is on top of a S_B step, a D_B step results. The rebonding is between the neighboring atoms of the intermediate lattice plane. The energy of rebonded double steps has been calculated in a series of papers [Aspnes and Ihm 1986; Chadi 1987]. The values $\varkappa_{DA} = 0.54$ eV/$a = 2.25 \times 10^{-5}$ erg cm^{-1} and $\varkappa_{DA} = 0.05$ eV/$a = 2.08 \times 10^{-6}$ erg cm^{-1} have been estimated [Chadi 1987].

In order to find the equilibrium structure of the steps we should estimate as before the corresponding works for kink formation. To this aim we follow a procedure analogous to that used above for the estimation of the specific step energies. In doing that we have to bear in mind that we have to preserve the integrity of the dimers. We consider first a completely smooth S_A step (Fig. 3.36(a)). We break a bond between two neighboring dimers and shift apart the two half-rows of atoms to form two single kinks. It is immediately seen that we spent an amount of work which is exactly equal to the work required to form a step S_A with length $2a$. Then the work

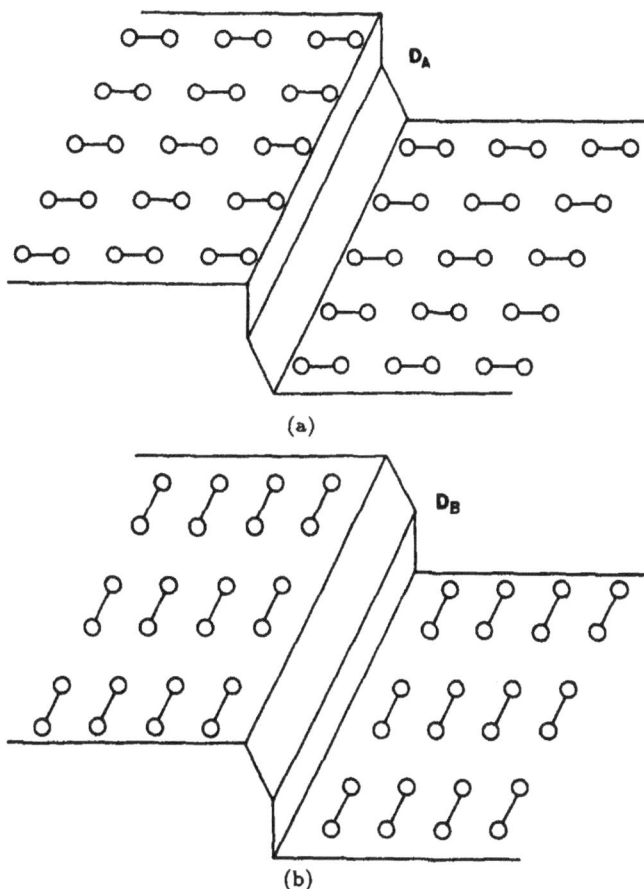

(a)

(b)

Fig. 3.35. Schematic view of (a) D_A (double A) and (b) D_B (double B) steps. The double steps can be thought of as consisting of single steps one on top of the other. In the case of the D_A step, the S_A step is on top of the S_B step and vice versa.

of formation of a single kink on the S_A step is $\omega_A = 2a\varkappa_{SB}/2 = a\varkappa_{SB}$. Applying the same procedure on the S_B step (Fig. 3.36(b)) we find $\omega_B = \varkappa_{SA}a$. It follows that the work of kink formation is greater for the step with lower specific edge energy, and vice versa [Van Loenen *et al.* 1990]. This leads in turn to the conclusion that the S_A steps will be smooth to much higher temperatures than the S_B steps. Then smooth and rough steps will alternate on a nonprimitive Si(001) surface (Fig. 3.37). Numerous STM investigations confirmed this conclusion (e.g., see Swartzentruber *et al.* [1990]).

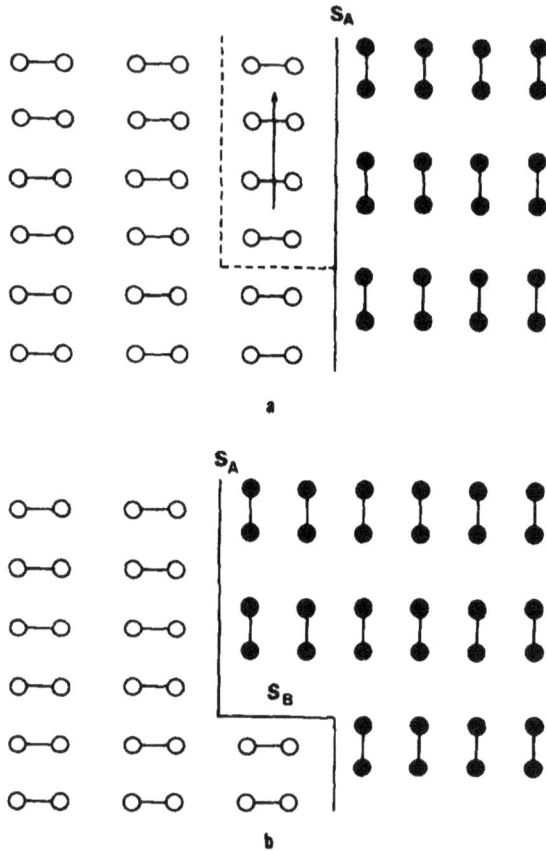

Fig. 3.36 (a) & (b). For the determination of the works of formation of kinks ω_A and ω_B along S_A and S_B steps, respectively. The atoms of the lower terraces are shown by filled circles. Imagine that in (a) we cleave the dimer row at the edge of a S_A step and shift to infinity the upper half-row as shown by the arrow. Two kinks are formed as a result, one of them being shown in (b). In fact a part of a S_B step with length $2a$ is formed, where $a = 3.84$ Å is the 1×1 interatomic spacing on the Si(001). The same in the case of a S_B step is shown in (c) and (d). A part of a S_A step with length $2a$ is formed in the latter case. Thus the side steps of the kinks are always equal to $2a$ in order to permit dimer formation at the lower terraces. Comparing (b) and (d) shows that the kinks along the S_A and S_B steps are equivalent although the steps differ. This is clear bearing in mind that the work to separate an atom from a kink position at 0 K is equal to the chemical potential of the bulk crystal taken with negative sign.

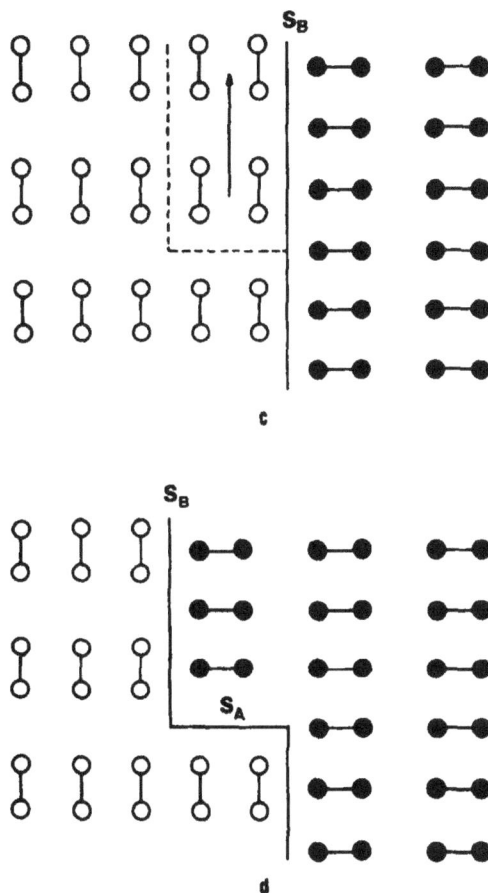

Fig. 3.36 (c) & (d) (*Continue*).

In other words, one of the edges of the kinks represents a part of a S_A step and the other edge a part of a S_B step, irrespective of whether the kink is on a S_A or S_B step. It is thus obvious that the detachment of a single atom from the kink position is no longer a repeatable step. One has to detach four atoms constituting two dimers in order to restore the initial state. It follows that the work spent to evaporate a complex of two dimers from a single kink is one and the same irrespective of the type of the step. Namely this work (per atom) taken with negative sign is equal to the chemical potential of the Si crystal at the absolute zero as shown in Chap. 1. Moreover, this leads to the conclusion that every kink should have a length being a multiple of $2a$ because of the way in which the

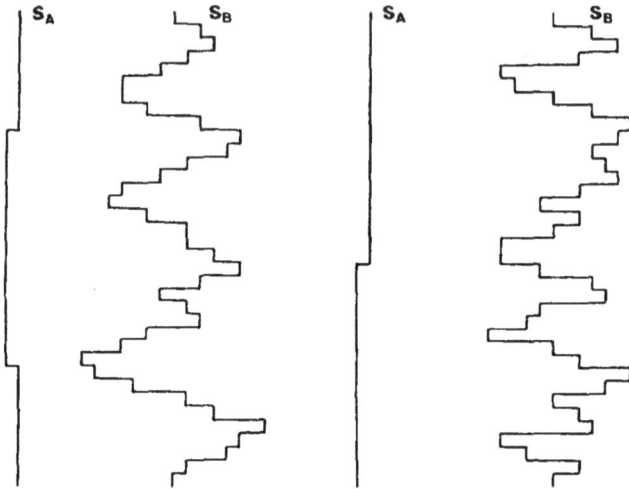

Fig. 3.37. Train of alternating smooth S_A and rough S_B steps.

step must terminate at the lower terrace and the latter has a periodicity of $2a$ normal to the step. The distance between two kinks should be a multiple of $2a$ for the same reason. These considerations were confirmed by STM observations of S_A and S_B steps [Swartzentruber *et al.* 1990]. These authors also showed that the kinks on the S_B step, which is very rough even at moderate temperatures, often have a length greater than $2a$ (but always a multiple of $2a$). This means that a statistics of kinks of amount greater than unity should be applied.

We can now calculate the density of kinks along the S_A and S_B steps as an illustration of the above considerations. To this aim we follow the ideal gas approximation (noninteracting kinks) of Burton, Cabrera and Frank [1951] although it was shown that there is considerable kink–kink interaction [Zhang, Lu and Metiu 1991c].

In analogy with Eq. (1.71) the sum of all kinks with arbitrary length r is

$$n_0 + \sum_{r=1}^{2L} n_{+r} + \sum_{r=1}^{2L} n_{-r} = n = 1/2a , \qquad (3.150)$$

where n_{+r} and n_{-r} are the numbers of positive and negative kinks of length r and n_0 is the number of the smooth parts. The summation is now carried out from $r = 1$ to $r = 2L$, where $2L$ is the mean distance between two steps in units of $2a$ of one and the same type A or B.

Equation (1.73) now becomes

$$\sum_{r=1}^{2L} r n_{+r} - \sum_{r=1}^{2L} r n_{-r} = \varphi/a ,$$ (3.151)

where φ is the angle between the step direction and the $\langle 110 \rangle$ azimuth.
As before (see (1.72))

$$\frac{n_+}{n_0} \frac{n_-}{n_0} = \eta^2 ,$$ (3.152)

where $n_\pm \equiv n_{\pm 1}$ and

$$\eta \equiv \eta_{A,B} = \exp\left(-\frac{\omega_{A,B}}{kT}\right) .$$ (3.153)

Burton, Cabrera and Frank [1951, Appendix C] showed that the following thermodynamical relation between multikinks of amount r and single kinks of amount $r = 1$ holds:

$$\frac{n_{\pm r}}{n_0} = \left(\frac{n_\pm}{n_0}\right)^r .$$ (3.154)

The latter is easy to understand bearing in mind that in order to form a kink of amount r, a kink of amount $r - 1$ must be formed before that. In other words, the probability of formation of a kink of amount r, P_r, is a product of the probability of formation of a kink of amount $r - 1$, P_{r-1}, and the probability of formation of a single kink, P_1, i.e. $P_r = P_{r-1}P_1$. Then by induction ($P_{r-2} = P_{r-1}P_1$, etc.) $P_r = P_1^r$.

The average spacing δ_0 between kinks of any amount is now given by (compare with (1.74))

$$\delta_0 = \left(\sum_{r=1}^{2L} n_{+r} + \sum_{r=1}^{2L} n_{-r}\right)^{-1} .$$ (3.155)

Solving the system (3.150), (3.151), (3.152) and (3.154) by summing a geometric series with $\varphi = 0$ gives

$$\delta_0(\text{A,B}) = a\left(1 + \frac{1}{\eta}\right) = a\left[1 + \exp\left(\frac{\omega_{A,B}}{kT}\right)\right] ,$$ (3.156)

which is a good approximation for wide enough terraces ($L > 10a$). Using the values taken by Van Loenen *et al.* [1990] for Monte Carlo simulation of the growth process, $\omega_A = 0.5$ eV and $\omega_B = 0.05$ eV for the mean kink spacings at $T = 750$ K, we obtain the values $\delta_0(\text{A}) \cong 2.3 \times 10^3 a$ and

$\delta_0(B) \cong 2a$. The values estimated by Chadi [1987], $\omega_A = 0.15$ eV and $\omega_B = 0.01$ eV, give smaller value for $\delta_0(A) \cong 12a$ and the same value for $\delta_0(B) \cong 2a$. In any case $\delta_0(A)$ is always greater than $\delta_0(B)$. It is worth noting, however, that Eq. (3.156) is only approximate as the ideal gas model (noninteracting kinks) has been used. More elaborate calculations including kink–kink interactions give more realistic results [Zhang, Lu and Metiu 1991c].

Following the same procedure as in Chap. 1 but solving Eqs. (3.150)–(3.154) for the Gibbs free energy of the S_B steps, one obtains

$$G_{SB} = \varkappa_{SB} - nkT \ln \left(\frac{1 + \eta_B - 2\eta_B^{2L+1}}{1 - \eta_B} \right) , \qquad (3.157)$$

where $\varkappa_{SB}$ is the energy of the straight step.

The Gibbs free energy of the S_A is obtained by replacing the index B by A. In the extreme case of steps far apart ($L \to \infty$) Eq. (3.157) turns into the one derived by Burton, Cabrera and Frank [1951]:

$$G_{SB} = \varkappa_{SB} - nkT \ln \left(\frac{1 + \eta_B}{1 - \eta_B} \right) . \qquad (3.157')$$

3.2.4.3. *Ground state of vicinal Si(100) surfaces*

The question of the lowest energy state of the Si(100) vicinal surfaces is very important as the latter are used as substrates for the growth of epitaxial films of GaAs and other III-V compounds. The latter are of utmost importance for potential device applications [Shaw 1989]. Obviously, the surface with single height steps necessarily leads to antiphase boundaries in the III-V epilayers [Kroemer 1986].

As seen, the S_A steps have the lowest edge energy. However, they unavoidably lead to the existence of S_B steps and the overall energy is $\varkappa_{SA} + \varkappa_{SB} = 0.16$ eV/$a = 6.66 \times 10^{-6}$ erg cm^{-1}. This value is three times higher than the energy of a D_B step but more than three times lower than that of a D_A step. It was concluded that the D_B steps are thermodynamically favored on a vicinal (001) surface of Si and a single domain 1×2 reconstruction should always dominate after sufficiently long annealing at high temperatures [Chadi 1987]. Such a surface is often called a primitive surface.

As shown in Fig. 3.31(b), terraces with 2×1 and 1×2 reconstructions alternate on the nonprimitive surface. Due to the displacements of the dimer atoms towards one another, such a reconstructed surface is under tensile stress $\sigma_=$ parallel to the dimer bonds and under compressive stress $\sigma_\perp$ in the perpendicular direction. The overall stress $\Delta\sigma = \sigma_= - \sigma_\perp$ is tensile. The stress rotates by 90° on the neighboring terrace and as a result tensile and compressive stresses alternate on a vicinal Si(100) surface, the period of alternation being given by the terrace width. It was first pointed out by Marchenko [1981] that on a surface with alternating stress domains (parquet-like surface) the stress relaxation lowers the surface energy. The decrease of the surface energy due to strain relaxation was found to depend logarithmically on the interstep distance [Marchenko 1981; Alerhand et al. 1988]. Obviously, in the case of a surface with double height steps (Fig. 3.35) all dimers strain the crystal in one and the same direction, the surface stress does not alter its sign and there is no strain relaxation. Then the difference between the Gibbs free energies of single height (SH) and double height (DH) stepped surfaces will be given by [Alerhand et al. 1990]

$$\Delta G = L^{-1}\left[\frac{1}{2}(G_{SA} + G_{SB} - G_{DB}) - \lambda_\sigma \ln\left(\frac{L}{\pi a}\right)\right], \qquad (3.158)$$

where $\lambda_\sigma = \Delta\sigma^2(1-\nu)/2\pi G$ (energy per unit length) originates from the anisotropy of the stresses, $\Delta\sigma$, and depends on the shear modulus G and the Poisson ratio ν of the bulk silicon [Marchenko 1981; Alerhand et al. 1988].

As discussed above, the work for kink formation is high for the S_A step and low for the S_B step. The same is valid for the double height steps. The D_B steps have high energy excitations, i.e. $\omega_{DB} = a\varkappa_{DA}$, and we can approximate G_{SA} and G_{DB} by the edge energies of the perfectly straight steps $\varkappa_{SA}$ and $\varkappa_{SB}$. We have to take the full expression (3.157) only for G_{SB}.

Then in the ideal gas approximation (neglecting for simplicity the term $2\eta^{2L+1}$ in (3.157)) (3.158) turns into

$$\Delta G \cong L^{-1}\left[\frac{1}{2}(\varkappa_{SA} + \varkappa_{SB} - \varkappa_{DB}) - \frac{kT}{2a}\ln\left(\frac{1+\eta_B}{1-\eta_B}\right) - \lambda_\sigma \ln\left(\frac{L}{\pi a}\right)\right].$$

The condition $\Delta G = 0$ determines the first order phase transition from nonprimitive (SH-stepped) to primitive (DH-stepped) surface. A critical terrace width L_c (or a critical tilt angle $\theta_c = \tan^{-1}(h/L_c)$; $h = a_0/4 = 1.36$ Å is the single step height) can be determined showing that the primitive surface will be energetically favored at $L < L_c$ ($\theta > \theta_c$), and the nonprimitive surface at $L > L_c$ ($\theta < \theta_c$). L_c is then given by

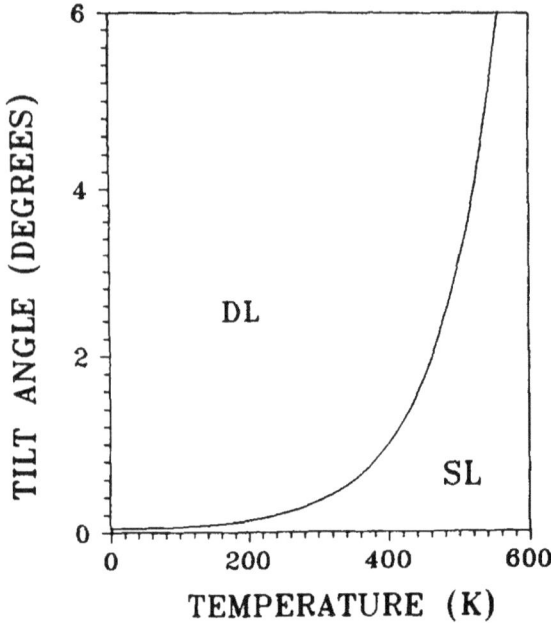

Fig. 3.38. Phase diagram of a vicinal Si(001) surface showing regions of stability of double layer (DL) and single layer (SL) stepped surfaces (after Alerhand et al. [1990]).

$$L_c = \pi a \exp \left[\frac{\varkappa_{SA} + \varkappa_{SB} - \varkappa_{DB}}{2\lambda_\sigma} - \frac{kT}{2a\lambda_\sigma} \ln \left(\frac{1 + \eta_B}{1 - \eta_B} \right) \right] . \qquad (3.159)$$

θ_c is plotted versus temperature in Fig. 3.38 with $\varkappa_{SA} + \varkappa_{SB} - \varkappa_{DB} = 110$ meV/a, $\lambda_\sigma = 11.5$ meV/a and $\omega_B = 10$ meV. As seen, at low temperatures the entropy term on the right-hand side of (3.159) goes to zero and $L_c \to L_{co} = 1440$ Å or $\theta_c \to \theta_{co} \cong 0.05°$. In general, the primitive surface is the ground state at high tilt angles, and vice versa, the single height stepped surface being energetically favored at low tilt angles. Equation (3.159) overestimates the result obtained by Alerhand et al. [1990]. More elaborate studies take into account the influence of strain relaxation on the step roughness [Alerhand et al. 1990], the corner energy of the kinks [Poon et al. 1990], etc. In particular, more elaborate evaluation of the specific edge energies [Poon et al. 1990] gives for θ_{co} a much higher value of about 1°. Pehlke and Tersoff [1991a] found that at the ground state the B type terraces are narrower than the A type terraces. The problem of the equilibrium structure of vicinal Si(100) surfaces is still a subject of intensive experimental and theoretical investigations. The interested reader is referred to the original papers [De Miguel *et al.* 1991; Barbier and Lapujoulade 1991; Barbier *et al.* 1991; Pehlke and Tersoff 1991b].

3.2.4.4. *Anisotropy of surface diffusion coefficient*

The anisotropy of the surface diffusion follows directly from the anisotropy of the crystal surface itself [Stoyanov 1989]. The first questions which arise are connected with the location of the adsorption sites on the reconstructed surface, with what happen to the dimers when adatoms appear on top of them or in their vicinity, etc. Then an energy surface should be constructed and the lowest energy path for surface diffusion should be determined. Brocks, Kelly and Car [1991a, b] found that the deepest minima for an adsorbed atom are located along the dimer rows between two neighboring dimers belonging to the same row (point M in Fig. 3.39). This site is favored by the fact that the bonds between the adatom and the nearest dimer atoms have the same length $a_0\sqrt{3}/4 = 2.35$ Å as the nearest neighbor spacing in the bulk silicon. The site B, which connects two dimers in adjacent rows, although looks very favorable, requires too long a bond of 2.49 Å and the dimers bond should be stretched also. Thus the energy of an adatom in site B is 1.0 eV higher than that in the deepest minimum M. The energy of the site H which is located between two adjacent dimers of one row is 0.25 eV higher than that of the site M, whereas the energy of the site D which is just on top of the dimer is 0.6 eV higher than the energy of the absolute minimum M. It was thus found that the lowest energy path of an adatom in a direction parallel to the dimer rows is D–H–M and the activation energy for surface diffusion is 0.6 eV. The activation energy for surface diffusion in a direction perpendicular to the dimer rows is greater than 1.0 eV. It was thus concluded that the direction of fast diffusion is parallel to the dimer rows. Miyazaki, Hiramoto and Okazaki [1991] found that an atom adsorbed on top of a dimer (site D) causes a little distortion of the dimer while that adsorbed at site B causes the dimer to break. They reached the same conclusion concerning the direction of fast diffusion as Brocks, Kelly and Car [1991a,b]. The activation energy along the path D–H–D parallel to the dimer rows was found to be 0.6 eV whereas in the perpendicular direction D–B–D the latter is somewhat larger (1.7 eV). Using a Stillinger–Weber interatomic potential [Stillinger and Weber 1987] Zhang, Lu and Metiu [1991a,b] (see also Lu, Zhang and Metiu [1991]) found even smaller activation energies for surface diffusion of about 0.3 eV along the dimer rows when the adatoms

diffuse on top of the dimers, 0.7 eV when the adatoms diffuse by the side
of the dimer string and 0.9 eV when the adatoms diffuse along the value
between the dimer rows. At the same time the latter authors found an
activation energy higher than 1.0 eV in a direction perpendicular to the
dimer rows. It follows that an adatom adsorbed on a dimer string can
quickly move to the end of the string and increase its length (Zhang, Lu
and Metiu 1991b). They also found that diffusion of dimers as entities is
highly improbable. Ashu, Matthai and Shen [1991] found the values 0.2 eV
and 2.8 eV for the activation energies for surface diffusion in directions
parallel and perpendicular to the dimer rows, respectively. Based on STM
measurements of the saturation island density, Mo *et al.* [1991] extracted
the values 0.67 ± 0.08 eV and 1×10^{-3} cm^2/sec for the activation energy for
fast diffusion and the pre-exponential factor. Excellent agreement with the
above values was found by Roland and Gilmer [1991, 1992a], the latter
authors having used the Stillinger–Weber potential for their study. In
addition they found that exchange between substrate atoms and adatoms
takes place even at low temperatures. This phenomenon gives additional
contribution to the surface diffusion.

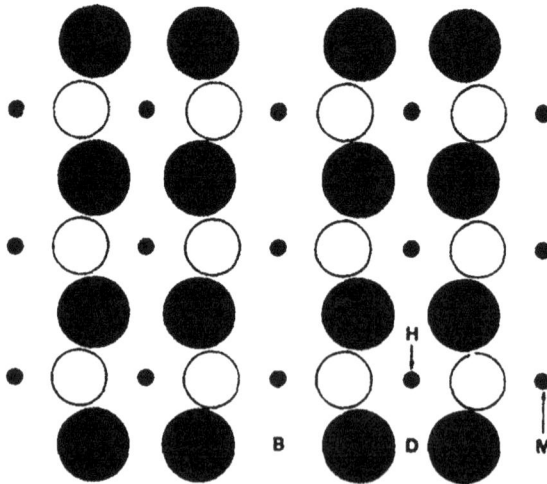

Fig. 3.39. Illustration of the directions of fast and slow surface diffusion of Si atoms on
Si(001). Top view of the topmost three layers of a (2×1) reconstructed Si(001) surface
is shown. The large filled circles represent the uppermost atoms, the medium-sized open
circles represent the second layer atoms and the small filled circles represent the third
layer atoms. The points denoted by B, D, M and H are explained in the text (after
Brocks, Kelly and Car [1991]).

The general conclusion is that, irrespective of the quantitative differences due to the different methods of calculations, the surface diffusion on the reconstructed Si(100) 2 × 1 surface is highly anisotropic and the direction of fast diffusion is parallel to the dimer rows. The values for the activation energy, although varying from 0.2 eV to nearly 0.7 eV in different studies, suggest that considerable surface diffusion takes place even at room temperature. What is much more important is that a critical temperature should exist below which only surface diffusion in one direction takes place and above which the diffusion in a direction normal to the dimer rows could become significant.

3.2.4.5. *Theory of 1D nucleation*

It follows from above that rough (S_B) and smooth (S_A) steps alternate on a vicinal double domain Si(001) surface. The S_B steps advance by direct incorporation of growth units to the kink sites in complete analogy with the normal growth of rough crystal faces. The growth of the S_A steps is more complicated. It requires the precursory formation of kinks. As thermally activated formation of kinks with sufficiently great enough density is inhibited another mechanism should obviously be involved. In analogy with the formation of 2D nuclei on a smooth defectless crystal surface one can think of the formation of 1D nuclei which represent finite atomic rows. Every row will thus give rise to two kinks. The theory of 1D nucleation has been treated by many authors [Voronkov 1970; Frank 1974; Zhang and Nancollas 1990].

Let us try to treat thermodynamically the problem of the formation of 1D nuclei just as we did in the case of 3D and 2D nuclei. To this aim we will use the atomistic approach suggested by Stranski and Kaischew [1934].

We consider first the formation of a 3D nucleus of a Kossel crystal with a cubic equilibrium shape. The nucleus consists of $N = n_3^3$ atoms, where n_3 is the number of atoms in the nucleus edge. The work for nucleus formation is given by Eq. (2.20), $\Delta G_3^* = N\bar\varphi_3 - U_N$, where $\bar\varphi_3 = 3\psi - 2\psi/n_3$ is the mean separation work (Eq. 1.60). The equilibrium vapor pressure of the nucleus is defined by Eq. (1.61), $\Delta\mu = \varphi_{1/2} - \bar\varphi_3 = 2\psi/n_3$. Substituting $\bar\varphi_3$ into Eq. (2.20) and bearing in mind that $U_N = 3n_3^3\psi - 3n_3^2\psi$ for the Gibbs free energy, one obtains $\Delta G_3^* = n_3^2\psi$. Applying the same procedure to a 2D nucleus with a square shape one correspondingly obtains $\Delta\mu = \psi/n_2$ and $\Delta G_3^* = n_2\psi$.

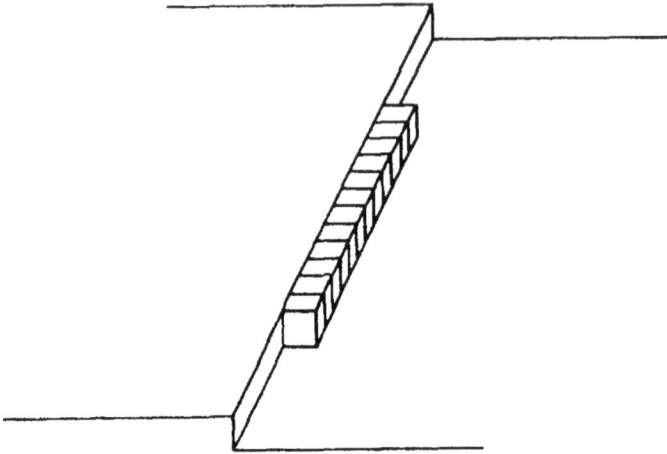

Fig. 3.40. Schematic view of 1D nucleus at the edge of a single step on the surface of a Kossel crystal.

We consider now a row of n_1 atoms in a step edge (Fig. 3.40). The mean separation work is calculated now by the detachment of the end atoms and is exactly equal to the separation work from a kink position, i.e. $\bar\varphi_1 = 3\psi = \varphi_{1/2}$. Then $\Delta\mu = \varphi_{1/2} - \bar\varphi_1 = 0$, i.e. the atomic row has the same chemical potential as the bulk crystal irrespective of its length. The potential energy is $U_N = 3n_1\psi - \psi$ and for the Gibbs free energy one obtains $\Delta G_1^* = \psi$. The results of the above calculations are summarized in Table 3.1. As seen the Gibbs free energy does not depend on the nucleus size, and hence we cannot define thermodynamically a critical size of the row of atoms. However, as shown by the authors mentioned above a critical 1D nucleus can be well defined kinetically. In what follows we will consider the problem of step advance by 1D nucleation following the approach of Voronkov [1970].

We consider the growth and dissolution of the row of atoms shown in Fig. 3.40 assuming that atoms attach to the row ends directly from the terraces. We rule out diffusion of atoms along the step edge. The diffusion of atoms to the kinks is fast enough and the step advances in a kinetic regime. Then a constant adatom concentration n_{st} or a constant supersaturation $\sigma = n_{st}/n_{se} - 1$ exists in the near vicinity of the step. We denote by $\omega^+ dt$ and $\omega^- dt$ the probabilities for attachment and detachment of adatoms to and from a kink position in a time interval dt. The corresponding frequencies ω^+ and ω^- are given by

Table 3.1. Gibbs free energies of formation and supersaturations for 3D, 2D and 1D nuclei of a Kossel crystal. n denotes the number of atoms in the edges of the 3D nucleus with a cubic shape, 2D nucleus with a square shape and 1D row of atoms. ψ denotes the work required to break the bond between the first neighbors.

Dimensionality of the nucleus	ΔG^*	$\Delta \mu$
3D	$n^2 \psi$	$2\dfrac{\psi}{n}$
2D	$n\psi$	$\dfrac{\psi}{n}$
1D	ψ	0

$$\omega^- = \nu \exp\left(-\frac{\Delta W + \Delta U}{kT}\right) , \qquad (3.160)$$

$$\omega^+ = \nu \frac{n_{st}}{N_0} \exp\left(-\frac{\Delta U}{kT}\right) , \qquad (3.161)$$

where $\Delta W = \varphi_{1/2} - \varphi_{des}$ is the work to transfer an atom from a kink position on the surface of the nearby terrace (Eq. (3.18)).

Bearing in mind that $\sigma = n_{st}/n_{se} - 1$ and $n_{se}/N_0 = \exp(-\Delta W/kT)$ (Eq. (3.18)) (3.161) can be rewritten in the form

$$\omega^+ = \omega^-(1 + \sigma) . \qquad (3.161')$$

As seen, at equilibrium $(\sigma = 0)$ $\omega^+ = \omega^-$ and $n_{st} = n_{se}$.

The kink performs random walk back and forth with a diffusion coefficient

$$D = a^2 \omega^- \qquad (3.162)$$

around a given constant position.

When not in equilibrium $(\sigma \neq 0)$ $\omega^+ > \omega^-$ and the rate of advance of the kink is given by

$$v_k = a(\omega^+ - \omega^-) = a\omega^- \sigma . \qquad (3.163)$$

At small supersaturations the probabilities of attachment and detachment of atoms are close. The kink can perform simultaneously random walk backward (dissolution of the row) and steady advance forward with a rate v_k. The direction of the random walk is opposite to the direction of advance. Then during a time t, n_b atoms will detach from the kink and n_f atoms will join the kink. The kink will shift backward by a distance

$$l = a(n_b - n_f) \cong 2(Dt)^{1/2} - v_k t \ .$$

As seen, the shift of the kink backward displays a maximum at some time $t_{\max} = D/v_k^2$. The maximum or the most probable shift of the kink backward is then

$$l_{\max} = \frac{D}{v_k} = \frac{a}{\sigma} \ . \tag{3.164}$$

Let the mean kink spacing be

$$\delta_0 = a \left(1 + \frac{1}{2} \exp(\omega/kT) \right)$$

as given by Eq. (1.73). If $a/\sigma \gg \delta_0$, the probability for the kink to encounter a neighboring kink with an opposite sign and to annihilate with it is very large. If this happens the atomic row will disappear. In other words, atomic rows smaller than a/σ will have a greater tendency to decay than to grow further.

If, however, the system is sufficiently far from equilibrium so that the supersaturation is large enough, $a/\sigma \ll \delta_0$ or

$$\sigma \gg a\rho_0 \ , \tag{3.165}$$

where $\rho_0 = 1/\delta_0$ is the equilibrium kink density, steady growth will prevail over random walk. An atomic row longer than a/σ most probably will not disappear, but will grow with a steady state rate v_k. Thus it is namely the quantity a/σ which plays the role of critical size of the 1D nucleus. The latter is defined solely on the basis of kinetic considerations.

It follows that when the inequality (3.165) is fulfilled the advance of the step will take place by formation and growth of 1D nuclei. Thus this inequality gives the lower limit of validity of the 1D nucleation mechanism of growth. Obviously, an upper limit should exist. In complete analogy with the growth of atomically smooth and defectless crystal face the upper limit will be defined by the condition of the kinetic roughness of the step. In the particular case of step advance the kinetic roughness is determined by the condition that every atom adsorbed at the step edge remains there for a sufficiently long time. Then each adsorbed atom will give rise to two kinks.

We denote the frequencies of adsorption and desorption of atoms at the step edge by w_a^+ and w_a^-, respectively. The adsorption frequency ω_a^+

should be nearly equal to the frequency w^+ of attachment to kink sites. The desorption frequency ω_a^- should be much greater than ω^- as the adsorbed atoms are more loosely bound to the steps than the atoms in the kink positions. If the supersaturation is not large, an atom adsorbed at the step will desorb with greater probability than attach new atoms on either side of it. In other words, when the frequency of desorption is comparable with the double frequency of attachment of atoms to kink positions, $\omega_a^- \cong 2\omega^+$, the step will by kinetically rough and the 1D nucleation mechanism will no longer be valid. The reverse condition, or the condition that the step is still smooth, is obviously $\omega_a^- \gg 2\omega^+$, or

$$\frac{2\omega^+}{\omega_a^-} \ll 1 \ . \tag{3.166}$$

The inequality (3.166) can be expressed through the equilibrium kink density and the supersaturation by using the condition for adsorption–desorption equilibrium. We denote by $\rho_{a,0}$ the equilibrium density of atoms adsorbed at the step. The condition of detailed balance reads $w_{a,0}^+ / a = w_a^- \rho_{a,0}$, or

$$\frac{\omega_{a,0}^+}{w_a^-} = a\rho_{a,0} \ .$$

Each adsorbed atom creates two kinks, one positive and one negative. Then the probability to find an adsorbed atom, $a\rho_{a,0}$, will be equal to the probability to find simultaneously one positive, $a\rho_0^+$, and one negative, $a\rho_0^-$, kink. Neglecting any possible energetic interaction between the kinks, the above results in $a\rho_{a,0} = (a\rho_0^+)(a\rho_0^-)$. Bearing in mind that $\rho_0 = \rho_0^+ + \rho_0^-$ and $\rho_0^+ = \rho_0^-$, one obtains $\rho_{a,0} = a(\rho_0/2)^2$. Then

$$\frac{\omega_{a,0}^+}{\omega_a^-} = \left(\frac{a\rho_0}{2}\right)^2 \ . \tag{3.167}$$

Excluding ω_a^- from (3.166) and (3.167) and bearing in mind that $\omega^+ \cong \omega_a^+ = \omega_{a,0}^+ (1 + \sigma)$, (3.166) turns into

$$\sigma \ll \frac{1}{(a\rho_0)^2} \ . \tag{3.168}$$

Thus the conditions (3.165) and (3.168) give the lower and upper limits of validity of the 1D nucleation mechanism of advance of single steps. Obviously, if $a\rho_0 \ll 1$ as in the case under study this mechanism of growth will be valid in a very wide interval of supersaturations. In the particular case of S_A steps on the vicinal surface of Si(001) at $T = 600$ K, $\delta_A = 870a$. Then $a\rho_0 = 1.15 \times 10^{-3}$ and the advance of the steps through 1D nucleation will take place from supersaturations as low as 1.15×10^{-3} up to supersaturations as high as 7.6×10^5.

We can now calculate the steady state rate of formation of 1D nuclei by using the classical approach of Becker and Döring [1935] described in Chap. 2. Equations (2.49') in this particular case will look like

$$J_0 = \omega_a^+ \frac{1}{a} - \omega_a^- \rho_a \,, \tag{3.169'}$$

$$J_0 = 2\omega^+ \rho_a - 2\omega^- \rho_2 \,, \tag{3.169''}$$

$$\vdots$$

$$J_0 = 2\omega^+ \rho_{n-1} - 2\omega^- \rho_n \,. \tag{3.169'''}$$

The first equation differs from all the others, on account of which we will solve the system beginning from the second equation (3.169''). Thus we will determine the density of the adsorbed atoms, ρ_a, and substitute it into the first equation (3.169') to obtain an expression for J_0.

The expression (2.50) now reads (from Eq. (3.169'') onwards)

$$J_0 = 2\omega^+ \rho_a \left[1 + \sum_{k=1}^{n} \left(\frac{\omega^-}{\omega^+} \right)^k \right]^{-1} \,.$$

The sum in the denominator represents that of a geometric series and can be easily found. The upper limit should be greater than the number of atoms in the critical nucleus $1/\sigma$. The ratio $\omega^-/\omega^+ = 1/(1 + \sigma)$ (Eq. (3.161')) is always smaller than unity as $\sigma > 0$. At large supersaturations the critical size $1/\sigma$ is small and the upper limit n should be a small number, and vice versa. At the same time, at large supersaturations the terms in the sum vanish faster than at small ones so that we will not make a large error if we extend the upper limit to infinity in both cases of large and small supersaturations. Then the sum in the denominator is equal to $1/\sigma$ and $J_0 = 2\omega^+ \rho_a \sigma/(1 + \sigma) = 2\omega^- \sigma \rho_a = 2(\omega^+ - \omega^-)\rho_a$, or

$$\rho_a = \frac{J_0}{2(\omega^+ - \omega^-)} \,.$$

This result can be immediately obtained from Eq. (3.169''') assuming $\rho_n = \rho_{n-1} = \rho_a = \text{const}$ [Voronkov 1970], i.e. the densities of the clusters do not depend on the cluster size as follows from the thermodynamic considerations given in the beginning of this section.

Substituting ρ_a into (3.169') gives for J_0 [Voronkov 1970]

$$J_0 = \frac{\omega_a^+}{a} \frac{2(\omega^+ - \omega^-)}{\omega_a^- + 2(\omega^+ - \omega^-)} . \qquad (3.170)$$

This expression can be easily simplified taking into account that $\omega_a^- \gg 2\omega^+$ (Eq. 3.166) and hence $\omega_a^- \gg 2(\omega^+ - \omega^-)$. Substituting $\omega_a^+ = \omega_{a,0}^+(1 + \sigma)$, $\omega_{a,0}^+ = w_a^-(a\rho_0/2)^2$ from Eq. (3.167) and $\omega^+ - \omega^- = \omega^-\sigma$ from Eq. (3.163) into Eq. (3.170) gives

$$J_0 = \frac{1}{2}a\omega^- \rho_0^2 \sigma(1 + \sigma) . \qquad (3.171)$$

Finally, making use of Eq. (3.160) and $\rho_0 = 1/\delta_0$ gives

$$J_0 = 2\frac{\nu}{a}\sigma(1 + \sigma) \exp\left(-\frac{\varphi_{1/2} - \varphi_{\text{des}} + 2\omega + \Delta U}{kT}\right) . \qquad (3.172)$$

As seen the steady state rate of 1D nucleation is a linear function of the supersaturation when the latter is much smaller than unity, but increases parabolically with it when $\sigma \gg 1$. Then at low temperatures, e.g., $T = 600$ K, with $\nu = 3 \times 10^{13} \text{sec}^{-1}$, $a = 3.84 \times 10^{-8}$ cm, $\sigma \cong 1 \times 10^3$, $\varphi_{1/2} = 4.33$ eV, $\varphi_{\text{des}} = 2.99$ eV [Roland and Gilmer 1991, 1992a], $\omega_A = 0.5$ eV and $\Delta U = 0.2$ eV, the steady state 1D nucleation rate is of the order of 7×10^5 cm^{-1}sec^{-1}. At $T = 1000$ K, $\sigma \cong 1 \times 10^{-3}$ and $J_0 \cong 2 \times 10^5$ cm^{-1}sec^{-1}.

3.2.4.6. Rate of step advance by 1D nucleation

As in the case of growth of a smooth and defectless crystal face by formation and lateral spreading of 2D nuclei we will consider separately the advance of infinitely long step and a step of finite length. In doing that we will follow exactly the same approach.

In the case of infinitely long step we assume that a row with length l is formed. Then the frequency of 1D nucleation of a new row of atoms next to the first one is $\bar{J}_0 = J_0 l$. The time elapsed from the nucleation of the first row to the nucleation of the second row is l/v_k. The latter is approximately inversely proportional to the nucleation frequency, or $l/v_k \cong 1/J_0 l$. Then $l = (v_k/J_0)^{1/2}$. The rate of step advance is given by $v = J_0 l a$, or [Voronkov 1970; Frank 1974]

$$v = a(J_0 v_k)^{1/2} . \qquad (3.173)$$

Making use of the expressions for J_0 and v_k, (3.171) and (3.163), gives for v

$$v = a^2 \rho_0 \omega^- \sigma (1 + \sigma)^{1/2} , \qquad (3.174)$$

or

$$v = 2a\nu\sigma(1 + \sigma)^{1/2} \exp\left(-\frac{\varphi_{1/2} - \varphi_{des} + \omega + \Delta U}{kT}\right) . \qquad (3.175)$$

On the other hand, $v = a\rho v_k = a^2 \rho \omega^- \sigma$, where ρ is the real kink density under conditions far from equilibrium. Comparing this expression with Eq. (3.174) gives for the kink density

$$\rho = \rho_0 (1 + \sigma)^{1/2} . \qquad (3.176)$$

It follows that at high temperatures (small supersaturations) when the equilibrium density of thermally activated kinks is large the kink density is close to the equilibrium one, i.e. the contribution of the 1D nucleation to the kink formation is negligible. The contribution of the 1D nucleation to the kink formation is significant at low temperatures.

The propagation of steps with a finite length is completely analogous to the layer-by-layer growth of finite crystal faces. The advance of a step with a finite length l in the row-by-row mode will be given by

$$v = J_0 la . \qquad (3.177)$$

Equation (3.177) is particularly important when the growth of a Si(001) surface through the formation of 2D islands is considered. As mentioned above the latter are surrounded by two smooth, S_A, edges and two rough, S_B, edges of finite length.

3.2.4.7. Growth of Si(001) vicinal by step flow

As follows from the above, the growth of a double domain vicinal Si(001) 2×1 surface is characterized by two fundamental properties: first, the nonequivalency of the steps, and, second, the anisotropy of the surface diffusion. As a consequence of the first property the alternating steps will propagate in general with different velocities and catch up with each other to form higher steps. The second factor leads to the conclusion that on B type terraces the atoms will diffuse predominantly in a direction perpendicular to the steps, while on the A type terraces the adatoms will diffuse in a direction parallel to the steps. It follows that the steps will in fact propagate only at the expense of the atoms diffusing to them on the B type terraces. The atoms on the A type terraces will not take part in the growth process at high enough temperatures. If the temperature is low enough the adatoms on the A type terraces will give rise to 2D nucleation and growth. Thus the S_A steps will propagate at the expense of the atoms diffusing to them on the lower terraces whereas the S_B steps will advance at the expense of the atoms diffusing on their upper terraces. In addition, Roland and Gilmer [1991] have found that the attachment of adatoms to the S_A steps from the above A type terrace is less probable than from the lower B type terrace. The reverse is valid for rebonded S_B steps. Note that at lower temperatures 2D nucleation will take place on the terraces and the growth will proceed by 2D nucleation mechanism. A critical temperature for transition from step flow growth to 2D nucleation growth should exist [Myers-Beaghton and Vvedensky 1990]. The problem of the growth of Si(001) vicinal surfaces has been studied in detail by many authors [Vvedensky *et al.* 1990a, b; Wilby *et al.* 1989] and the interested reader is referred to their papers. A Monte Carlo simulation with video animation to visualize the results was performed by Wilby *et al.* [1991]. We will consider in this section the growth of Si(001) at high temperatures to avoid 2D nucleation on A type terraces following the analysis given by Stoyanov [1990].

We consider a double domain (nonprimitive) vicinal Si(100) surface on which S_A and S_B alternate (see Fig. 3.37). The initial interstep spacing is denoted by λ. The beginning of the coordinate system is at the middle point of a B type terrace so that the S_A and S_B steps are located at $x = -\lambda/2$ and $x = \lambda/2$, respectively.

The S_B steps are rough and propagate with a rate

$$v_B = a^2 \beta_B n_{se} \sigma_B \ . \tag{3.178'}$$

The S_A steps are smooth and propagate through 1D nucleation with a rate

$$v_A = a^2 \beta_A \sigma_A (1 + \sigma_A)^{1/2} \cong a^2 \beta_A \sigma_A \ . \tag{3.178''}$$

In the above equations, $\beta_{A,B}$ and $\sigma_{A,B} = n_{A,B}/n_{se} - 1$ are the corresponding kinetic coefficients and supersaturations in the vicinities of the steps. $n_{A,B}$ are the corresponding adatom concentrations.

In the case of complete condensation (the re-evaporation is strongly inhibited) the diffusion on the B type terraces is governed by the diffusion equation

$$\frac{d^2 n_s(x)}{dx^2} + \frac{\mathcal{R}}{D_s} = 0 \ . \tag{3.179}$$

The solution of the equation subject to the boundary conditions $x = -\lambda/2\,n_s = n_A$ and $x = \lambda/2\,n_s = n_B$ reads

$$n_s(x) = \frac{\mathcal{R}}{2D_s}\left(\frac{\lambda^2}{4} - x^2\right) + \frac{x}{\lambda}(n_B - n_A) + \frac{1}{2}(n_B + n_A)\,. \tag{3.180}$$

Bearing in mind that $v_{A,B} = a^2 D_s(dn_s/dx)_{x=\pm\lambda/2}$ for the rates of growth one obtains

$$v_A = \frac{\mathcal{R}\lambda}{2N_0} - \frac{D_s n_{se}}{\lambda N_0}(\sigma_A - \sigma_B)\,, \tag{3.181'}$$

$$v_B = \frac{\mathcal{R}\lambda}{2N_0} + \frac{D_s n_{se}}{\lambda N_0}(\sigma_A - \sigma_B)\,. \tag{3.181''}$$

Comparing (3.178') with (3.181''), and (3.178'') with (3.181'), we find σ_A and σ_B and in turn for v_A and v_B we obtain

$$v_A = \frac{\mathcal{R}\lambda}{2N_0}\frac{M_A(2 + M_B)}{M}\,, \tag{3.182'}$$

$$v_B = \frac{\mathcal{R}\lambda}{2N_0}\frac{M_B(2 + M_A)}{M}\,, \tag{3.182''}$$

where $M_{A,B} = \beta_{A,B}\lambda/D_s$ and $M = M_A + M_B + M_A M_B$.

The ratio of the step velocities reads

$$\frac{v_A}{v_B} = \left(\frac{\rho_{A,0}}{\rho_{B,0}} + \frac{\beta_A\lambda}{2D_s}\right)\left(1 + \frac{\beta_A\lambda}{2D_s}\right)^{-1}\,. \tag{3.183}$$

It follows that the rates ratio depends on the ratio of the crystallization rate β_A and the rate of diffusion D_s/λ. When $\beta_A \gg D_s/\lambda$, both steps will propagate in a diffusion regime with equal rates. The supersaturations σ_A and σ_B will be equal and $v_A = v_B = \mathcal{R}\lambda/2N_0$. As a result, the Si(001) vicinal will grow with single height steps. In the reverse case, $\beta_A \ll D_s/\lambda$, the propagation rates relate as the corresponding equilibrium densities of the thermally activated kinks $\rho_{A,0}/\rho_{B,0} \ll 1$. The rate of growth of the S_B steps is much greater than the rate of growth of the S_A steps. Then the former will catch up with the latter and D_B steps will be formed. Roland and Gilmer [1992b] treated the growth of Si(001) vicinal by step flow in detail using Eq. (3.173) and found that the S_B steps always propagate with higher velocity than the S_A steps.

Let us now try to define the conditions of growth with single and double height steps. The ratio $\beta_A\lambda/2D_s$ reads

$$\frac{\beta_A \lambda}{2D_s} = \frac{\lambda}{a} \exp\left(\frac{\varphi_{sd} - \omega_A - \Delta U}{kT}\right) , \qquad (3.184)$$

where $\lambda/a = \sqrt{2}/4 \tan\theta \gg 1$, θ being the tilt angle.

The condition for the diffusion regime of growth, $\beta_A \lambda/2D_s \gg 1$, and hence of single height steps will always be fulfilled when $\varphi_{sd} - \omega_A - \Delta U > 0$ because $\lambda \gg a$. Only in the reverse case when $\varphi_{sd} - \omega_A - \Delta U < 0$ should a transition from single height steps to double height steps be observed.

A critical temperature for transition from single to double steps can be defined from the condition $\beta_A \lambda/2D_s = 1$:

$$T_{tr} = -\frac{\varphi_{sd} - \omega_A - \Delta U}{k \ln(4 \tan\theta/\sqrt{2})} . \qquad (3.185)$$

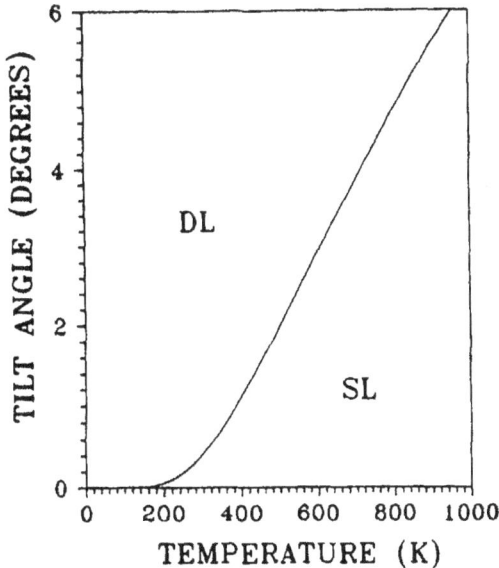

Fig. 3.41. Diagram of a vicinal Si(001) surface θ–T showing regions of formation of double layer (DL) and single layer (SL) stepped surfaces during growth. The S_A steps grow by 1D nucleation mechanism.

As discussed above, the activation energy for surface diffusion in a direction parallel to the dimer rows was calculated to vary in the interval 0.3–0.65 eV. The value of 0.15 eV was estimated for ω_A [Chadi 1987] but a higher value of 0.5 eV has been taken by Van Loenen *et al.* [1990]. The kinetic barrier ΔU should be connected with the attachment of a single atom to the kink site, bearing in mind that the repeatable step consists of attachment and detachment of building units containing four atoms. A value of the order 0.2–0.4 eV seems reasonable. Thus the difference $\varphi_{sd} - \omega_A - \Delta U$ could vary between -0.6 eV and $+0.3$ eV. A value of -0.1 eV seems reasonable. The critical temperature T_{tr} is plotted against the tilt angle θ in Fig. 3.41. As seen, single steps should be observed at high temperatures. At low temperatures, double steps should be observed as a result of the kinetic regime of growth of the S_A steps. Note that this result which is based on a kinetic treatment should obviously be considered as complementary to the equilibrium considerations of Alerhand *et al.* [1990] (Fig. 3.38). Following the same approach but using Eq. (3.177) for the rate of advance of the S_A steps one can study the dynamical evolution of the step density on Si(001) as was done in the previous section [Markov 1992].

3.3. Kinematic Theory of Crystal Growth

The layer growth is often realized as a lateral propagation of monomolecular (monoatomic) steps, which is not always the case. Just the opposite, the propagation of thicker steps, is frequently observed. As discussed briefly in Sec. 3.2 the rate of advance of such steps should be lower than that of monomolecular or elementary steps. This is easy to understand, but nevertheless we will illustrate it by a simple example.

The example consists in a comparison of the rates of advance of two elementary steps due to 2D nuclei formed one over the other (Fig. 3.26) and the double step which is formed when the upper step catches up with the lower step. In other words, we consider a pyramid of growth as shown in Fig. 3.26.

The solution of the diffusion problem (3.58) now reads

$$\Psi(r) = \Psi(\rho_2) \frac{I_0\left(\dfrac{r}{\lambda_s}\right)}{I_0\left(\dfrac{\rho_2}{\lambda_s}\right)} \quad \text{for } r < \rho_2 , \tag{3.186'}$$

$$\Psi(r) = A I_0\left(\frac{r}{\lambda_s}\right) + B K_0\left(\frac{r}{\lambda_s}\right) \quad \text{for } \rho_2 < r < \rho_1 , \tag{3.186''}$$

$$\Psi(r) = \Psi(\rho_1) \frac{K_0\left(\dfrac{r}{\lambda_s}\right)}{K_0\left(\dfrac{\rho_1}{\lambda_s}\right)} \quad \text{for } r > \rho_1 , \tag{3.186'''}$$

where

$$A = \frac{\Psi(\rho_1)K_0\left(\frac{\rho_2}{\lambda_s}\right) - \Psi(\rho_2)K_0\left(\frac{\rho_1}{\lambda_s}\right)}{I_0\left(\frac{\rho_1}{\lambda_s}\right)K_0\left(\frac{\rho_2}{\lambda_s}\right) - I_0\left(\frac{\rho_2}{\lambda_s}\right)K_0\left(\frac{\rho_1}{\lambda_s}\right)}, \qquad (3.187')$$

$$B = -\frac{\Psi(\rho_1)I_0\left(\frac{\rho_2}{\lambda_s}\right) - \Psi(\rho_2)I_0\left(\frac{\rho_1}{\lambda_s}\right)}{I_0\left(\frac{\rho_1}{\lambda_s}\right)K_0\left(\frac{\rho_2}{\lambda_s}\right) - I_0\left(\frac{\rho_2}{\lambda_s}\right)K_0\left(\frac{\rho_1}{\lambda_s}\right)}. \qquad (3.187'')$$

Following Burton, Cabrera and Frank [1951] and making use of the relation $I_1(z)K_0(z) + I_0(z)K_1(z) = 1/z$ and the approximations valid for $\rho_{1,2} > \lambda_s$, $I_0(z) \cong (\pi z/2)^{1/2}\exp(z)$ and $K_0(z) \cong (\pi/2z)^{1/2}\exp(-z)$, we find

$$v(\rho_1) = v_\infty\left(1 - \frac{\rho_c}{\rho_1}\right)\frac{1 - \frac{\Psi(\rho_2)}{\Psi(\rho_1)}\sqrt{\frac{\rho_2}{\rho_1}}\exp\left(-\frac{\lambda}{\lambda_s}\right)}{1 - \exp\left(-2\frac{\lambda}{\lambda_s}\right)}, \qquad (3.188')$$

$$v(\rho_2) = v_\infty\left(1 - \frac{\rho_c}{\rho_2}\right)\frac{1 - \frac{\Psi(\rho_1)}{\Psi(\rho_2)}\sqrt{\frac{\rho_1}{\rho_2}}\exp\left(-\frac{\lambda}{\lambda_s}\right)}{1 - \exp\left(-2\frac{\lambda}{\lambda_s}\right)}, \qquad (3.188'')$$

where now $\lambda = \rho_1 - \rho_2$ is the step spacing. $\Psi(\rho_2) > \Psi(\rho_1)$, $\rho_1 > \rho_2$ and the rate of propagation of the upper island is smaller than that of the lower one. This holds for small enough radii ρ_1 and ρ_2. At large enough sizes of ρ_2 and ρ_1 when $(\rho_1/\rho_2)^{1/2} \cong 1$, $\Psi(\rho_1)/\Psi(\rho_2) \cong 1$ and $v(\rho_1) \cong v(\rho_2)$. Expanding the exponents in power series to the linear term at $\lambda \to 0$ we find that the rate of advance of a double step is exactly twice smaller than that of a single elementary step (Eq. 3.62).

When considering the propagation of a step with arbitrary thickness $h > a$ in the case of growth from vapors, one has to take into account direct incorporation of atoms from the vapor phase to the step in addition to surface diffusion [Chernov 1984]. The surface diffusion flux per unit length of the step is (the diffusion gradient is approximated by $(n_s - n_{se})/\lambda_s$, see Eq. (3.33))

$$j_s \cong 2\frac{D_s}{\lambda_s}(n_s - n_{se}) = 2\lambda_s\frac{P - P_0}{\sqrt{2\pi m k T}}.$$

The flux from the vapor phase directly onto the step per unit length of the step is

$$j_v = h \frac{P - P_0}{\sqrt{2\pi m k T}} \; .$$

Bearing in mind that the total number of atoms required to shift the step by one atomic spacing a is $a^2 h / v_c = h/a$, the step velocity is equal to

$$v_\infty = \frac{j_s + j_v}{h N_0} a = \left(1 + \frac{2\lambda_s}{h}\right) v_c \frac{P - P_0}{\sqrt{2\pi m k T}} \; .$$

In other words, the factor $1 + 2\lambda_s/h$ should be added to the expressions for the rate of step advance in vapors to account for the step height. Obviously, when $h \gg \lambda_s$ the step should be considered as a separate crystal face and its growth does not depend on the step height. In the opposite case $h \ll \lambda_s$, the rate of the step advance is inversely proportional to the step height.

In growth from solutions (and melts) the boundary condition (3.65′) should read $C(r = h/\pi) = C_{st}$ and the height a of the elementary step should be replaced by h everywhere. Then v_∞ becomes dependent on the step thickness. Figure 3.42 illustrates the decrease of the rate of step advance with increasing step height for the particular case of solution growth (Eq. (3.77)).

Once a double step is formed it can catch up with other elementary steps and grows thicker, becoming a bunch of steps or a macrostep. On the other hand, elementary steps can leave the bunch of elementary steps and the macrostep can dissipate. Thus macrosteps and elementary steps usually coexist making the detailed description of the processes of growth very complex. As has been discussed in Chap. 1 vicinal surfaces can break up into closely packed facets under the influence of impurity atoms adsorbed on them. The facets should grow through formation of 2D nuclei if they are larger than the size of the 2D nucleus [Chernov 1961]. In order to overcome the difficulties connected with the complicated relief of the crystal surfaces Frank [1958b], Cabrera and Vermilyea [1958] developed the so-called kinematic theory of crystal growth (see also Bennema and Gilmer [1973]).

In order to illustrate the essence and the consequences of the kinematic theory we will use an example whose mathematical treatment served as a basis of Frank's considerations. This is the model of road traffic developed by Lighthill and Whitham [1955].

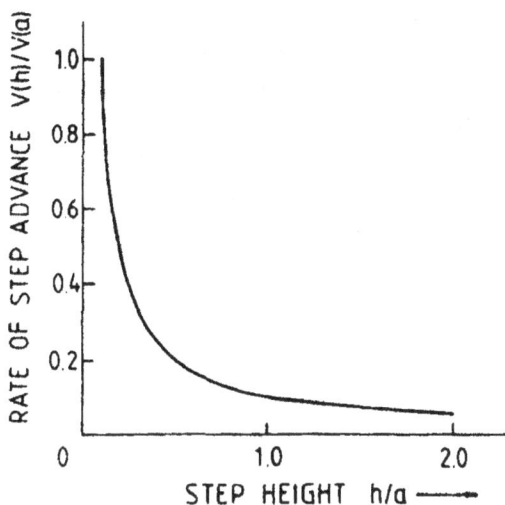

Fig. 3.42. Dependence of the rate of advance $v(h)$ of a step of height h relative to the rate of advance of a single height step, $v(a)$, on the step height in units of a. The curve represents the case of solution growth after Eq. (3.77) in which the step height a is replaced by h.

We consider a road and cars moving along it. The cars cannot outstrip each other but can catch up with each other. In fact the same is also true of the behavior of the single elementary steps on a vicinal two-dimensional crystal surface (prismatic or cylindrical). The speed of the cars depends on their proximity just like the rate of advance of steps in a train depends on their spacing (see Eq. (3.50)). In other words, we assume that the speed of the cars depends on their local density only. When the cars (steps) are equidistant all cars move with one and the same speed v. We denote by ρ the density of the cars (cars per mile) which is just equal to the reciprocal of the distance between them. Obviously the local car density is analogous to the slope of the vicinal hillock p. As the system is discrete the local car density cannot be determined at a point. That is why we take the average over a large distance neglecting any small fluctuations of speed and density. Imagine now an arbitrary car drops accidentally its speed and is caught up with the car behind. The pair of cars (a double step) moves together and its speed is lower than the speed of a single car (the cars cannot overtake each other). Then more and more cars catch up with them thus forming a pack or a "wave" of cars which moves with a speed of its own, denoted by c. Assuming now the flux of the cars along the road is constant (the number of cars per unit time entering the road is equal to the number of cars per unit time leaving the road), waves and nearly empty spacings will alternate. Plotting the local car density along the road we find a wavy line. That is why the speed c of the wave has been called by Lighthill and Whitham a "kinematic wave velocity." The same result would be obtained if by some reason an arbitrary car increases its speed and catches up with the car in front. Thus the kinematic wave velocity can in general be larger or smaller than the speed v of the single cars. If $v > c$, the front cars of the wave break off and leave the wave, whereas cars from behind catch up with the wave. When $v < c$ the wave catches up with the front cars but the back cars drop behind and leave the wave. Thus the wave does not consist of one and the same cars but continuously exchanges them. Thus a particular car will join a kinematic wave, then leave it and join the next one, and so on. In between the waves its speed will be higher than that of the waves. In some cases the shape of the wave could display a discontinuity in the sense of a sharp edge which can be either behind or in front of the wave. The edge divides the wave into two regions with different density. Such a wave is called a "kinematic shock wave" or simply a shock wave and it will move with a speed determined by the difference of the densities of cars on either side of the edge and their respective speeds.

Precisely the same can occur on vicinal crystal surfaces or on the sides of growth hillocks due to screw dislocations. Besides, external factors like the hydrodynamic conditions in solutions (the direction of solution flow above the crystal face with respect to the direction of step advance) can affect the formation or dissipation of the kinematic waves, thus smoothing or roughening the crystal face [Chernov and Nishinaga 1987; Chernov 1989].

The growing crystal face is represented by the surface

$$z = z(x, y, t) \,, \tag{3.189}$$

where the axis z is normal to the crystal surface and is on average parallel to the growing singular crystal plane.

The rate of growth of the crystal face is

$$R = \frac{\partial z}{\partial t} \,. \tag{3.190}$$

The real profile $z(x, y, t)$ deviates locally from $z = 0$ and the slopes p and q at the particular points x and y of the crystal surface are

$$p = -\frac{\partial z}{\partial x} \quad \text{and} \quad q = -\frac{\partial z}{\partial y} \,. \tag{3.191}$$

Assuming $z(x, y, t)$ is an analytic function, i.e. neglecting the discrete character of the system,

$$\frac{\partial^2 z}{\partial x \partial t} = \frac{\partial^2 z}{\partial t \partial x}, \quad \frac{\partial^2 z}{\partial y \partial t} = \frac{\partial^2 z}{\partial t \partial y}, \quad \text{and} \quad \frac{\partial^2 z}{\partial x \partial y} = \frac{\partial^2 z}{\partial y \partial x},$$

then

$$\frac{\partial p}{\partial t} + \frac{\partial R}{\partial x} = 0, \quad \frac{\partial q}{\partial t} + \frac{\partial R}{\partial y} = 0, \quad \text{and} \quad \frac{\partial p}{\partial y} + \frac{\partial q}{\partial x} = 0 . \tag{3.192}$$

In fact Eqs. (3.192) represent the law of conservation of the elementary steps. If $p = h/\lambda$, where h is the step height and λ is the step spacing, the local density of the steps is $\rho_{st} = 1/\lambda = p/h$. From (3.15) it follows that $J_{st} = v/\lambda = R/h$ is the flux of steps which pass over a point of the crystal surface. Then

$$\frac{1}{h}\frac{\partial p}{\partial t} + \frac{1}{h}\frac{\partial R}{\partial x} = \frac{\partial\left(\frac{p}{h}\right)}{\partial t} + \frac{\partial\left(\frac{R}{h}\right)}{\partial x} = \frac{\partial \rho_{st}}{\partial t} + \frac{\partial J_{st}}{\partial x} = 0 .$$

The basic equations (3.192) are unusable if some simplifying assumptions were not made. We assume that the growth rate depends on x and y <u>only</u> through the local average slopes p and q, i.e. $R(x, y) = R(p, q)$. Second, when determining the slope p or q we take the average of a region sufficiently wider than the step spacing, i.e. in this respect the theory neglects any microscopic change in the surface relief. Nevertheless, some very valuable consequences can be drawn at the expense of the approximation used. In the analysis to follow we will consider for simplicity the two-dimensional case accepting $q = 0$.

With the above approximation (3.192) becomes

$$\frac{\partial p}{\partial t} + \frac{\partial R}{\partial x} = \frac{\partial p}{\partial t} + \frac{\partial R}{\partial p}\frac{\partial p}{\partial x} = 0 . \tag{3.193}$$

We then denote $\partial R/\partial p = c(p)$ and

$$\frac{\partial p}{\partial t} + c(p)\frac{\partial p}{\partial x} = 0 \tag{3.193'}$$

or

$$\frac{dx}{dt} = -c(p) . \tag{3.193''}$$

It follows that $c(p)$ is the x component of the velocity of a point of the surface relief with a slope p. In other words, we have regions with a constant slope which are called *kinematic waves*. In fact the kinematic waves represent bunches of elementary steps divided by regions with wider terraces (Fig. 3.43(b)), and $c(p)$, which is a function of the slope only, represents the velocity of the bunches as a whole. As $c(p)$ is a function of p only, x is a linear function of time.

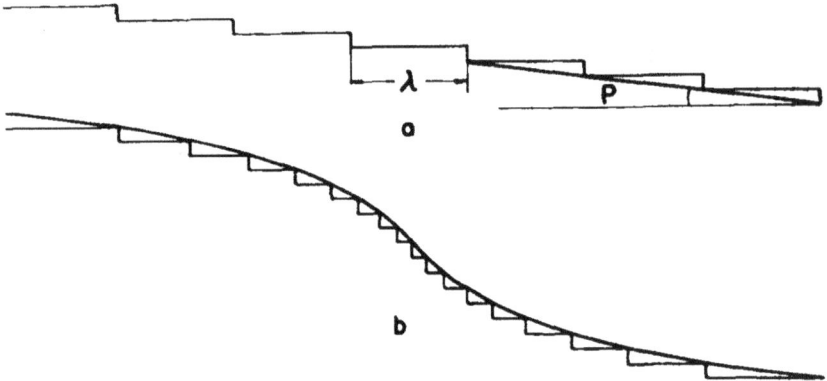

Fig. 3.43. Schematic view of a kinematic wave (b) as compared with a train of equidistant steps (a).

Let us follow the change of the relief of the growing crystal surface with time (Fig. 3.44) and find the trajectories of the waves or the lines which connect points with one and the same slope p or step density $1/\lambda = p/h$. The slope dz/dx at $p = $ const (z and x are coordinates of points on the crystal surface) is given by ($R = pv$, Eq. (3.15))

$$\left(\frac{dz}{dx}\right)_p = \frac{\partial z}{\partial t}\left(\frac{dt}{dx}\right)_p + \frac{\partial z}{\partial x} = \frac{R}{c} - p = \frac{pv}{c} - p = p\frac{v-c}{c} . \qquad (3.194)$$

Thus the slope of the wave trajectory is proportional to the relative difference of the velocities of the elementary steps and the kinematic wave. When $v > c$ the slope $(dz/dx)_p > 0$ (Fig. 3.44(a)), and vice versa (Fig. 3.44(b)). As seen in the latter case when the wave moves faster than the train of more distant steps the wave tends to leave the vicinal and to disappear from the crystal face.

Figure 3.45 illustrates the same phenomenon in the space of the growth rate R (or the flux $J_{st} = R/h$) versus the step density p. Figure 3.45(a)

(a)

(b)

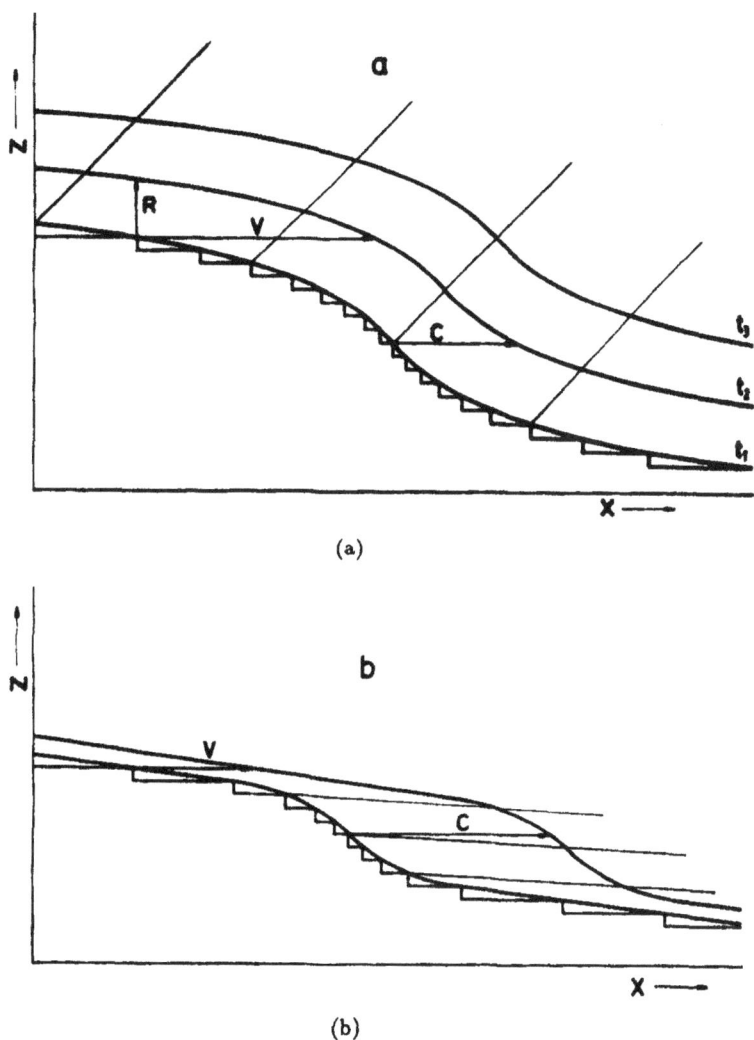

Fig. 3.44. Time behavior of kinematic waves: (a) $v > c$ and (b) $v < c$, where v and c are the velocities of the single steps far from the wave and the wave, respectively. The latter are shown by arrows denoted by v and c. The straight lines connect points of the same densities of steps.

gives the dependence of R on p in the diffusion regime of growth under clean conditions (no impurities adsorbed). The growth is proportional to p but with increasing step density the step velocity decreases according to Eq. (3.48) and R deviates downwards from the straight line. Hence, the

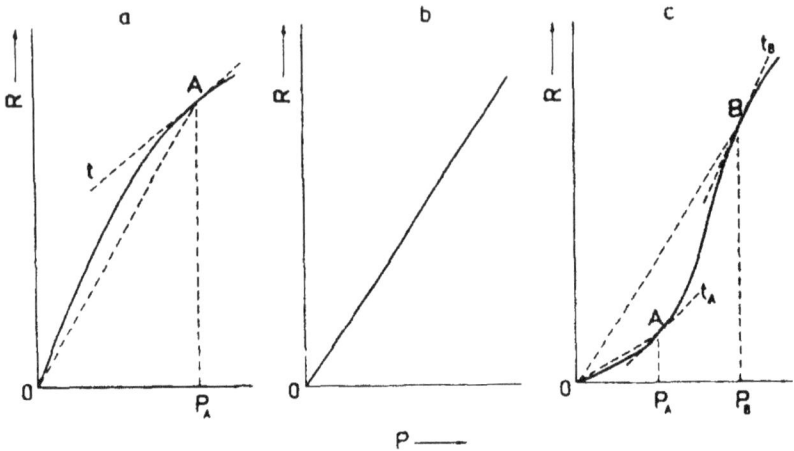

Fig. 3.45. Three possible variations of the rate of growth versus the step density. (a) The growth occurs in a diffusion regime in an absence of impurities. The growth rate $R = pv$ is determined by the overlapping of the diffusion fields (the hyperbolic tangent in the BCF theory) and deviates downwards from the straight line at large p. The curvature of the $R(p)$ dependence is everywhere negative ($d^2R/dp^2 < 0$). The rate of advance of the shock wave, c_{sh}, with slopes $p = p_A$ and $p \cong 0$ on both sides of the edge is given by the slope of the chord represented by the dashed line. It lies between the slopes of the tangents at $p = 0$ and $p = p_A$ (the latter being given by the tangent t at the point A). (b) The growth takes place in a kinetic regime in an absence of impurities. The rate of step advance v is independent of the step density and R is a linear function of p. The curvature of the $R(p)$ dependence is equal to zero. The single steps and the shock waves propagate with equal rates, i.e. $v = c$. (c) The growth takes place in a diffusion regime in the presence of impurities. The concentration of impurity atoms is large at small p, i.e. on wide terraces, and vice versa. Then the curvature of the $R(p)$ dependence is positive at small p and negative at large p. The rate of propagation of the shock wave $c < v$ at smaller p, but $c > v$ at larger p. The slopes of the dashed lines OA and OB give the rate of propagation of the shock waves, c, whereas the slopes of the tangents denoted by t_A and t_B give the rate of propagation of the single steps, v.

second derivative d^2R/dp^2 is negative everywhere. In the case of kinetic regime of growth under clean conditions, the step velocity does not depend on the step density up to very high value of the latter ($\cong 1$); R is a linear function of p as shown in Fig. 3.45(b) and $d^2R/dp^2 = 0$. In the intermediate case (kinetic regime at small p but overlap of the diffusion fields at large p), R initially increases linearly with p and then gradually deviates from the straight line. Then $d^2R/dp^2 = 0$ for small p, and $d^2R/dp^2 < 0$ for larger p. Figure 3.45(c) illustrates the case where impurity atoms are adsorbed on the terraces between the steps [Frank 1958]. At small p (wide terraces) there is time enough for an adsorption equilibrium to be established and there the concentration of the impurity atoms is high. The latter inhibits strongly the propagation of the steps. At large p (narrow terraces) there is not enough time for considerable adsorption to take place and the propagation of the steps is faster although they are nearer to each other. Then the $R(p)$ dependence has a positive curvature ($d^2R/dp^2 > 0$) at small p but a negative one ($d^2R/dp^2 < 0$) at large p. When $d^2R/dp^2 < 0$ (Fig. 3.45(a), purely diffusion regime of growth) on the whole crystal surface, the rate of advance of the single steps, $v = R/p$, is always greater than the rate of advance of the step bunch, $c = dR/dp$, the slope of the trajectory $(dz/dx)_p$ is always positive, and the kinematic waves will be present on the crystal surface. Under conditions of a kinetic regime of growth (Fig. 3.45(b)), $c = v$ all over the crystal surface and the slope of the trajectory is $(dz/dx)_p = 0$. If there are regions where $d^2R/dp^2 > 0$ as shown in Fig. 3.45(c), $v < c$, $(dz/dx)_p < 0$ and the kinematic waves tend to leave the vicinals.

We consider now the formation of shock waves, or waves with sharp edges. Figure 3.46 shows the step density, or $p/h = -(1/h)dz/dx$, in a kinematic wave as a function of x at different times. At some initial moment t_1 the step density represents a symmetric bell-shaped curve. Trajectories far from the wave are parallel. Near the wave the trajectories are no longer parallel because their slopes are directly proportional to the step density according to Eq. (3.194). Therefore the trajectories become more and more steep from the rear end; they are steepest at the maximum and after passing the latter they again tend to take the initial slope. As a result, the trajectories in the rear part of the wave will intersect each other after some time. The step density will have the shape shown by the curve denoted by t_2. Going back to the representation in coordinates (z, x) a discontinuity or a sharp edge will appear on the crystal surface as shown in Figs. 3.47(b) and (c). This is called a shock wave to distinguish it from the usual kinematic wave (Fig. 3.47(a)).

The rate of advance c_{sh} of the shock wave is easy to find. It follows from Eq. (3.194):

$$p_1(v_1 - c_{sh}) = p_2(v_2 - c_{sh})$$

or

$$c_{sh} = \frac{p_2 v_2 - p_1 v_1}{p_2 - p_1} = \frac{R_2 - R_1}{p_2 - p_1} \ . \tag{3.195}$$

Thus the rate of advance of the shock wave is determined by the differences of the growth rates and the slopes of the crystal surface on the two sides of the edge. In fact, the shock wave represents a boundary

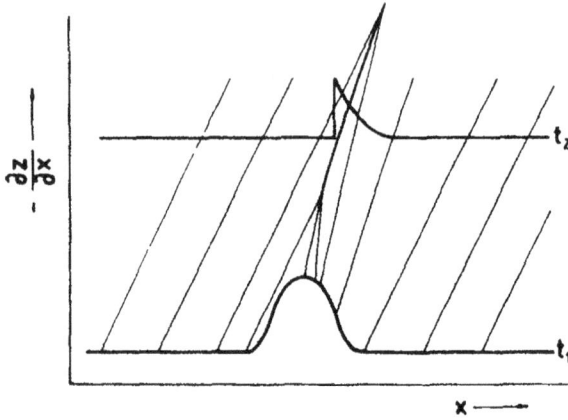

Fig. 3.46. The transformation in time $(t_2 > t_1)$ of a kinematic wave into a shock wave. For details see text.

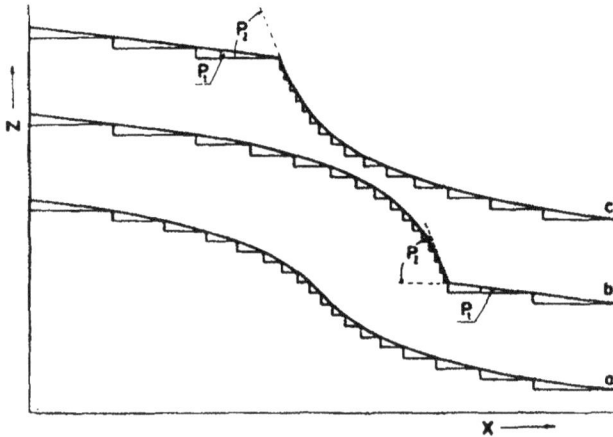

Fig. 3.47. Kinematic shock waves with the sharp edge (b) in front of the wave and (c) behind the wave. A usual kinematic wave is given for comparison in (a). p_1 and p_2 denote the slopes on both sides of the edge.

between two regions with different step densities or, in other words, between two kinematic waves.

Figure 3.47 shows a geometrical construction illustrating the interrelation between the velocities of the kinematic waves constituting the shock wave and the shock wave itself in the diffusion regime of growth (see Fig. 3.45(a)). The velocities of the two kinematic waves are given by the slopes of the tangents, $(dR/dp)_{p=p_1}$ and $(dR/dp)_{p=p_2}$. The velocity of the shock wave, $c_{sh} = (R_2 - R_1)/(p_2 - p_1)$, is given by the chord connecting the points $R_1(p_1)$ and $R_2(p_2)$. Its slope is obviously in between the slopes $(dR/dp)_{p=p_1}$ and $(dR/dp)_{p=p_2}$. Then the velocity of the shock wave has a value in between the velocities of the kinematic waves that constitute it. It is evident that in the kinetic regime of growth (linear dependence of R on p, Fig. 3.45(b)) the velocities of the kinematic waves as well as of the shock wave do not depend on the step density and are equal to each other.

As mentioned above, the retardation (or acceleration) of elementary steps which leads to bunching or formation of kinematic waves is due primarily to impurities [Frank 1958] or accidental local changes (fluctuations) of the supersaturation. In the first case the bunches are usually stabilized by the impurities and should move faster than the isolated steps [Van der Eerden and Müller-Krumbhaar 1986]. As discussed above, this is due to the fact that on the wide terraces between the isolated steps an adsorption–desorption equilibrium of the impurities is established and the concentration of the impurities there is higher compared with that on the narrow terraces which divide the steps constituting the bunch. Under clean conditions where the rate of advance of the elementary steps depends on the overlap of the diffusion fields the rate of advance of the bunch should be smaller than that of the isolated steps.

3.4. A Classical Experiment in Crystal Growth

The predictions of crystal growth theories have been the subject of numerous experimental verifications in different media (vapors, solutions and melts). Many accurate experiments have been carried out and, as shown above, most of the theoretical conclusions have been confirmed. In this chapter we describe one of the most elegant and precise experiments in crystal growth—the electrocrystallization of silver in aqueous solutions. In doing this we do not mean to underestimate the fine experimental work of many other investigators [Chernov 1989; Neave, Joyce and Dobson 1984; Neave et al. 1985; Wolf et al. 1985; Keshishev, Parshin and Babkin 1981; Avron et al. 1980; etc.].

In the case of electrocrystallization of metals, the rate of growth is given by the electric current flowing through the electrolytic cell, the amount of material deposited is given by the amount of the electricity and the supersaturation is given by the overpotential. The electrical quantities and then the parameters characterizing the process of growth can be measured with great accuracy. A particular advantage of the experiments of this kind is that a single crystal face with a definite crystallographic orientation can be produced and put in contact with the electrolyte solution. As will be shown below screw-dislocation-free crystal faces can be prepared as well as faces with definite number of dislocations. Thus the spiral growth of a single crystal face as well as the 2D nucleation growth, both layer-by-layer and multilayer, can be studied in one and the same system under well-defined conditions.

The electrolytic cell for the preparation of single crystal faces both dislocation-free and with one or more dislocations is shown in Fig. 3.48 [Kaischew, Bliznakow and Scheludko 1950; Budewski and Bostanov 1964; Budewski *et al.* 1966; see also Kaischew and Budewski 1967]. The cathode represents a glass tube containing the seed crystal which ends with a capillary with an inner diameter of approximately 200 μm. In some cases a capillary with rectangular cross section (100 μm $\times$400 μm) has been used for measuring of the rate of advance of a single monoatomic step or a train of steps (see Fig. 3.57). The bottom of the cell is a plane-parallel glass window which permits microscopic observation of the front face of the growing crystal. The cell is filled with 6N aqueous solution of $AgNO_3$ acidified with HNO_3. Special measures are usually taken in the process of preparation to make the solutions as pure as possible. After switching on the electric current the seed crystal begins to grow and fills the capillary. A single crystal filament is formed which has the same crystallographic orientation as the seed crystal. Thus single crystal faces with (100) and (111) orientations are obtained [Budewski *et al.* 1966]. When alternating current is superimposed onto the direct current of growth the silver filament begins to grow thicker and fills in the whole cross section of the capillary and the area of the front face becomes equal to the opening of the capillary.

The filament inherits defects (screw dislocations) from the seed crystal which can be detected easily by the following procedure. The crystal face is initially smoothed by applying a low current. Then a high current pulse is applied which results in the appearance of distinct pyramids of growth at the emergency points of the screw dislocations. Figure 3.49 shows a (100) face with several square-based growth pyramids revealed by this method [Budewski *et al.* 1966]. Triangular pyramids of growth (not shown) is the result in the case of (111) single crystal faces. Burger's vectors of the screw dislocations are usually inclined with respect to the growing crystal face (in most cases $\frac{1}{2}\langle 110 \rangle$ dislocations are detected). When the filament is carefully grown the emergency points of the screw dislocations inherited

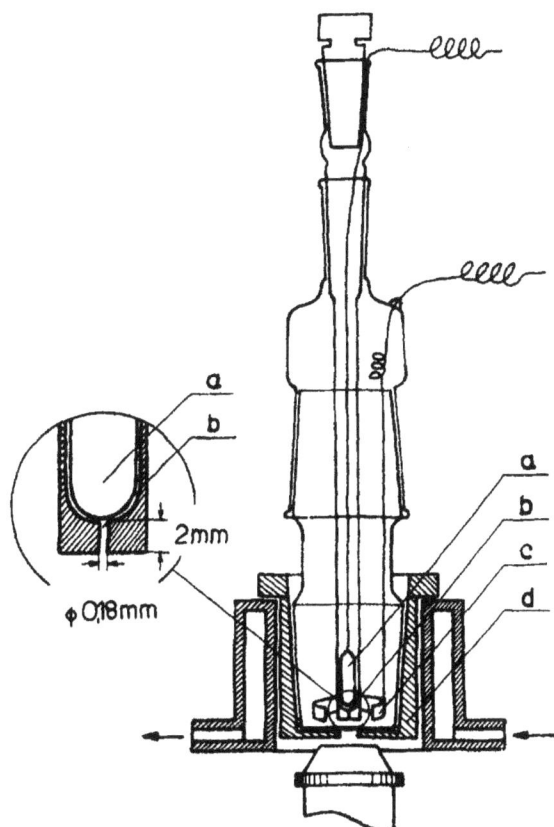

Fig. 3.48. Electrolytic cell for the investigation of the growth of Ag single crystal faces: (a) Ag seed crystal, (b) capillary, (c) silver anode, (d) brass block. The inset shows an enlargement of the end of the capillary. (E. Budewski and V. Bostanov, *Electrochim. Acta* **9**, 477 (1964). By permission of Pergamon Press Ltd. and courtesy of V. Bostanov.)

from the seed crystal leave the front face and appear on the side faces of the filament. Defectless single crystal faces are thus prepared.

The first convincing evidence of growth through 2D nucleation has been obtained on perfect faces prepared by the above method. A constant current density $i = 0.5$ mA cm^{-2} has been applied and the overpotential measured. It has been found that the latter oscillated from zero to a maximum value of about 10 mV (Fig. 3.50) [Budewski *et al.* 1966]. Increasing the current density has led to a decrease of the period of oscillation but the product of the current density and the period of oscillation remained constant.

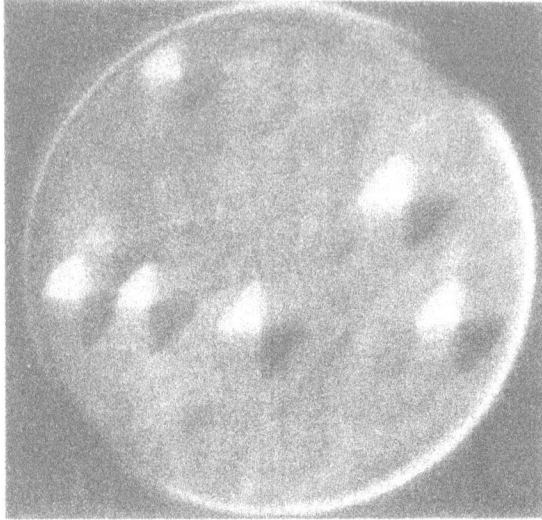

Fig. 3.49. Pyramids of growth obtained by applying a short overpotential pulse, showing
the emergency points of screw dislocations on Ag(001). (E. Budewski, V. Bostanov, T.
Vitanov, Z. Stoynov, A. Kotzeva and R. Kaischew, *Electrochim. Acta* 11, 1697 (1966).
By permission of Pergamon Press Ltd. and courtesy of V. Bostanov.)

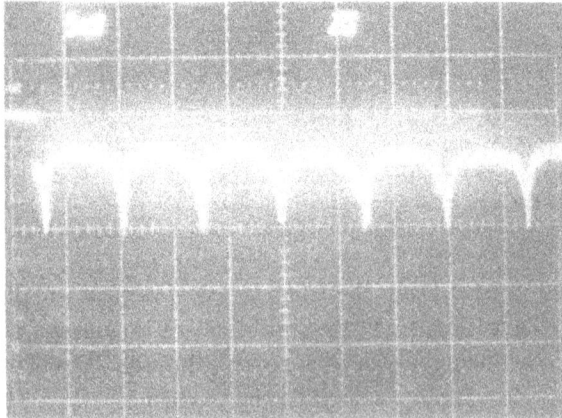

Fig. 3.50. Oscillations of the overpotential when a constant current is applied on the cell.
The product of the current and the period of oscillations gives the amount of electricity to
complete one monolayer. (E. Budewski, V. Bostanov, T. Vitanov, Z. Stoynov, A. Kotzeva
and R. Kaischew, *Electrochim. Acta* 11, 1697 (1966). By permission of Pergamon Press
Ltd. and courtesy of V. Bostanov.)

The latter is exactly equal to the amount of electricity required for the completion of a monolayer, $itS = 3.83 \times 10^{-8}$ C, where i is the constant current density, $S = 2 \times 10^{-4}$ cm^2 is the area of the opening of the capillary and t is the period of oscillation. The amount of electricity required for the completion of a monolayer of one square centimeter is $zeN_0 \cong 1.92 \times 10^{-4}$ C cm^{-2}, where $N_0 = 1.2 \times 10^{15}$ cm^{-2} (for Ag(100)) is the atom density of the corresponding crystal face, $e = 1.6 \times 10^{-19}$ C is the elementary charge of an electron and $z = 1$ is the valency of the neutralizing ions. Then $itS = zeN_0 S$. This behavior of the overpotential which is a measure of the difficulties accompanying the electrodeposition process can be easily explained if one assumed that each oscillation is due to the formation and lateral propagation of one 2D nucleus. At the initial moment the crystal face does not offer growth sites and the overpotential increases up to a critical value of about 10 mV necessary for 2D nucleation to take place. Once a 2D nucleus is formed, it begins to grow, offering more and more kink sites along its periphery, crystallization becomes easier, and the overpotential rapidly drops to zero when the edges surrounding the growing monolayer island reach the walls of the capillary. Then the formation of a new 2D nucleus is required for further growth and the process is repeated. Therefore, the amplitude of the oscillations should be equal to the critical supersaturation for 2D nucleation and the value 36% has been estimated for it ($\sigma_c = ze\eta_c/kT$, where $\eta_c = 10$ mV is the critical overpotential), in good agreement with the prediction of the theory of layer-by-layer growth in solutions.

When the overpotential is fixed slightly above the critical value spontaneous oscillations of the current are observed (Fig. 3.51) [Bostanov *et al.* 1981]. The oscillations appear through irregular intervals of time, thus reflecting the random character of the nucleation process [Toschev, Stoyanov and Milchev 1972; Toschev 1973]. However, the mean number of oscillations, averaged over a longer period of time, remains one and the same. Ascribing the formation of a 2D nucleus to each oscillation it becomes obvious that the mean number of the oscillations per unit time gives the steady state nucleation rate.

In another experiment a constant overpotential lower than the critical one was applied on the electrolytic cell. No current was detected except for a very low capacitive current. The cell under these conditions was cut off. Then short potentiostatic pulses higher than the critical overpotential ($\cong 9-10$ mV) were superimposed on the constant overpotential and a current was detected to flow through the cell. The latter increased with

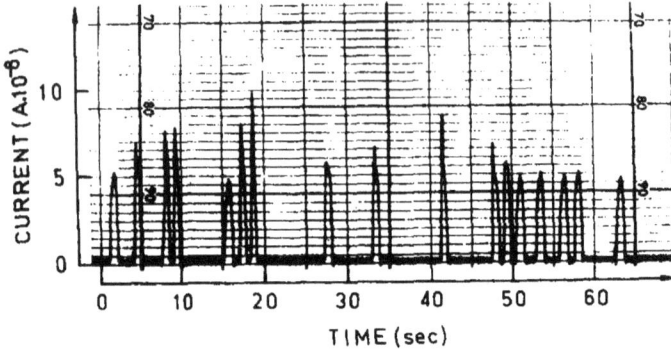

Fig. 3.51. Oscillations of the current at a constant overpotential slightly higher than the critical overpotential for 2D nucleation. The variation of the time elapsed between consecutive peaks reflects the random character of the nucleation process. (V. Bostanov, W. Obretenov, G. Staikov, D. K. Roe and E. Budewski, *J. Crystal Growth* **52**, 761 (1981). By permission of Elsevier Science Publishers B.V. and courtesy of V. Bostanov.)

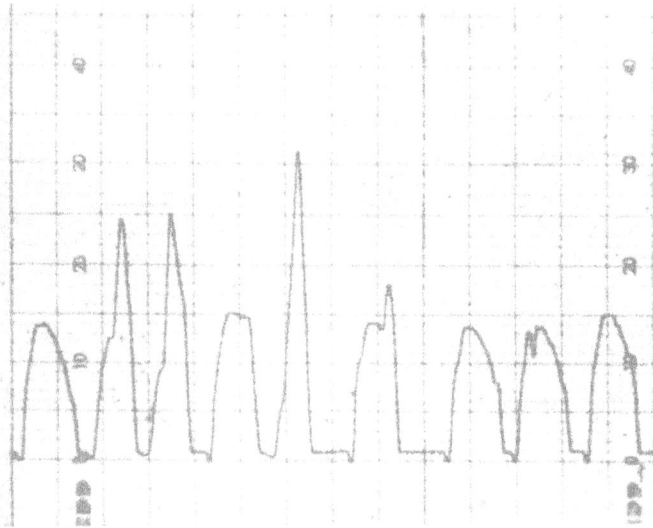

Fig. 3.52. Oscillations of the current obtained by superposition of high short potentiostatic pulses over a constant overpotential lower than the critical one for 2D nucleation. The shape of the curves reflects the site on the electrode surface on which the 2D nucleus is formed (e.g., the narrow high peak in the middle is originated by a nucleus formed at nearly the center of the electrode). The areas under the curves are equal to each other and to the amount of electricity to complete one monolayer. (E. Budewski, V. Bostanov, T. Vitanov, Z. Stoynov, A. Kotzeva and R. Kaischew, *Electrochim. Acta* **11**, 1697 (1966). By permission of Pergamon Press Ltd. and courtesy of V. Bostanov.)

time, displayed a maximum and dropped again to zero. No current had been detected until a new pulse was applied (Fig. 3.52) [Budewski *et al.* 1966]. The current–time curves had different shape but the integral of the current or the amount of electricity remained constant and equal again to that required for the completion of a monolayer. The difference of the shapes of the current–time curves is easily explained by the different locations of the nucleation event. Obviously the potentiostatic pulse provokes the formation of a 2D nucleus which then grows to cover the crystal face completely. The current is proportional to the length of the steps surrounding the growing island and goes to zero when the latter reach the walls of the capillary.

The experimentally observed current transients have been compared with the time variation of the edge length of a monolayer island calculated numerically under the assumption of a different location of the nucleation event on the crystal face. This led to the conclusion that in most cases only one 2D nucleus has been formed or, in other words, an artificial layer-by-layer growth has taken place. Then the expression for the layer-by-layer growth rate (3.111) can be compared with the experimental observations. To this aim the height and duration of the nucleation pulse have been adjusted in such a way that nucleation took place only in 50% of the pulses applied. Then the pulse duration τ can be treated as the time necessary for the formation of one nucleus. Neglecting the nonsteady state effects one can accept that the reciprocal of τ is just equal to the rate of 2D nucleation, $J_0(2D)$.

In the particular case of electrolytic nucleation the flux of atoms to the critical nucleus is given by

$$\omega^* = 4l^* \frac{i_{\rm st}}{ze} = \frac{8\varkappa a^2}{ze\eta} \frac{i_{\rm ost}}{ze} \exp\left(\frac{(1-\alpha)ze\eta}{kT}\right) \;,$$

where $i_{\rm st} = i_{\rm ost}\exp[(1-\alpha)ze\eta/kT]$ is the cathodic part of the current per unit length of the step (A cm^{-1}), $i_{\rm ost}$ is the exchange current density per unit step length ($i_{\rm s} = i_{\rm st}/a$ is the current density per unit area in A cm^{-2}) and $\alpha \cong 0.5$ is the so-called transfer coefficient.

The factor of Zeldovich reads

$$\Gamma = \frac{(ze\eta)^{3/2}}{8\varkappa a(kT)^{1/2}} \;.$$

Then for the nucleation rate of square-shaped nuclei one obtains

$$J_0(2D) = A\sqrt{\eta}\exp\left(\frac{(1-\alpha)ze\eta}{kT}\right)\exp\left(-\frac{4\varkappa^2 a^2}{kTze\eta}\right) \;,$$

where A $(cm^{-2}sec^{-1}V^{-1/2})$ is a constant given by

$$A = aN_0 \frac{i_{ost}}{ze} \left(\frac{ze}{kT}\right)^{1/2} .$$

With $N_0 = 1.2 \times 10^{15}$ cm^{-2}, $i_{ost} = 5.5 \times 10^{-6}$ A cm^{-1}, $T = 318$ K, $e = 1.6 \times 10^{-19}$ C, $A = 7.1 \times 10^{21}$ cm^{-2}sec^{-1}V$^{-1/2}$, and with $\eta = 0.01$ V for the pre-exponent, one obtains $K_1 = 7.1 \times 10^{20}$ cm^{-2}sec^{-1}.

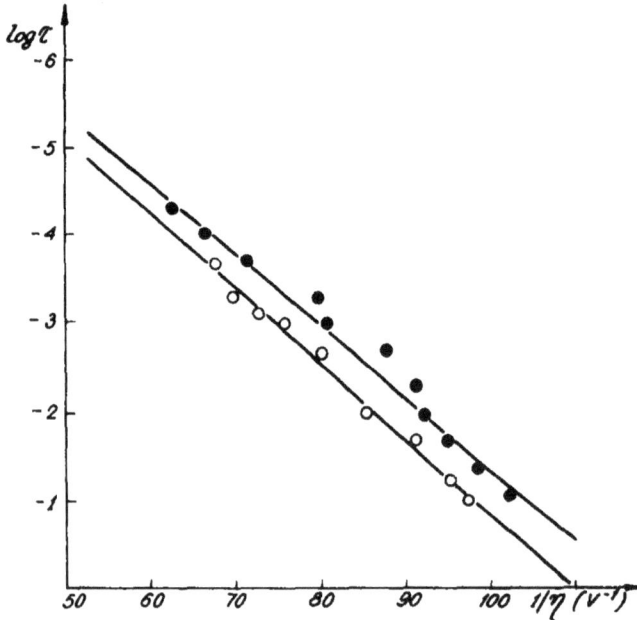

Fig. 3.53. Logarithmic plots of the time of appearance of one 2D nucleus versus the reciprocal of the overpotential. The open and filled circles are from two series of measurements under slightly different conditions. The plots are in fact equivalent to the plots of the steady state nucleation rate versus the reciprocal of the overpotential if one neglects the time lag for transient nucleation. The straight line shows the validity of the classical theory of nucleation. (E. Budewski, V. Bostanov, T. Vitanov, Z. Stoynov, A. Kotzeva and R. Kaischew, *Electrochim. Acta* **11**, 1697 (1966). By permission of Pergamon Press Ltd. and courtesy of V. Bostanov.)

Figure 3.53 shows the plot of $\ln \tau$ vs $1/\eta$ [Budewski *et al.* 1966]. As seen a straight line is obtained as required by the theory. The value 1.9×10^{-13} J cm^{-1} has been found for the specific edge energy $\varkappa$ from the slope, and $A = 1 \times 10^{19}$ cm^{-2}sec^{-1}V$^{-1/2}$ from the intercept of the line on the ordinate. Approximately the same value ($\varkappa = 2 \times 10^{-13}$ J cm^{-1}) has been found for the specific edge energy in the case of nucleation on the (111) face. Using the relations $\sigma_{100} = 4\varkappa_{10}/d_{100}$ and $\sigma_{111} = 2\varkappa_{11}/d_{111}$ following the first neighbor model (d_{100} and d_{111} are the interplanar spacings of the (100) and (111) planes, respectively) for the fcc lattice [Markov and Kaischew 1976], we find $\sigma_{100} = 372 \text{ erg cm}^{-2}$ and $\sigma_{111} = 170 \text{ erg cm}^{-2}$ for the Ag/AgNO$_3$ (aq. $6N$) boundary.

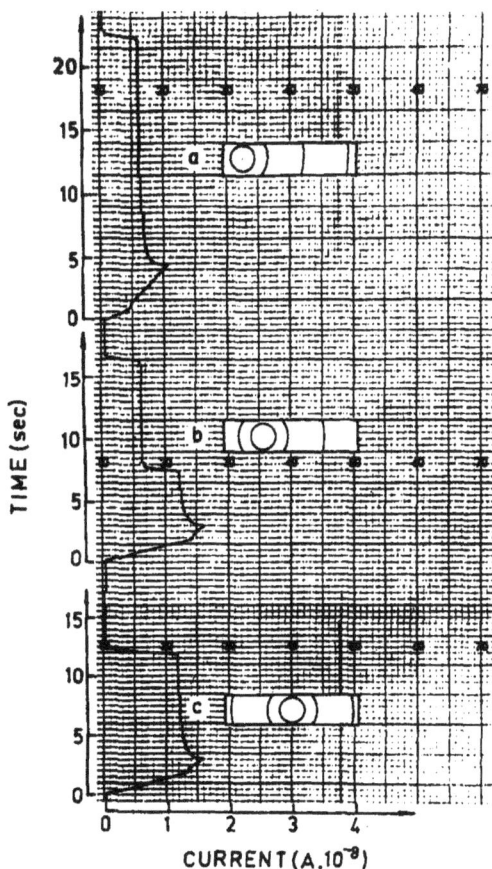

Fig. 3.54. The current of growth of single height steps versus time in a rectangular capillary (see Fig. 3.57). The inset below each curve shows the site of the nucleation event: (a) at the very end of the capillary, (b) at approximately one third from the end of the capillary, (c) at the middle of the capillary. (V. Bostanov, G. Staikov and D. K. Roe, *J. Electrochem. Soc.* **122**, 1301 (1975). By permission of Electrochemical Society Inc. and courtesy of V. Bostanov.)

The rate of step advance has been measured directly as a function of the supersaturation by using the rectangular capillary mentioned above (see Fig. 3.57) [Bostanov, Staikov and Roe 1975]. Figure 3.54 shows the current–time curves which follow the application of a short potentiostatic pulse. As seen the shape of the transients depends again on the location of the nucleation event. It displays a plateau with single or double height when the nucleus was formed at the very end or the middle of the capillary, respectively. It follows first that the current is directly proportional to the total step length L. Second, a linear dependence of the plateau current on the overpotential is established as shown in Fig. 3.55 [Bostanov, Staikov and Roe 1975]. Besides, the current is proportional to the length of the advancing step. The latter depends on the orientation of the seed crystal and, in turn, of the 2D nucleus with respect to the capillary edge. Thus in the case of $\langle 100 \rangle$ direction of the step advance the current is greater by $\sqrt{2}$ than the current which flows when the direction of the step advance is $\langle 110 \rangle$.

In order to shift the step with a length L by one atomic spacing we have to add L/a atoms or zeL/a coulombs of electricity. The time elapsed will be $t = zeL/ia$, where $i(A)$ is the current of growth. Then the rate of step advance will be $v_\infty = a/t$ or $v_\infty = i/qL$, where $q = zeN_0$ is the amount of electricity required for the completion of a monolayer. The current i is given by the well-known expression in electrochemistry:

$$i = i_0 \left[\exp\left(\frac{\alpha ze\eta}{kT} \right) - \exp\left(-\frac{(1-\alpha)ze\eta}{kT} \right) \right] ,$$

which gives, in fact, the net flux of atoms $j_+ - j_-$ to the propagating step. At small values of the overpotential ($\eta \ll kT/ze$) the latter turns into $i = k\eta$, where $k = i_0 ze/kT$ and i_0 is the exchange current. Then $v_\infty = k\eta/qL$, where obviously the ratio $\beta_{st} = k/qL$ is just the kinetic coefficient of the step in the case of electrocrystallization. In other words,

$$v_\infty = \beta_{st}\eta ,$$

where

$$\beta_{st} = \frac{k}{qL} = \frac{ze}{kTqL}i_0 .$$

k is simply the slope of the current–overpotential curves and from Fig. 3.55 one finds $k = 2\times10^{-6}$ ohm^{-1}. Bearing in mind that $q = 1.92\times10^{-4}$ C cm^{-2} and $L = 1 \times 10^{-2}$ cm, we find $\beta_{st} = 1$ cm sec^{-1}V^{-1}. Step advance rates are then of the order of 1×10^{-3} cm sec^{-1}. The exchange current is then equal

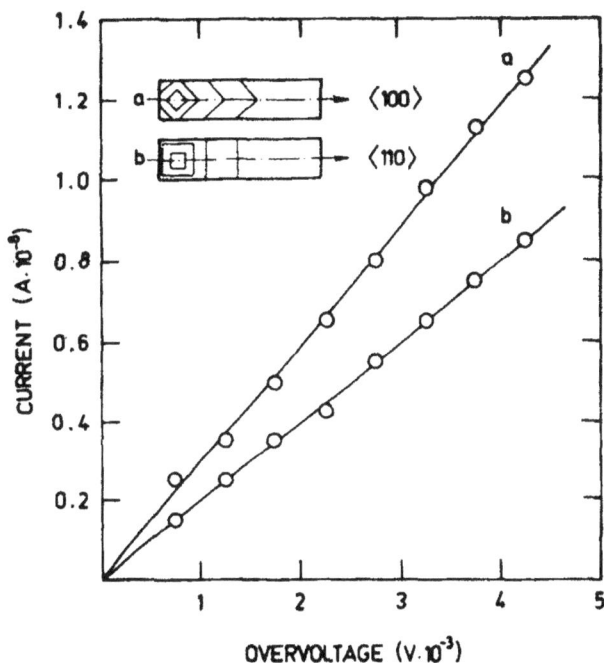

Fig. 3.55. The current of growth of single height steps versus the overpotential in a rectangular capillary (see Fig. 3.57). Curves a and b are for nuclei oriented along the $\langle 100 \rangle$ and $\langle 110 \rangle$ directions as shown in the insets. (V. Bostanov, G. Staikov and D. K. Roe, *J. Electrochem. Soc.* **122**, 1301 (1975). By permission of Electrochemical Society Inc. and courtesy of V. Bostanov.)

to $i_0 = 5.5 \times 10^{-8}$ A, or $i_{ost} = 5.5 \times 10^{-8}/1 \times 10^{-2} = 5.5 \times 10^{-6}$ A cm^{-1} per unit length of the step.

One can express the kinetic coefficient in the usual form (Eq. (3.26))

$$\beta_{st} = a\nu \frac{a}{\delta_0} \frac{ze}{kT} \exp\left(-\frac{\Delta U}{kT}\right) \ ,$$

and estimate the energy barrier for crystallization. In the particular case of electrocrystallization it includes the energy of desolvation as well as the transition through the electric double layer.

First, one has to evaluate the roughness factor a/δ_0 through Eq. (1.74). In order to do that one has to find the energy to break a first neighbor bond, ψ. To account for the situation in electrolytic solutions it is better if one uses the value of the specific edge energy found from the experiment.

Assuming for simplicity the existence of monoatomic kinks only, Eq. (1.79) gives

$$\varkappa = G_{st} = -nkT \ln \eta - nkT \ln(1 + 2\eta) \cong \frac{\psi}{2a} - \frac{2kT}{a}\eta$$

from where

$$\frac{\varkappa a}{kT} = \frac{\psi}{2kT} - 2\exp\left(-\frac{\psi}{2kT}\right) \ .$$

The same result is obtained if we come out from Eq. (1.80) valid for polyatomic kinks, bearing in mind the simplification $(1+\eta)/(1-\eta) \cong 1+2\eta$ valid for small η.

Solving numerically the above equation with $\varkappa = 1.9 \times 10^{-13}$ J cm^{-1}, $a = 2.889 \times 10^{-8}$ cm and $T = 318$ K, we find $\psi/2kT = 1.64$ and $\delta_0 \cong 4a$ [Budewski 1983]. Bearing in mind that $\beta_{st} = 1$ cm sec^{-1}V^{-1} for ΔU one obtains the value 9 kcal mole^{-1} which is typical for aqueous solutions.

At this point it is interesting to compare the possible contribution of the surface diffusion processes in the lateral spreading of the steps. Dividing the exchange current i_{ost} per unit step length by the atomic spacing $a = 2.889 \times 10^{-8}$ cm we find the value $i_{os} = 190$ A cm^{-2} for the exchange current per unit area [Vitanov, Popov and Budevski 1974]. Vitanov *et al.* [1969] (see also Vitanov, Popov and Budevski [1974]) carried out impedance measurements on dislocation-free (100) Ag crystal faces at $T = 318$ K in $6N$ AgNO$_3$ and found that the exchange current density due to adsorbed atoms is $i_{0,ad} = 0.06$ A cm^{-2}. Comparison with the above value $i_{os} = 190$ A cm^{-2} shows that the surface diffusion contribution to the step advance does not exceed 0.03%. Estimations of the rate of step advance assuming surface diffusion supply of growth units show that the latter should be 60 times smaller than that found experimentally [Bostanov, Staikov and Roe 1975; Bostanov 1977; see also Budewski 1983].

The rate of step advance has been measured independently by using quite a different method. A small number of screw dislocations have been left on the crystal surface. The monoatomic steps are invisible while growth pyramids with apparently smooth sides are observed during growth. The step spacing is a function of the overpotential applied, and increasing the latter leads to an increase of the slope of the pyramids. Assume now that we grow the crystal face at a low value of the overpotential. Comparatively flat pyramids of growth are observed. When a short overpotential pulse of higher amplitude is superimposed on the lower one a stripe with higher slope will result. The latter is visible under the microscope. In fact, this stripe is an artificially produced kinematic wave. If now high overpotential pulses are applied regularly at equal intervals of time a train of kinematic waves will be formed (Fig. 3.56) [Budewski, Vitanov and Bostanov 1965]. As shown in the previous section the velocity c of the kinematic waves is equal to the rate of advance v of the elementary steps in a kinetic regime of growth (Fig. 3.45(b)). Then, measuring the velocity of the kinematic waves, we find in fact the rate of advance of the elementary steps. The velocity of the kinematic waves is easily estimated from the distance between the stripes and the frequency of the high overpotential pulses. The value $\beta_{st} \cong 1$ cm sec^{-1}V^{-1} has been found, confirming once again that the electrolytic growth of silver in aqueous solutions takes place under conditions of kinetic regime and the surface diffusion plays a negligible role [Bostanov, Russinova and Budewski 1969; see also Budewski 1983].

If higher pulses are superimposed on the overpotential of growth a macrostep instead of an elementary step is produced. With the help of the Nomarski differential contrast technique steps thicker than, say, $10-15$ atomic diameters can be observed directly in the rectangular capillary mentioned above (Fig. 3.57) [Bostanov, Staikov and Roe 1975]. Thus the rate of propagation of such steps can be measured directly and knowing the current the thickness can be estimated. It was found, quite unexpectedly, that the velocity of the macrostep was the same as that of the elementary step up to thicknesses of the order of 100 Å. The slope of the macrostep has been estimated by analyzing the decay of the current–time curve when the step reached the end of the capillary [Bostanov, Staikov and Roe 1975]. The values 0.01275 for the slope and 160 Å ($\cong$ 55 interatomic spacings) for the interstep distance have been found. The macrosteps can be considered as kinematic waves and, as before, their velocity should be equal to that of the elementary steps when the crystal grows in a kinetic regime.

We now have all the information we need to make predictions concerning the mechanism of growth through 2D nucleation. The lateral size of the crystal plane ($L = 2 \times 10^{-2}$ cm) becomes smaller than $(v_\infty/J_0)^{1/3}$ at $\eta \geq 7$ mV. Then layer-by-layer growth should be observed at overpotentials smaller than 7 mV. At higher overpotentials multilayer growth should take place. Indeed, increasing the overpotential beyond 8 mV leads to current transients of the kind shown in Fig. 3.58 [Bostanov *et al.* 1981]. As seen the curve reproduces fairly well what has been theoretically predicted.

The steady state current density at large times is given in this case by

$$i_{st} = qb(J_0 v_\infty^2)^{1/3} \,,$$

where b is a constant of the order of unity.

Fig. 3.56. Pyramids of growth on Ag(111). The darker stripes are obtained by periodical superposition of higher overpotential pulses and represent trains of steps with larger densities (larger slopes), or in other words, artificially produced kinematic waves. (E. Budewski, T. Vitanov and V. Bostanov, *Phys. Status Solidi* 8, 369 (1965). By permission of Akademie Verlag GmbH and courtesy of V. Bostanov.)

Fig. 3.57. Micrograph of the opening of the rectangular capillary with a macrostep photographed at different times to measure its rate of advance. (V. Bostanov, G. Staikov and D. K. Roe, *J. Electrochem. Soc.* 122, 1301 (1975). By permission of Electrochemical Society Inc. and courtesy of V. Bostanov.)

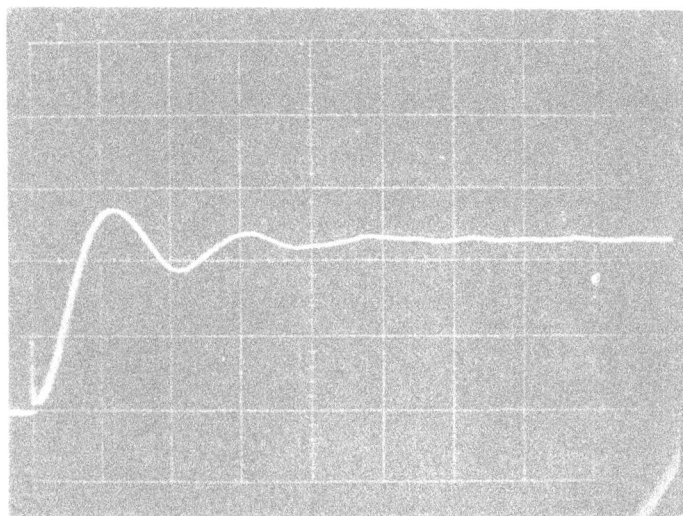

Fig. 3.58. The current of growth versus time at an overpotential considerably higher than the critical overpotential. Several oscillations are clearly visible. At a longer time the current reaches a constant value which corresponds to steady state growth. This is the first experimental recording of the oscillations of the rate of multilayer growth. (V. Bostanov, W. Obretenov, G. Staikov, D. K. Roe and E. Budewski, *J. Crystal Growth* **52**, 761 (1981). By permission of Elsevier Science Publishers B.V. and courtesy of V. Bostanov.)

The interpretation of the experimental results in coordinates $\log\{i_{st}\eta^{-5/6}\exp[-(1-\alpha)ze\eta/kT]\}$ vs $1/\eta$ (see Eq. (3.114)) gives a straight line (Fig. 3.59), in good qualitative agreement with the theory [Bostanov *et al.* 1981]. From the slope and the intercept of the straight line the values $\varkappa = 2.0 \times 10^{-13}$ J cm^{-1} and $A = 2 \times 10^{18}$ cm^{-2}sec^{-1}V$^{-1/2}$ have been estimated for the specific edge energy and the pre-exponential constant, respectively, in good agreement with the results from the study of the layer-by-layer growth.

In the case where the crystal face is not defectless but contains several screw dislocations, polygonized pyramids of growth are usually visible (Fig. 3.60) [Bostanov, Russinova and Budewski 1969; see also Budewski 1983], which represent hillocks due to growth spirals. No critical supersaturation should be overcome in order for the growth to take place and the current density is a parabolic function of the overpotential (Fig. 3.61) as required by the theory of Burton, Cabrera and Frank [1951] [Bostanov, Russinova and Budewski 1969].

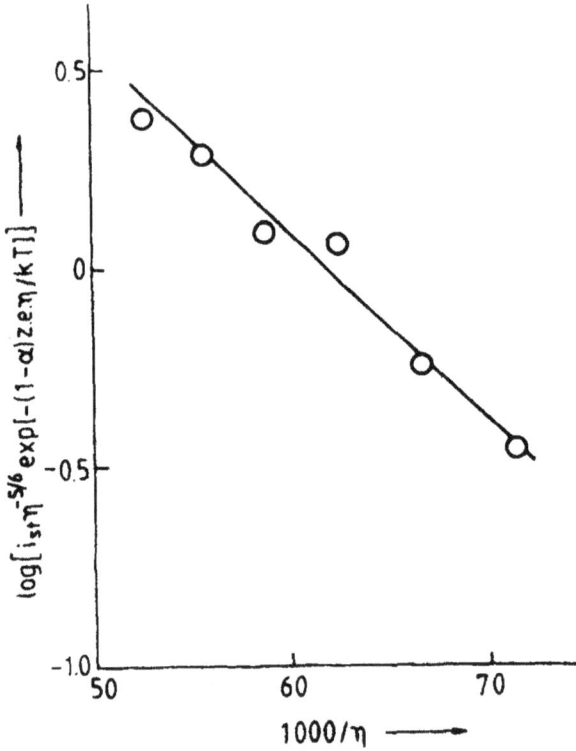

Fig. 3.59. Plot of the logarithm of the steady state current of growth versus the reciprocal of the overpotential according to Eq. (3.114). (V. Bostanov, W. Obretenov, G. Staikov, D. K. Roe and E. Budewski, *J. Crystal Growth* **52**, 761 (1981). By permission of Elsevier Science Publishers B.V. and courtesy of V. Bostanov.)

Assuming the growth rate is given by (3.15) and the slope of the growth pyramid is given by

$$p = \frac{a}{19\rho_c} ,$$

one obtains for the growth rate

$$R = \frac{aq\beta_{st}}{19\varkappa}\eta^2 .$$

The rate of growth is given by $R = d/t$, where d is the thickness of a monolayer and $t = q/i$ is the time required to deposit a monolayer. Then the current density should be proportional to the square of the supersaturation:

(a)

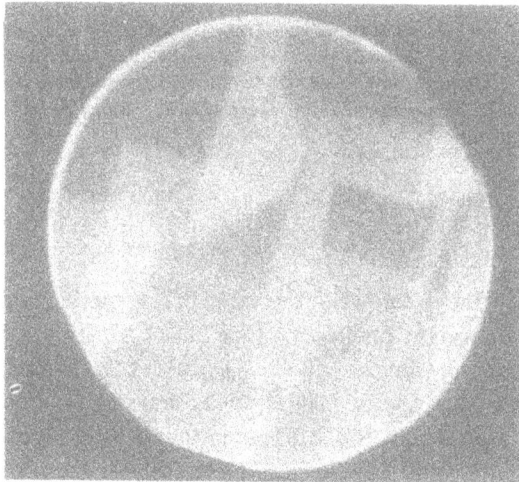

(b)

Fig. 3.60. Pyramids of growth around the emergency points of screw dislocations on (a) Ag(111) and (b) Ag(100). As seen the pyramids are well polygonized. (V. Bostanov, R. Russinova and E. Budewski, *Comm. Dept. Chem.* (Bulg. Acad. Sci.) **2**, 885 (1969). Courtesy of V. Bostanov.)

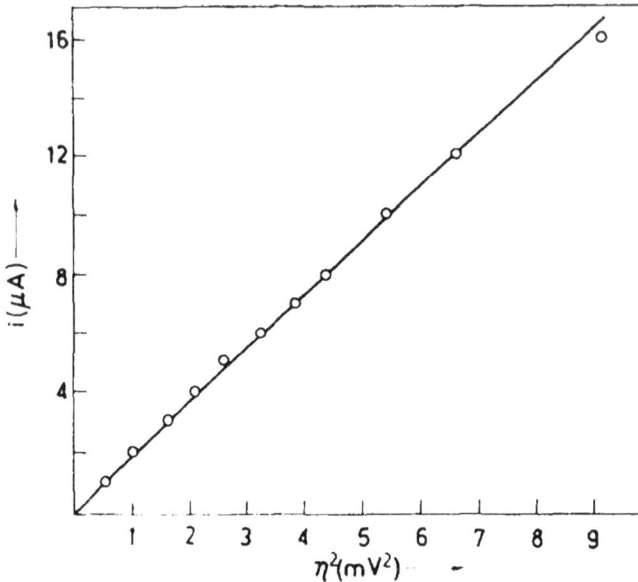

Fig. 3.61. Plot of the growth current versus the square of the overpotential of a Ag(001) crystal face with screw dislocations. The straight line confirms the validity of the theory of Burton, Cabrera and Frank [1951] for small supersaturations. (V. Bostanov, R. Russinova and E. Budewski, *Comm. Dept. Chem.* (Bulg. Acad. Sci.) **2**, 885 (1969). Courtesy of V. Bostanov.)

$$i = \frac{aq^2 \beta_{st}}{19 \varkappa d} \eta^2 \; .$$

Obviously, one can calculate the value of the specific edge energy $\varkappa$ from the slope of the straight line i vs η^2 (Fig. 3.61). Bostanov, Staikov and Roe [1975] estimated the value $\varkappa = 2.4 \times 10^{-13}$ J cm^{-1}, in good agreement with the value found from a study of 2D nucleation.

The latter shows that the expression found by Cabrera and Levine for the step spacing is valid for the case of electrolytic growth. At high enough supersaturations the step spacing should become so small that the diffusion fields around the step should overlap and the parabolic dependence of the growth rate on the supersaturation should change gradually to a linear one. It is then interesting to calculate the step spacing $\lambda = 19 \varkappa a^2 / z e \eta = 19 \varkappa / q \eta$ at the highest overpotential used at which a parabolic dependence is still observed. The value 793 Å is obtained at the highest overpotential, $\eta = 3 \times 10^{-3}$ V, applied. The latter is equal to 275 interatomic spacings.

Investigations [Bostanov, Staikov and Roe 1975] on the rate of advance in a rectangular capillary of macrosteps which represent, in fact, trains of monoatomic steps, have shown that the rate of advance of such a step train is just the same as that of a monoatomic step when the overall thickness of the macrostep does not exceed 80 Å. Moreover it has been found that the interstep spacing is 160 Å. We conclude that the parabolic law of growth should be observed approximately at least up to $\eta = 15$ mV.

CHAPTER 4

EPITAXIAL GROWTH

4.1. Basic Concepts and Definitions

The oriented growth of a crystalline material on the single crystal surface of a different material is called epitaxy (or "ordered on" from the Greek words $\epsilon\pi\iota$ — on and $\tau\alpha\xi\iota\sigma$ — in order). The term has been coined by Royer [1928] more than half a century ago. A typical example of epitaxial overgrowth is shown in Fig. 4.1 in the case of deposition of copper on the (111) surface of silver [Markov, Stoycheva and Dobrev 1978]. As seen the truncated triangular copper crystallites are lying with their (111) faces on the (111) Ag surface. What is not immediately evident from the micrograph is that the $\langle 110 \rangle$ direction of the copper crystallites is parallel to the $\langle 110 \rangle$ direction of the silver substrate. This parallelism of directions, which we call *epitaxial orientation*, is usually described in terms of the Miller indices of crystal planes and directions. In our particular case it is $(111)\langle 110 \rangle_{Cu} \| (111)\langle 110 \rangle_{Ag}$ and we say that the copper deposit is in *parallel* epitaxial orientation with the silver substrate. Although the parallel epitaxial orientation is frequently observed particularly in the very important cases of deposition of semiconductor compounds one on top of the other, this is not always the case. For example, when PbS or PbTe are deposited onto the (100) surface of MgO the epitaxial orientation is $(100)_d\langle 110 \rangle_d \| (100)_s\langle 100 \rangle_s$. This means that the crystal planes in contact with each other are (100) for both the substrate and the deposit, but the $\langle 110 \rangle$ direction of the deposit coincides with the $\langle 100 \rangle$ direction of the substrate [Honjo and Yagi 1969, 1980].

Fig. 4.1. Electron micrograph of three-dimensional copper crystallites eletrodeposited at constant overpotential on Ag(111) surface. The crystallites are in a parallel epitaxial orientation $(111)[110]_{Cu} \parallel (111)[110]_{Ag}$ with the substrate. The Ag(111) substrate is prepared by evaporation of Ag on mica in conventional vacuum. (I. Markov, E. Stoycheva and D. Dobrev, *Commun. Dept. Chem.* [Bulg. Acad. Sci.] **3**, 377 (1978).)

In general the epitaxial orientation depends on the temperature. Massies and Linh [1982a, b and c] deposited Ag on the As side $(00\bar{1})$ of GaAs and established that at temperatures lower than 200°C the Ag plane in contact with GaAs$(00\bar{1})$ is (110), i.e. the Ag $\langle 111 \rangle$ direction is parallel to the direction $\langle \bar{1}10 \rangle$ of the As dangling bonds. The epitaxial orientation is $(011)\langle 111 \rangle_{Ag} \parallel (00\bar{1})\langle \bar{1}10 \rangle_{GaAs}$. The (011) plane of fcc metals consists of parallel rows of atoms whose spacing is equal to the lattice parameter (4.086 Å of Ag). Along the rows, the atom spacing is equal to the first neighbor distance 2.889 Å for Ag. The lattice parameter of GaAs is $a_0 = 5.6531$ Å and the atom spacing in the (100) plane is $a_0/\sqrt{2} = 3.9973$ Å. So across the rows the lattice misfit is compressive and is equal to +2.22%. The lattice misfit is defined as the relative difference of the unit atom spacings, $f = (b-a)/a$, where b and a are the atom spacings of the deposit

and the substrate, respectively. Along the rows, the relative difference of
the unit atom spacings is very large, in absolute value -27.7%, but with
the opposite sign. However, it is easy to realize that four-atom spacings
of the silver nearly coincide with three-atom spacings of the GaAs. Then
the lattice misfit can be expressed as the relative difference of the multiple
atom spacings, $f = (4b - 3a)/3a = -3.63\%$ [Matthews 1975b]. In other
words, along the rows the silver bonds are stretched out, whereas across
the rows they are compressed. At temperatures higher than 200°C the
silver deposit is in parallel epitaxial orientation with $(00\bar{1})$ GaAs, i.e. the
epitaxial orientation is $(001)\langle 010\rangle_{Ag}\|(00\bar{1})\langle 010\rangle_{GaAs}$, and the lattice misfit
in both orthogonal directions is equal to -3.63%, i.e. the silver bonds
are stretched out in both directions. As will be shown below such an
orientation is connected with the lower energy due to the anharmonism
of the interatomic bonding.

Another interesting example of epitaxial orientation is established in
the deposition of Cu on Ag (001) by Bruce and Jaeger [1977, 1978a, b].
Both bulk metals have one and the same fcc lattice, but the thin Cu
films have a bcc lattice, the epitaxial orientation being $(001)\langle 1\bar{1}0\rangle_{bcc\ Cu}\|$
$(001)\langle 010\rangle_{fcc\ Ag}$. No evidence for the existence of bcc Cu in nature has been
found.

The last two examples show that the epitaxial orientation is determined
by the condition for the minimum of the free energy of the system. We know
however that the bcc lattice of copper has higher energy than the natural
fcc one. It follows that the energy of the epitaxial interface between bcc Cu
and fcc Ag overcompensates the energy difference between the bcc and fcc
lattices of Cu. We can conclude that the structure and hence the energy
of the epitaxial interface play a significant role in determining the epitaxial
orientation.

The parallelism of the contact planes is often called *fibre* or *texture
orientation* whereas the parallelism of the crystallographic directions at
the contact plane is called *azimuthal orientation*. Thus epitaxy means
simultaneous realization of texture and azimuthal orientations [Gebhardt
1973]. Very often only texture orientation takes place, the deposit being
azimuthally misoriented. Here we will not discuss this case.

In general, epitaxy does not require parallelism of low index crystallographic directions. A nonzero angle between these directions is possible
provided that it is the same for all islands (2D or 3D) of the deposit. This is
the case, for example, of Pb deposition on (111) Ag surface which is rotated
about $+4°$ and $-4°$ from the parallel orientation around the normal of the

(111) surface [Takayanagi 1981; Rawlings, Gibson and Dobson 1978]. In such cases, which while appearing more exotic are not that rare, we have to look for a parallelism of higher index crystallographic directions. For more details the reader is referred to Stoyanov [1986].

There exist terms in the specialized literature such as homoepitaxy, autoepitaxy, heteroepitaxy, etc. They are used sometimes quite arbitrarily. Thus homoepitaxy and autoepitaxy are often confused. This is the reason why we will define them here more rigorously in terms of the chemical potentials of the substrate and deposit materials.

One way to understand epitaxial growth, perhaps the best way, is to compare it with crystal growth, or in other words, with the growth of a single crystal film on the surface of the same material [Stranski and Kuleliev 1929; Stranski and Krastanov 1938]. What differentiate epitaxial growth from crystal growth are the nature and strength of the chemical bonds of both the substrate and deposit crystals on one hand, and the crystal lattices and/or the lattice parameters on the other. In other words, both crystals differ *energetically and geometrically*. If both crystals do not differ simultaneously energetically and geometrically, which means that they are identical, we have the usual crystal growth. Strictly speaking, this means that the chemical potentials of the substrate, μ_s, and the deposit, μ_d, are precisely equal. Obviously, terms like autoepitaxy or homoepitaxy for the description of this case are irrelevant. Epitaxial growth takes place only when the chemical potentials of the deposit and the substrate crystals differ.

Let us consider for example the case of deposition of Si on Si single crystal, the latter being doped with B [Sugita, Tamura and Sugawara 1969]. The nature and strength of the chemical bonds in both substrate and deposit crystals are one and the same and we can neglect the effect of the dopant on the strength of the chemical bonds. However, due to the presence of the dopant in the substrate crystal its lattice parameter is different (smaller) from that of the pure silicon and the deposit film should be compressed to match the substrate. Then the chemical potential of the strained deposit differs from that of the large deposit crystal. The difference of the chemical potentials is just given by the strain energy per atom. Thus we have a case in which both the crystals have different chemical potentials ($\mu_s \neq \mu_d$) and the difference is due solely to the difference of the lattice parameters, the nature and strength of the chemical bonds remaining practically the same. We will call this case *homoepitaxy*.

There are cases, such as deposition of $In_x Ga_{1-x} As$ on (100)InP with $x = 0.47$, when the lattice parameters coincide exactly and the lattice misfit is equal to zero. The difference of the chemical potentials is in this case due to the difference in strength of the chemical bonds (different bond strengths mean different works of separation from half-crystal positions and different equilibrium vapor pressures). In the general case, however, the two materials differ geometrically as well. Then the strain energy per atom due to lattice misfit is added to the difference in bond strengths of the two materials. This case is known as heteroepitaxy.

Summarizing we distinguish in general two cases:

1. Homoepitaxy — when the difference of the chemical potentials of the substrate and deposit crystals is due mainly to the lattice misfit;

2. Heteroepitaxy — when the difference of the chemical potentials of the substrate and deposit crystals is due mainly to the difference in strength of the chemical bonds irrespective of the value of the lattice misfit.

When both substrate and deposit crystals do not differ in any way and their chemical potentials are exactly equal, we have crystal growth which we have just considered in the previous chapter. Some investigators call this case *autoepitaxy* but we will restrain ourselves from using this term.

The influence of the bonding across the interface and the lattice misfit on the occurrence of epitaxy was noted for the first time in the famous rules of Royer [1928]. On the basis of experimental observations of epitaxial growth of ionic crystals one on top of the other he formulated the following rules:

(i) The crystal planes in contact must have one and the same symmetry and close lattice parameters, the difference of the latter being no greater than approximately 15%. The lattice misfit should be considered in a more generalized sense — one has to compare not only the unit but also the multiple lattice parameters.

(ii) Both crystals must have one and the same nature of the chemical bonds.

(iii) When ionic crystals grow one upon the other the alternation of ions with opposite signs across the interface should be preserved. Although he mentioned the importance of the bonding in both materials he emphasized more on the effect of the difference of the lattice parameters. The importance of lattice misfit was noted even earlier by Barker [1906, 1907, 1908] who concluded that oriented growth of alkali halide crystals one upon the other is more likely to occur when their molecular volumes are approximately equal. More details concerning the early works on epitaxy can be found in the excellent historical review of Pashley [1975].

When dealing with epitaxial problems one has to bear in mind the following. The epitaxial orientation of the deposit depends on the structure of the crystal planes in contact and the nature of the bonding across the epitaxial interface. In other words, it does not depend on the process of growth. Indeed, as stated at the beginning of this chapter the phenomenon of epitaxy by definition does not refer to growth at all. On the other hand, the kinetics of growth is just the same as outlined in Chap. 3 of this book. Thus, in the case of vapor deposition it includes the same processes of adsorption and desorption, surface diffusion of adatoms and incorporation into kink sites. In deposition from solutions, the bulk diffusion of the growth species should be accounted for, etc. It follows that when considering epitaxy we could treat the problems of the equilibrium structure of the epitaxial interface, which is intimately connected with the epitaxial orientation, separately from the problems of the growth kinetics of the epitaxial films and the problems connected with it.

Since the time Royer [1928] formulated his rules, the epitaxial growth of thin films has been developed to the basis for the fabrication of numerous modern devices. Thus, microelectronic devices are fabricated by epitaxial deposition of materials, varying from such "simple" ones as elementary semiconductors (Si, Ge) to binary compounds (GaAs, CdTe), and even to ternary and quaternary alloys such as $In_xGa_{1-x}As$ and $In_xGa_{1-x}As_yP_{1-y}$. Bubble memory devices are prepared by epitaxial deposition of ferromagnetic garnets such as $Y_xGd_{3-x}Ga_yFe_{5-y}O_{12}$ on the surface of nonmagnetic garnets, e.g., $Gd_3Ga_5O_{12}$, in high temperature solutions. Varying the values of x and y, one can change smoothly the crystallographic parameters (e.g., the lattice parameter) and the physical properties (e.g., width of the forbidden energy gap in semiconductors).

Research in epitaxial growth is inseparable from the surface analytical methods that are employed and the development of the corresponding tools. At about the same time as vacuum techniques were developed, the electron diffraction methods, RHEED and LEED (that is, "Reflection High Energy Electron Diffraction" and "Low Energy Electron Diffraction"), arose from the work of Thomson and Reid [1927] and Davisson and Germer [1927]. In addition, X-Ray Diffraction (XRD), X-Ray Topography (XRT), Replica Electron Microscopy (REM), Transmission Electron Microscopy (TEM), Scanning Electron Microscopy (SEM), Auger Electron Spectroscopy (AES) and so on allowed quite an accurate characterization of the epitaxial films. REM and SEM investigations give information concerning the surface morphology of the growing deposit. TEM micrographs of the cross sections of the substrate–deposit system reveal the structure of the epitaxial interface [Gowers 1987]. The *in situ* measurement of the variation of the RHEED intensity of the specular beam as a function of time gives the possibility to follow the growth and determine the thickness of the growth front, the concentration of the dopant, etc. [Neave *et al.* 1983; Neave, Joyce and Dobson 1984; Sakamoto *et al.* 1987]. In particular, a combination of LEED and AES has made it possible to obtain information on the initial stages of the epitaxial deposition. AES allows very small fractions of a monolayer of the deposited material to be detected on the substrate surface. It is, in addition, a powerful analytical method for detecting any impurities on the surface of the substrate prior to deposition. On the other hand, LEED gives a very accurate picture of the geometric disposition of the substrate atoms, and of the adsorbed atoms of the deposit as well. Combining further AES and LEED with other methods like work function measurements, Thermal Desorption Spectroscopy (TDS), the mechanism of growth of epitaxial films can be followed from the very beginning (fraction of a monolayer) to the formation of a continuous film [Bauer *et al.* 1974, 1977]. A new powerful method, Scanning Tunneling Microscopy (STM), has been recently invented, which allows the surface of the growing deposit to be visualized to the resolution of single atoms [Binnig *et al.* 1982a, b; Binnig and Rohrer 1983]. The structure of the growing surface, monoatomic steps, 2D islands and their edges can thus be observed and analysed [Hamers, Tromp and Demuth 1986; Swartzentruber *et al.* 1990].

A great variety of methods for epitaxial deposition of different materials have been invented so far. Chemical Vapor Deposition (CVD), Liquid Phase Epitaxy (LPE), Atomic Layer Epitaxy (ALE), Molecular Beam Epitaxy (MBE), Metal Organic Chemical Vapor Deposition (MOCVD), and such combinations as Low Pressure Metal Organic Chemical Vapor Deposition (LP-MOCVD) and Metal Organic Molecular Beam Epitaxy (MOMBE) are among the most widely used at present [Farrow *et al.* 1987]. That is why the list of the epitaxial systems studied up to 1975 contains approximately 6000 entries [Grünbaum 1975]. Review papers, monographs [Pashley 1956, 1965, 1970; Kern, LeLay and Metois 1979; Honjo and Yagi 1980; Vook 1982; van der Merwe 1979; Markov and Stoyanov 1987; Matthews 1975], etc., and whole volumes of journals such as *Surface Science* and the *Journal of Crystal Growth* are devoted to different aspects of epitaxial growth and characterization of epitaxial films.

4.2. Structure and Energy of Epitaxial Interfaces

4.2.1. *Boundary region*

The interface represents the region between two bulk phases. The surface of a crystal in contact with its vapors or melt is also considered as an interface. The epitaxial interface is the boundary between two single crystals — the overgrowth crystal and the substrate crystal, the former being in epitaxial orientation with the latter.

In principle, the boundary between two single crystals which are characterized by their bulk properties can have different structures depending on the nature of the chemical bonds, the crystal lattices and lattice parameters, the chemical properties of both materials, etc. Mayer [1971] classified the boundaries into five groups:

(i) *Layers of ordered adatoms or adions.* For example, the layer between the bulk K deposit and W(100) substrate consists of one monolayer of negatively charged W ions, one monolayer of positively charged K ions and two monolayers of K dipols [Mayer 1971].

(ii) *Layers in which the difference between the lattice parameters is accommodated by periodic strains due to misfit dislocations* (Fig. 4.5(d)). The latter were predicted theoretically by Frank and van der Merwe [1949] and found experimentally in a multitude of systems [Matthews 1961, 1963].

(iii) *Pseudomorphic layers in which the deposit is homogeneously strained to fit exactly the periodicity of the substrate* (Fig. 4.5(c)). Such layers have been detected in many epitaxial systems, e.g., Ni on Cu (111) [Gradmann 1964, 1966], Ge on GaAs (100) [Matthews and Crawford 1970], etc. (for a review, see Pashley [1970]; Matthews [1975]). The concept of pseudomorphism (or forced isomorphism) was introduced by Finch and Quarrell [1933, 1934] to explain the experimental observations of epitaxial growth of ZnO on Zn. Frank and van der Merwe [1949] found theoretically that a critical misfit exists under which the overgrowth should be pseudomorphous with the substrate. Beyond this misfit the interface should be resolved in a sequence of misfit dislocations. The value 14% was estimated for the critical misfit, in tempting agreement with the experimental finding of Royer [1928].

(iv) *Layers due to interdiffusion or consisting of alloys, solid solutions, metastable phases, etc.* Thus an intermetallic compound Au_2Pb is formed upon deposition of Pb on Au (100) [Green, Prigge and Bauer 1978].

(v) *Layers consisting of chemical compounds between the substrate and deposit crystals.* A typical example is the formation of $NiSi_2$ on Si(111)

[Jentzsch, Froitzheim and Theile 1989]. Stoichiometric metal silicides are in fact often observed in the deposition of metals on Si(111) and Si(001) (for a review, see Chen and Tu [1991]).

The considerations in this chapter will be confined to the accommodation of the lattice misfit by homogeneous strain (HS) (pseudomorphism) or by periodic strain (misfit dislocations), all other phenomena like alloying, interdiffusion or chemical reactions between both partners being ruled out. We then consider the epitaxial interface as a geometric plane dividing two crystals which in general differ energetically and geometrically. We assume further that the structure of the interface is such that it minimizes the energy of the system.

4.2.2. Models of epitaxial interfaces

Various models have been invented for the theoretical description of the equilibrium (minimum energy) structure of the epitaxial interfaces.

The *coincidence lattice model* developed by Bollmann [1967, 1972] considers the two lattices of the substrate and deposit crystals as rigid. The interatomic forces are assumed to have spherical symmetry and the model is valid for materials with metallic or van der Waals bonds. Two lattice points, one from each crystal, are brought into coincidence and the density of lattice points which are in perfect or near coincidence serves as a measure of registry of the two lattices (Fig. 4.2). The azimuthal orientation with maximum coincident points is considered as the minimum energy (ground state) orientation.

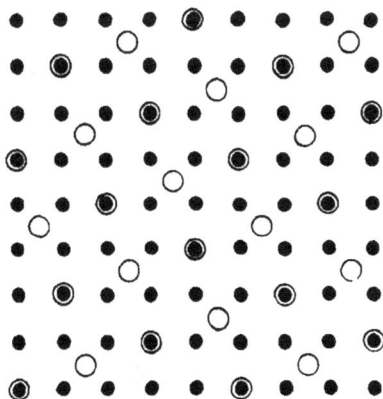

Fig. 4.2. Illustration of the coincidence lattice model of epitaxial interfaces. The atoms of the two planes in contact are denoted by open and filled circles.

The *ball-and-wire model* [Hornstra 1958] was developed initially to describe dislocations in diamond. It assumes anisotropic chemical bonds such as the directional covalent bonds, and is applicable to semiconductor materials. Holt [1966] extended it further to describe the structure of the semiconductor heterojunctions and found that dangling (unsaturated) bonds should exist at the interface originating from the material with the smaller lattice parameter (Fig. 4.3). Depending on the surface polarity of the adjacent crystal planes the dangling bonds can act as acceptors or donors [Holt 1966]. Oldham and Milnes [1964] showed that the dangling bonds are expected to constitute deep energy levels in the forbidden energy gap and thus play the role of recombination centers. When the density of the dangling bonds is too high they can create a conducting band in the forbidden energy gap and alter significantly the properties of the heterojunction [Holt 1966; Sharma and Purohit 1974].

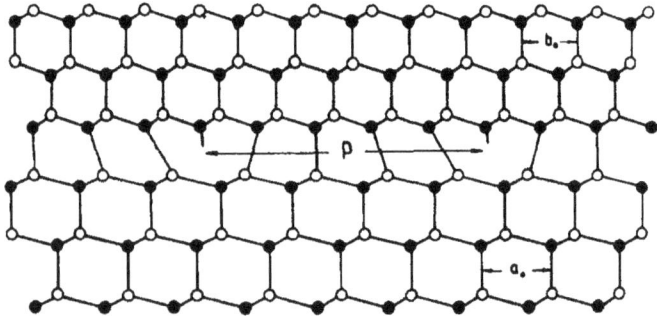

Fig. 4.3. Illustration of the ball-and-wire model of epitaxial interfaces between two crystals with diamond lattice. The misfit dislocations represent unsaturated dangling bonds spaced at an average distance p (after Holt [1966]).

In the *variational approach* developed by Fletcher [1964, 1967], Adamson [Fletcher and Adamson 1966] and Lodge [Fletcher and Lodge 1975] the positions of the atoms of adjacent crystal planes have been varied to find the minimum of the energy. A pairwise interatomic potential has been used for this purpose. The calculations were carried out in the reciprocal space. It has been found that the energy has a minimum value when as many as possible lattice points of either side of the interface coincide as in the coincidence lattice model. If elastic displacements of the atoms are allowed (restricted for simplicity to crystal planes in contact with each other) the resulting structure of the interface is very much like the one in the misfit dislocation model developed by Frank and van der Merwe [1949].

The *Volterra dislocation model* was developed on the basis of the dislocation theory of low angle grain boundaries [Brooks 1952, Matthews 1975]. The concept of edge type dislocations for accommodation of the lattice misfit was introduced explicitly in the model, both substrate and deposit crystals having been considered as elastic continua. The most general definition of edge dislocations given by Volterra [1907] at the beginning of the century (see also Hirth and Lothe [1968]) has been used. The energy of the interface is represented as a sum of the energy of the dislocations and the energy of the residual homogeneous strain. Minimization of the energy with respect to the homogeneous strain allows the calculation of the equilibrium strain and the critical thickness for pseudomorphous growth. The advantage of the model is its mathematical simplicity. The arbitrariness in the choice of the quantities determining the energy of the cores of the dislocations is, however, a shortcoming of the model. More details will be given below.

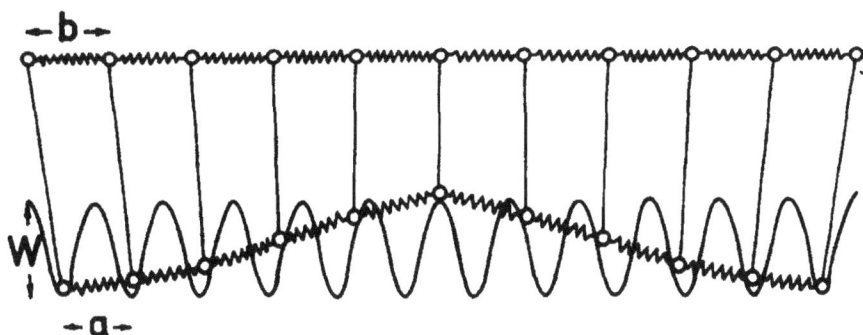

Fig. 4.4. Illustration of the one-dimensional misfit dislocations model of Frank and van der Merwe [1949a]. The deposit is simulated by a chain of atoms connected by elastic springs of length b and force constant γ. The rigid substrate exerts a periodic potential with period a and amplitude W. The figure shows the atomic chain before and after being put in contact with the substrate. In the latter case 11 atoms of the chain are distributed over 12 potential troughs of the substrate, thus forming a misfit dislocation.

The *misfit dislocation model* of Frank–van der Merwe [1949a, b] (for a review, see van der Merwe [1974]) deals with a linear chain of atoms connected with elastic springs subject to an external periodic potential exerted by a rigid substrate (Fig. 4.4). The misfit dislocations appear naturally as a result of the mathematical analysis of the model.

In the present chapter the dislocation models of Frank and van der Merwe [1949] and van der Merwe [1950] will be described in more detail as they are the most well known and most exploited. Besides, the Volterra approach of Matthews will also be described in more detail. The reader interested in the other models mentioned above is referred to the original papers and the review papers of van der Merwe [1974] and Woltersdorf [1981].

4.2.3. *Misfit dislocations*

We will confine our considerations in this chapter only to the case where the deposit crystal possesses a crystal face with the same symmetry as the substrate surface.

In general when two geometrically dissimilar crystals (A and B) join each other in such a way that two particular crystal planes come into contact at the interface, the only physical reality is that the atoms in the adjoining crystal halves in the near vicinity of the boundary between them are displaced from their ideal positions which they should occupy if the foreign crystal was replaced by the same crystal. Two lateral forces act on each atom. The first is the force exerted by the neighboring atoms of the same crystal which tends to preserve its natural crystal lattice and keep the interatomic distances equal to their natural bond lengths. The second is the force exerted by the atoms of the adjoining crystal which tends to force the atoms to occupy the lattice sites of the foreign crystal. One can distinguish several limiting cases. First, when the interfacial bonding ψ_{AB} is very weak compared with the bonds strengths, ψ_{AA} and ψ_{BB}, both crystals tend to preserve their natural lattices. In such a case the difference of the periodicities of the two adjoining crystal lattices degenerates into a vernier of misfit (Fig. 4.5(a)). A special case is when the lattice parameters a and b are multiples of each other, i.e. $ma = nb$, where m and n are small integers such that $m = n+1$ (Fig. 4.5(b)). Then every mth atom of A coincides with every nth atom of B and we arrive at the coincidence lattice model outlined above. In the other limiting case in which ($\psi_{AB} \gg \psi_{BB}$ and $\psi_{AB} \simeq \psi_{AA}$) the crystal B is forced to adopt the lattice of A, or in other words, the crystal B is homogeneously strained to fit the crystal A, we say then that B is pseudomorphous with A (Fig. 4.5(c)). However, the elastic strains and in turn the energy of B increase linearly with the thickness. That is why beyond some critical thickness the pseudomorphous growth becomes energetically unfavored and the homogeneous strain should be replaced by a periodical strain which can attenuate with increasing film thickness. Thus, misfit dislocations with a lower energy are introduced at the interface to accommodate the lattice misfit. Obviously, the smaller the natural misfit the greater will be the critical thickness for pseudomorphous growth. In the intermediate case ($\psi_{AB} \simeq \psi_{AA} \simeq \psi_{BB}$), the interfacial forces are not strong enough to produce a pseudomorphous layer with considerable thickness and the lattice misfit will be accommodated by misfit dislocations (MDs), or in other words, by periodical strain (Fig. 4.5(d)), from the beginning of the growth process.

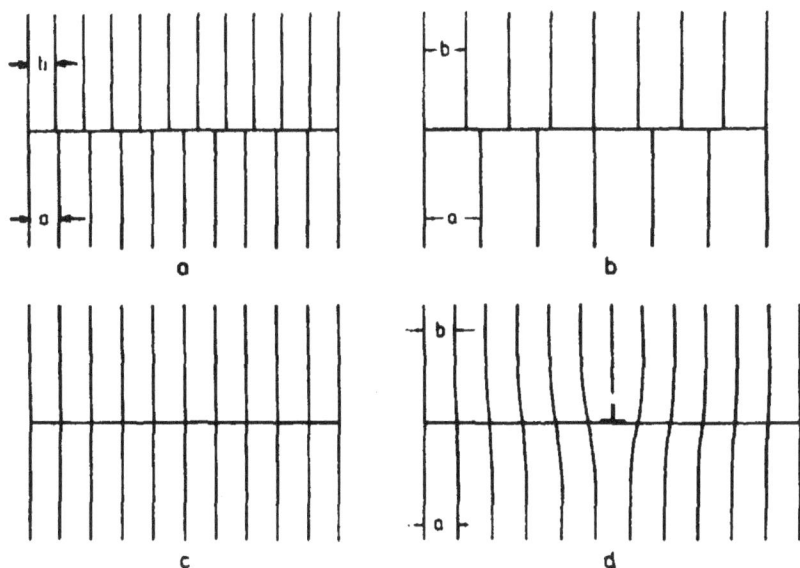

Fig. 4.5. Four possible modes of misfit accommodation: (a) vernier of misfit, (b) coincidence lattices, (c) homogeneous strain, (d) misfit dislocations (after van der Merwe, Woltersdorf and Jesser [1986]).

Misfit dislocations are a convenient concept for the description of lattice distortions in the vicinity of the epitaxial interface. They represent atomic planes in excess in the material with the smaller atomic spacing (Fig. 4.5(d); see also Fig. 4.11). Their fundamental feature is the local strain with opposite sign in the cores of the dislocations. If the atomic spacing of the overgrowth is smaller than that of the substrate the chemical bonds in between the MDs will be stretched out but the bonds in the cores of the dislocations will be compressed, and vice versa. Thus an interface resolved in a sequence of MDs is characterized by a periodical elastic strain

with a period equal to the dislocation spacing. That is why the concept
of the MDs is applicable only when the bonding across the interface is
strong enough to ensure the appearance of local strains. It follows that
the periodic distortions of both lattices (substrate and deposit) lead to
an almost perfect match of the crystal planes in contact in some areas
of the epitaxial interface. These areas are separated by stripes in which
the two lattices are out of registry. Figure 4.6 is an illustration of an
epitaxial interface between PbS and PbSe resolved in a sequence of misfit
dislocations [Böttner, Schießl and Tacke 1990]. The situation is similar,
in a topological sense, to that existing in a single crystal containing edge
dislocations, from where the term "dislocation" was borrowed. Unlike edge
dislocations, however, misfit dislocations are not linear defects of the crystal
lattices themselves and their equilibrium density does not tend to zero with
decreasing temperature.

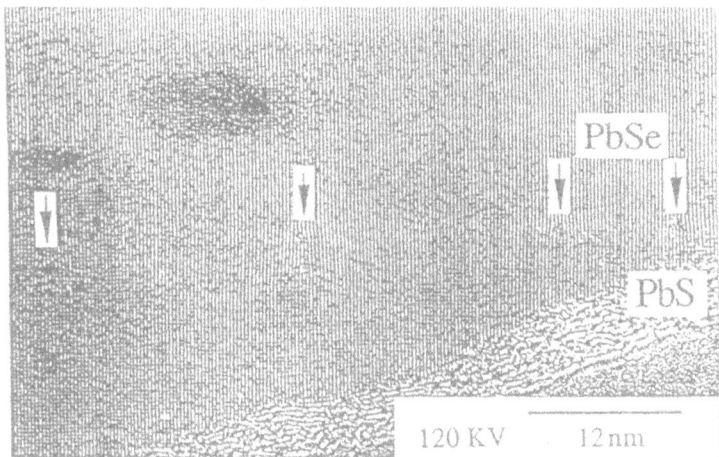

Fig. 4.6. High resolution TEM micrograph of a PbS/PbSe interface. The misfit dislo-
cations are shown by arrows. (H. Böttner, U. Schießl and M. Tacke, *Superlattices and
Superstructures* **7**, 97 (1990). By permission of Academic Press Ltd. and courtesy of H.
Böttner.)

In fact there are two separate, though similar, misfit dislocation models
which are solved exactly: the model of overgrowth with monolayer thickness, which is in fact the famous one-dimensional model of Frank and van
der Merwe [1949], and the model of fairly thick deposit developed later by
van der Merwe [1950, 1963a, b]. We consider first the monolayer model

in more detail although it can be taken only as a first approximation to a growing epitaxial film. However, it is very illustrative and helpful for a deeper understanding of the more realistic models of epitaxial interfaces. Then the model of thick overgrowth will be briefly out lined. The misfit dislocation models have been reviewed comprehensively in a series of papers by van der Merwe [1973, 1975, 1979] and others [Matthews 1975b].

4.2.4. *Frank–van der Merwe model of thin overlayer*

The Frank and van der Merwe model [1949] recently gained prominence not only in the field of epitaxy but also in various other fields, a common feature of which is the competing periodicities. Thus it provided the grounds of the theory of commensurate–incommensurate phase transitions in physisorbed layers [Villain 1980] and in layer compounds [McMillan 1976], the alignments of cholesteric liquid crystals in a magnetic field [de Gennes 1968], etc. (for a review, see Bak 1982). The treatment is based on an earlier model of Frenkel and Kontorova [1939] who considered the "worm-like motion" of edge dislocations in crystals to explain the plastic flow of the latter. That is why the one-dimensional model is also known as the model of Frenkel and Kontorova.

In this chapter the original one-dimensional model of Frank and van der Merwe [1949] will be considered first. Then it will be generalized for a two-dimensional monolayer overgrowth, and after that it will be applied to the case of thickening overgrowth although it is inadequate for quantitatively describing this situation [van der Merwe, Woltersdorf and Jesser 1986]. The influence of the anharmonicity and nonconvex characters of the more realistic interatomic potentials will be discussed at the end of the chapter.

4.2.4.1. *Interatomic potentials*

Pairwise interatomic potentials with simple analytical form are often used in direct lattice calculations in solid state physics. The potentials of Morse and Lennard-Jones (the 6-12 Mie potential) are the common choices [Kaplan 1986].

The potential of Morse [1929],

$$V(r) = V_0 \left\{ \left[1 - \exp\left(-\omega \frac{r - r_0}{r_0} \right) \right]^2 - 1 \right\}, \qquad (4.1)$$

where V_0 is the energy of dissociation, r_0 is the equilibrium atom spacing and ω is a constant which governs the range of action of the interatomic forces, was originally suggested for the evaluation of the vibrational energy levels in diatomic molecules. By varying ω we shift the repulsive and attractive branches of the potential in opposite directions so that the degree of anharmonicity remains practically the same. Girifalco and Weiser [1959] adjusted the constants of the Morse potential to fit the lattice parameters, the cohesive energies and the elastic properties of a series of metals, and found a value for ω varying around 4.

The potential of Morse does not behave well at small and large atom spacings. At $r = 0$, the potential does not go to infinity but has a finite value. The exponential dependence is not believed to describe well the atom attraction at $r > r_0$. In this respect the inverse power Mie potential

$$V(r) = V_0 \left[\frac{n}{m-n} \left(\frac{r_0}{r} \right)^m - \frac{m}{m-n} \left(\frac{r_0}{r} \right)^n \right] \tag{4.2}$$

is much more flexible than the Morse potential. The repulsive and attractive branches are governed by two independent parameters, m and n ($m > n$). The Mie potential with $m = 12$ and $n = 6$ which is known as the Lennard-Jones potential [Lennard-Jones 1924] describes satisfactorily the properties of the noble gases.

A generalized Morse potential

$$V(r) = V_0 \left(\frac{\nu}{\mu - \nu} \exp[-\mu(r - r_0)] - \frac{\mu}{\mu - \nu} \exp[-\nu(r - r_0)] \right) \tag{4.3}$$

has been recently suggested [Markov and Trayanov 1988]. It has all the shortcomings of the Morse potential except that the repulsive and attractive branches are governed by two independent parameters μ and ν ($\mu > \nu$). An advantage of both Morse potentials, particularly for solving interface problems, is that they are expressed in terms of strains $r - r_0$ which makes the mathematical formulation of the problems and the calculation of the strains, stresses and strain energy easier. If we put $\mu = 2\omega/r_0$ and $\nu = \omega/r_0$ into (4.3) it turns into the Morse potential. It is worth noting that the 6-12 Lennard-Jones potential is practically indistinguishable from the one expressed by (4.3) with $\mu = 18$ and $\nu = 4$. The generalized Morse potential with $\mu = 4$ and $\nu = 3$ is plotted in Fig. 4.7.

The pairwise potentials counted above have two fundamental properties. First they are anharmonic in the sense that the repulsive branch is steeper than the attractive one, and second they have an inflection point r_i beyond

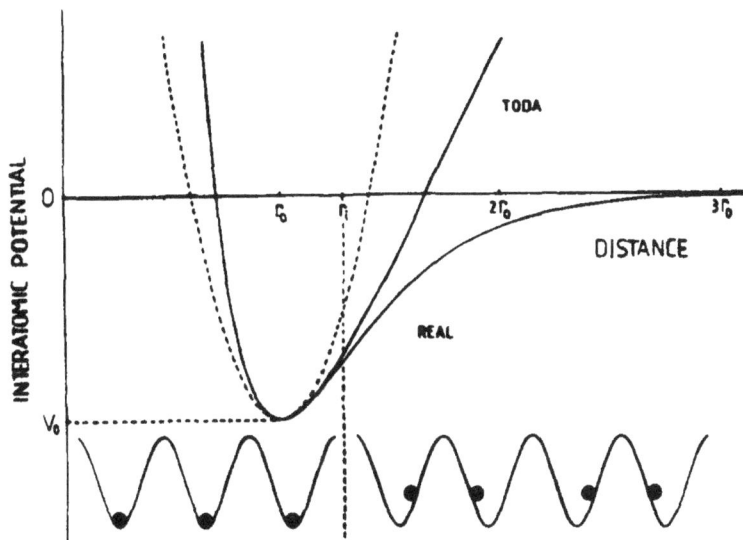

Fig. 4.7. The generalized Morse potential (Eq. (4.3)) (shown as real) with $\mu = 4$ and $\nu = 3$ and the Toda potential (Eq. (4.4)) with $\alpha = 2$ and $\beta = 6$. The parameters α, β, μ and ν are chosen in such a way that the repulsive branches of both potentials coincide up to the third digit and have one and the same harmonic approximation given by the dashed line. For misfits smaller than $f_i = (r_0 - r_i)/a$ (Eq. (4.84)), where r_i denotes the inflection point of the real potential, the atoms of the chain are equidistant as shown below to the left. For misfits larger than f_i, the distorted state shown below to the right is the ground state. The latter consists of alternating short, strong and long, weak bonds. (I. Markov and A. Trayanov, *J. Phys.: Condens. Matter* 2, 6965 (1990). By permission of IOP Publishing Ltd.)

which they become nonconvex. In order to distinguish the influence of the anharmonicity from that of the nonconvex character one can use the well-known potential of Toda [Toda 1967]

$$V(r) = V_0 \left(\frac{\alpha}{\beta} \exp[-\beta(r - r_0)] + \alpha(r - r_0) - \frac{\alpha}{\beta} - 1 \right). \qquad (4.4)$$

which is shown in Fig. 4.7 with $\alpha = 2$ and $\beta = 6$. By varying α and β but keeping their product constant we can go smoothly from the harmonic approximation ($\alpha \to \infty$, $\beta \to 0$, $\alpha\beta = $ const) to the hard sphere limit ($\alpha \to 0$, $\beta \to \infty$, $\alpha\beta = $ const). It has no inflection point (or has an inflection point at infinity) and can be used to study the effect of anharmonicity in its pure form on the equilibrium structure of the epitaxial interfaces [Milchev and Markov 1984; Markov and Milchev 1984a, b, 1985]. It is immediately seen that expanding the second exponential term of the generalized Morse potential (4.3) in Taylor series to the linear term results in the Toda potential with $\alpha = \mu\nu/(\mu - \nu)$ and $\beta = \mu$.

Expanding any of the above potentials in Taylor series to the parabolic term gives the harmonic approximation (the dashed line in Fig. 4.7), which for the generalized Morse potential (4.3) reads

$$V(r) = \frac{1}{2}\mu\nu V_0(r - r_0)^2 - V_0 , \qquad (4.5)$$

where the product $\gamma = \mu\nu V_0$ gives the elastic modulus. The force between the neighboring atoms, $F(r) = \gamma(r - r_0)$, satisfies Hooke's law exactly. Obviously, the harmonic approximation can be used for small deviations from the equilibrium atom separation, i.e. for small strains $r - r_0$. This is equivalent to small misfits in interface problems. The force constant γ is a measure of the bonding between the overgrowth atoms.

Let us analyze more closely the above pairwise potentials. Figure 4.8(a) demonstrates the variation with the atom spacing of the first derivative, or the force exerted on one atom by its neighbor. As seen the force goes linearly to infinity in the harmonic case. This means that increasing the atom spacing leads to an increase of the force which tends to keep the atoms together. The Toda force, however, goes to a constant value at large atom separations. This means that in applying a force greater than the maximum one the corresponding bond can break up and both atoms can be separated from each other. The same is valid for the potentials (4.1)–(4.3) (henceforth to be referred to as real potentials). The force displays a maximum — the theoretical tensile strength of the material.

Figure 4.8(b) demonstrates the variation of the second derivative of the pairwise potentials which in fact determines the sign of the curvature. In the harmonic case the second derivative is constant and positive. In the Toda case it is a decreasing function of the atom separation and goes asymptotically to zero but remains always positive. Only in the case of the real potentials the curvature changes its sign from positive to negative at the inflection point $r = r_i$. In other words, the real potentials become nonconvex at $r > r_i$. As shown by Haas [Haas 1978, 1979] the nonconvex character of the real potential results in distortion or polymerization of the chemical bonds in expanded chain (or epilayers); long, weak and short, strong bonds alternate (Fig. 4.7). The driving force of such a distortion is the energy difference between the distorted and undistorted structures. It is easy to show that the mean energy of the distorted (dimerized) structure

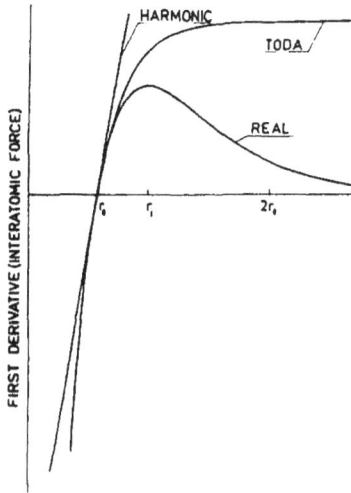

(a)

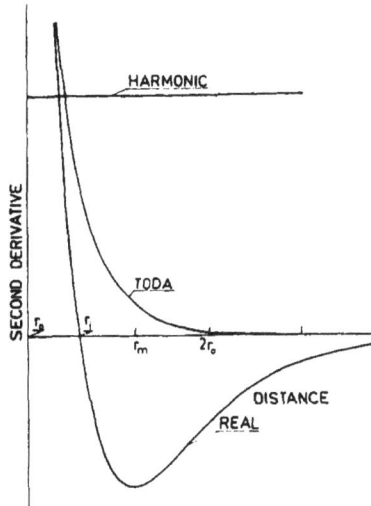

(b)

Fig. 4.8. (a) First and (b) second derivatives of the harmonic, Toda and real potentials. The force acting between the atoms (the first derivative) goes to infinity for the harmonic potential but is finite for the other potentials. The maximal force is in fact the theoretical tensile strength of the material. The second derivative which determines the sign of the curvature is positive for both the harmonic and Toda potentials but becomes negative beyond the inflection point in the case of the real pairwise potentials.

$[V(r + u) + V(r - u)]/2 < V(r)$ for a curve with a negative curvature, $[V(r + u) + V(r - u)]/2 = V(r)$ for a straight line (zero curvature) and $[V(r + u) + V(r - u)]/2 > V(r)$ for a curve with a positive curvature. It follows that when applying the harmonic potential or the real potential at misfits smaller than that corresponding to the inflection point, the ground state will be the undistorted structure. The distorted structure will be the ground state in epilayers expanded beyond the inflection misfit when a real potential is adopted.

4.2.4.2. *Interfacial interactions*

A single atom moving on a single crystal surface should feel a two-dimensional periodic potential relief. It is convenient to represent it in the form [Frank and van der Merwe 1949b]

$$V(x, y) = \frac{1}{2}W_x \left(1 - \cos 2\pi \frac{x}{a_x}\right) + \frac{1}{2}W_y \left(1 - \cos 2\pi \frac{y}{a_y}\right), \qquad (4.6)$$

where a_x and a_y are the atom spacings (or more correctly the spacings between the neighboring potential troughs) and W_x and W_y are the overall amplitudes (the depths of the potential troughs) in the two directions x and y. A potential relief of this kind should be exerted for example by the (110) face of a fcc crystal.

In the case of a face with quadratic symmetry $a_x = a_y = a$ and $W_x = W_y = W$, the relief (4.6) simplifies to

$$V = \frac{1}{2}W \left(1 - \cos 2\pi \frac{x}{a}\right) + \frac{1}{2}W \left(1 - \cos 2\pi \frac{y}{a}\right). \qquad (4.7)$$

If one assumes a corrugation in one direction only, (4.6) simplifies further to the potential field (Fig. 4.4)

$$V(x) = \frac{1}{2}W \left(1 - \cos 2\pi \frac{x}{a}\right) \qquad (4.8)$$

introduced initially by Frenkel and Kontorova [1939].

It is generally believed that the interfacial potential has a flatter crest than is represented by a single sinusoide. That is why Frank and van der Merwe [1949c] suggested a refined potential of the form

$$V = \frac{1}{2}W \left(1 - \cos 2\pi \frac{x}{a}\right) + \frac{1}{2}w \left(1 - \cos 4\pi \frac{x}{a}\right),$$

where the maximum flattening is achieved when $w/W = 1/4$.

The amplitude W is related to the substrate–deposit bond strength by

$$W = g\varphi_d , \qquad (4.9)$$

where φ_d is the desorption energy of an overlayer atom from the substrate surface and g is a constant of proportionality varying from 1/30 for long range van der Waals forces to approximately 1/3 for short range covalent bonds [van der Merwe 1979]. In fact, W is the activation energy for surface diffusion and g gives the relation between the activation energies for surface diffusion and desorption.

In some cases the rather unrealistic parabolic model has been employed which replaces the smooth sinusoide (4.8) by a sequence of parabolic arcs

$$V(x) = \text{ const } x^2 \qquad (|x| \leq a/2)$$

with sharp crests between them [van der Merwe 1963a; Stoop and van der Merwe 1973]. The parabolic model permits linearization of the mathematical problem and makes it possible to obtain exact analytical solutions [Markov and Karaivanov 1979] in order to illustrate some properties of the system.

A smooth potential, known as biparabolic potential, consisting of parabolic segments

$$V(x) = \frac{1}{2}\lambda x^2, \qquad |x| \leq \frac{1}{4}a ,$$

$$V(x) = -\frac{1}{2}\lambda \left(x - \frac{1}{2}a \right)^2 + \frac{1}{16}\lambda a^2, \qquad \frac{1}{4}a \leq x \leq \frac{1}{2}a ,$$

$$V(x) = -\frac{1}{2}\lambda \left(x + \frac{1}{2}a \right)^2 + \frac{1}{16}\lambda a^2, \qquad -\frac{1}{2}a \leq x \leq -\frac{1}{4}a ,$$

was constructed by Stoop and van der Merwe [1973] and in a more general form by Kratochvil and Indenbom [1963]. The different interfacial potentials are shown in Fig. 4.9.

4.2.4.3. 1D model of epitaxial interface

The overgrowth is simulated by a chain of atoms connected by purely elastic (Hookean) springs (Eq. 4.5) as a substitute of the real interatomic forces (Fig. 4.4). The springs are characterized by their natural length $b(\equiv r_0)$ and force constant γ. The chain is subject to an external periodic potential field (4.8) exerted by a rigid substrate. The assumption of the substrate rigidity

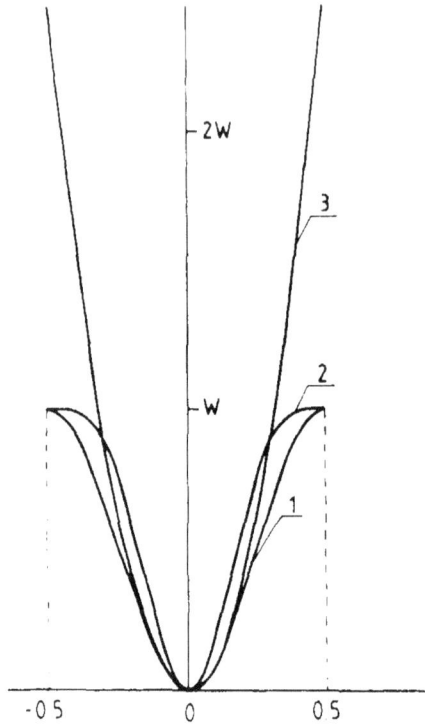

Fig. 4.9. Comparison of several interfacial potentials: curve 1 — the sinusoidal potential (Eq. (4.8)), curve 2 — the refined sinusoidal potential, curve 3 — the parabolic potential.

could be considered to reflect the real situation in the case of sufficiently thin overgrowth (not more than a few monolayers). In the case of thick deposit this assumption is not valid anymore and elastic strains in both substrate and deposit should be allowed [van der Merwe 1950].

As b is not equal to a, the atoms will not sit at the bottoms of the potential troughs but will be displaced. In fact, our task is to find the atoms' displacements. Then the energy can be calculated and in turn the ground state of the system can be easily found. To do this we have to analyze the forces acting upon every atom. As mentioned above two forces act on each atom: first a force exerted by the neighboring atoms, and second a force exerted by the substrate. The first force tends to preserve the natural spacing b between the atoms, whereas the second one tends to place all the atoms at the bottoms of the corresponding potential troughs of the substrate. As a result of the competition between the two forces, the atoms will be spaced in general at some compromise distance $\bar{b}$ in between b and a. In the case of $\bar{b} = a$ the natural misfit will be accommodated by homogeneous strain and the overgrowth will be pseudomorphous with the substrate. At the other extreme, $\bar{b} = b$, the deposit preserves on average its own atomic spacing and the natural misfit is accommodated entirely by misfit dislocations. It follows that in the intermediate case, $a < \bar{b} < b$, part of the natural misfit defined as

$$f = \frac{b - a}{a} \tag{4.10}$$

will be accommodated by misfit dislocations

$$f_{\mathrm{d}} = \frac{\bar{b} - a}{a} \tag{4.11}$$

and the remaining part

$$f_{\mathrm{e}} = \frac{\bar{b} - b}{b} = \frac{a}{b}(f_{\mathrm{d}} - f) \tag{4.12}$$

by homogeneous strain. In other words, the natural misfit appears in the general case as a sum of the homogeneous strain and the periodical strain due to the misfit dislocations, i.e.

$$f \simeq f_{\mathrm{d}} + |f_{\mathrm{e}}| . \tag{4.13}$$

In order to find the forces acting on the atoms and in turn the atoms' displacements we have to write an expression for the potential energy of the system. The force exerted by the neighboring atoms depends on the distance between them. To find it we choose the origin of the coordinate system at an arbitrary point to the left of the atoms under consideration (Fig. 4.10). Without loss of generality we can place the origin at the bottom of an arbitrary potential trough. Then the distances from the origin to the $(n+1)$th and nth atoms will be $X_{n+1} = (n+1)a + x_{n+1}$ and $X_n = na + x_n$, respectively, where x_n and x_{n+1} are the displacements of the atoms from the bottoms of the potential troughs with the same numbers. Then the distance $\Delta X_n = X_{n+1} - X_n$ between the $(n + 1)$th and the nth atoms is

$$\Delta X_n = x_{n+1} - x_n + a = a(\xi_{n+1} - \xi_n + 1) ,$$

where $\xi_n = x_n/a$ is the relative displacement of the nth atom from the bottom of the nth potential trough. The strain of the bond between the atoms will be

$$\varepsilon(n) = \Delta X_n - b = a(\xi_{n+1} - \xi_n - f) . \tag{4.14}$$

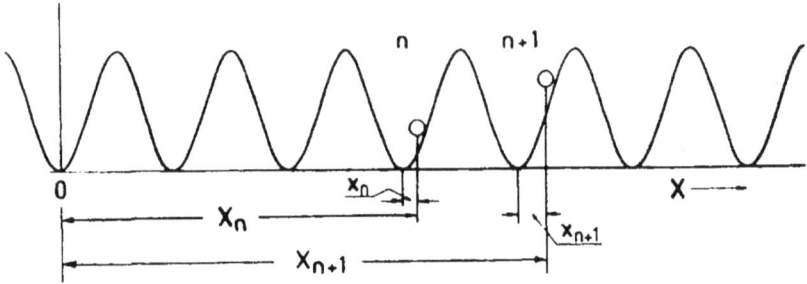

Fig. 4.10. For the determination of the atom displacements X_n and x_n and the atomic spacing $\Delta X_n = X_{n+1} - X_n$ in the 1D model of Frank and van der Merwe [1949a].

Bearing in mind (4.8) the potential energy of a chain consisting of N atoms reads

$$E = \frac{1}{2}\gamma a^2 \sum_{n=0}^{N-2} (\xi_{n+1} - \xi_n - f)^2 + \frac{1}{2}W \sum_{n=0}^{N-1} (1 - \cos 2\pi\xi_n) , \qquad (4.15)$$

where the first sum gives the strain energy of the system while the second sum accounts for the interaction across the interface.

We assume for definiteness that $b > a$ and $f > 0$ (compression of the chain). The analysis is valid for the opposite case $b < a$ and $f < 0$ (expansion of the chain), which follows from the symmetric (Hookean) shape of the interatomic potential. The only difference consists in the fact that at positive misfits the atom (corresponding to atomic plane in the 3D case) in excess is in the substrate. In the 1D case this is equivalent to an empty potential trough as shown in Fig. 4.11(a). At negative misfits the atom (plane) in excess is in the overgrowth which is equivalent to a pair of atoms in one trough (Fig. 4.11(b)). In the harmonic approximation adopted both configurations are symmetric and have one and the same energy. As shown below, this is not the case when a more realistic interatomic potential is adopted.

The derivative of the potential energy E with respect to the displacement ξ_n gives the overall force acting on the nth atom. At equilibrium this force is equal to zero and $dE/d\xi_n = 0$ appears as the condition for equilibrium. Minimizing (4.15) with respect to ξ_n leads to the following set of recurrent equations:

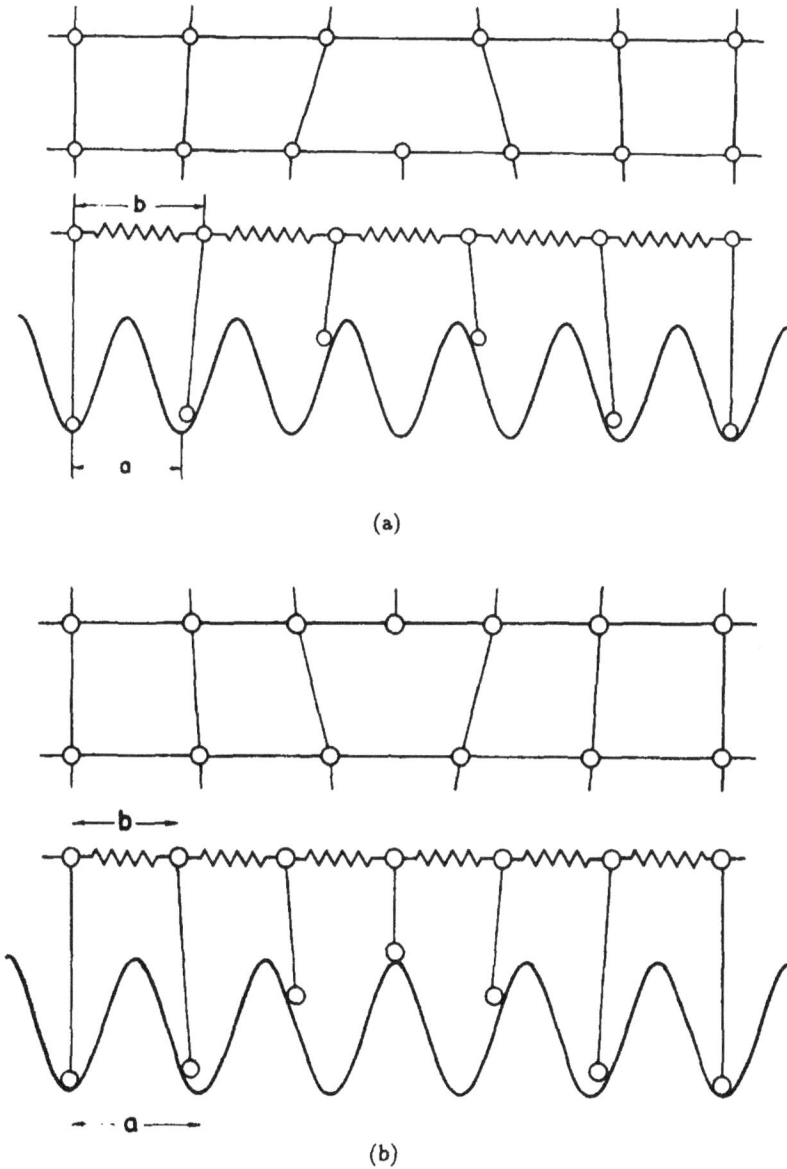

Fig. 4.11. Structure of misfit dislocations in the chain model of Frank and van der Merwe at (a) positive ($b > a$) and (b) negative ($b < a$) misfits. In (a) the misfit dislocation represents an empty potential trough (light wall) which corresponds to an excess atomic plane in the substrate. In (b) the dislocation represents two atoms in one trough (or three atoms in two troughs, heavy wall) which corresponds to an excess atomic plane in the overgrowth.

$$\xi_1 - \xi_0 - f = \frac{\pi}{2l_0^2} \sin 2\pi\xi_0 \, ,$$

$$\xi_{n+1} - 2\xi_n + \xi_{n-1} = \frac{\pi}{2l_0^2} \sin 2\pi\xi_n \, , \qquad (4.16)$$

$$\xi_{N-1} - \xi_{N-2} - f = -\frac{\pi}{2l_0^2} \sin 2\pi\xi_{N-1} \, ,$$

where

$$l_0 = \left(\frac{\gamma a^2}{2W} \right)^{1/2} \qquad (4.17)$$

is a parameter which accounts for the ratio of the forces between the overgrowth atoms and across the interface. The equations (4.16) can be solved numerically and the atoms' displacements can thus be found. However, another procedure can be used to find an analytical solution for ξ_n.

Assuming that the displacements vary slowly with the atom number we can approximate the discrete quantities ξ_n by continuous variables $\xi(n)$ and replace the differences $\xi_{n+1} - \xi_n$ by the derivative $d\xi(n)/dn$. Expanding it in a Taylor series and neglecting higher order differentials result in a differential equation of second order [Frank and van der Merwe 1949]:

$$\frac{d^2\xi(n)}{dn^2} = \frac{\pi}{2l_0^2} \sin 2\pi\xi(n) \, . \qquad (4.18)$$

This is the continuum approximation of the problem. It replaces the real discrete chain of atoms by an elastic continuum (a rubber cord). Although some details are usually lost in this procedure it has the merit of giving rise to an analytical solution. In fact Eq. (4.18) is the pendulum equation but in this particular case it is known as the static sine–Gordon equation or simply the sine–Gordon equation [Barone *et al.* 1971; Scott, Chu and McLaughlin 1973; Villain 1980].

The integration of the sine–Gordon equation can be carried out in two stages. First, we find the first integral by using the following simple procedure. We multiply both sides of (4.18) by $d\xi$ and reorganize the left-hand side to obtain

$$d\xi \frac{d^2\xi(n)}{dn^2} = d\xi \frac{d}{dn} \left(\frac{d\xi}{dn} \right) = \frac{d\xi}{dn} d \left(\frac{d\xi}{dn} \right) = \frac{\pi}{2l_0^2} \sin 2\pi\xi(n) d\xi \, .$$

The integration then gives

$$\left(\frac{d\xi}{dn} \right)^2 = -\frac{\cos 2\pi\xi(n)}{2l_0^2} + C \, ,$$

where C is the integration constant. To find it we assume that in general the solution $\xi(n)$ crosses zero at some angle so that $d\xi/dn = \omega$ at $\xi = 0$. Then $C = (1 + 2\omega^2 l_0^2)/2l_0^2$ and

$$\left(\frac{d\xi}{dn}\right)^2 = -\frac{\cos 2\pi\xi(n)}{2l_0^2} + \frac{1 + 2\omega^2 l_0^2}{2l_0^2} \ .$$

Making use of the relations $\cos 2\pi\xi = 2\cos^2 \pi\xi - 1$ and $\omega^2 l_0^2 = 1/k^2 - 1$ one finally obtains

$$\frac{d\xi}{dn} = \frac{\sqrt{1 - k^2 \cos^2 \pi\xi(n)}}{kl_0} \ . \tag{4.19}$$

Single dislocations. We consider first for illustration the limiting case of a single dislocation assuming $\omega = 0$ and $k = 1$. Then the integration of the resulting equation

$$\frac{d\xi(n)}{dn} = \frac{\sin \pi\xi(n)}{l_0}$$

subject to the boundary condition $n = 0$, $\xi(n) = 1/2$ gives

$$\xi(n) = \frac{2}{\pi} \arctan\left[\exp\left(\frac{\pi n}{l_0}\right)\right] \ . \tag{4.20}$$

The atomic displacements as a function of the atom number according to Eq. (4.20) are plotted in Fig. 4.12 (curve 1). As seen the solution has the form of a single wave. On the left-hand side, $n \to -\infty$, the displacements approach zero, which means that the atoms lie at the bottoms of their respective troughs. On the right-hand side, $n \to +\infty$, the displacements approach unity, i.e. the atoms lie at the bottoms of the neighboring potential troughs. In other words, N atoms are distributed over $N + 1$ (or $N - 1$ at negative misfit) potential troughs. This is equivalent to one missing atomic plane (or a plane in excess) in the overgrowth with respect to the substrate if we imagine the one-dimensional model under consideration as a cross section of two crystal halves. (Note that the overgrowth crystal has as many atomic planes as its own structure demands). It is known in the literature as a *misfit or interface dislocation* [Frank and van der Merwe 1949a] or a *soliton* [Scott, Chu and McLaughlin 1973; Villain 1980]. As will be shown below this is the shape the dislocations possess when they are far apart and do not interact with each other.

As seen in Fig. 4.12 atoms which are in marked disregistry with the substrate potential troughs (or atoms) occupy a region with a width l_0 measured in number of atoms. The atoms to the left and to the right of this region are in a good fit, and those far enough from it are in a perfect fit,

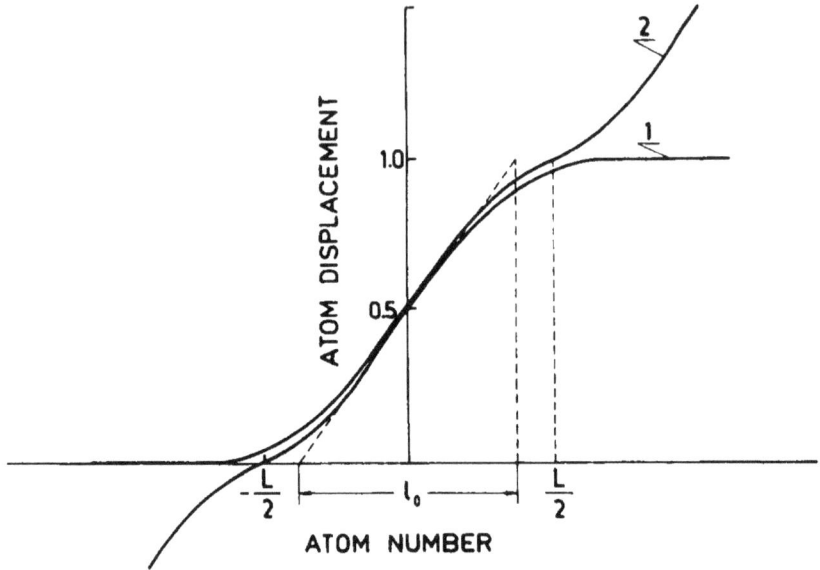

Fig. 4.12. Dependence of the atom displacement on the atom number. Curve 1 represents a single misfit dislocation (single soliton) given by Eq. (4.20). Curve 2 represents a misfit dislocation in a sequence of interacting dislocations given by Eq. (4.23). As seen, the latter crosses zero and unity under an angle ω which determines the boundary condition for finding the first integral and the value of the modulus k of the elliptic integrals $K(k)$ and $E(k)$. The greater the dislocation density, the greater the angle ω and the smaller than unity the modulus k. Thus the latter is a measure of the dislocation density. In the case of a single dislocation or dislocations far apart, $w = 0$ and $k = 1$. l_0 denotes the width of a single dislocation. The width of a dislocation in a sequence of dislocations is given by $l = kl_0 < l_0$. The reciprocal of the length L gives the density of the misfit dislocations according to Eq. (4.25).

with the substrate potential. Thus l_0 gives the width of single isolated misfit dislocations which do not interact with each other. Within the framework of the harmonic approximation under consideration the dislocation width does not depend on the natural misfit, but on the energetic parameters γ and W only. As will be shown below it becomes a steep function of the misfit in the more realistic model with anharmonic interactions [Markov and Trayanov 1988].

It is of interest to consider the elastic strains of the consecutive springs. The latter can be written in the continuum approximation in the form

$$\varepsilon(n) = a(\xi_{n+1} - \xi_n - f) \cong a\left(\frac{d\xi}{dn} - f\right) = a\left(\frac{1}{l_0 \cosh{(\pi n/l_0)}} - f\right) . \quad (4.21)$$

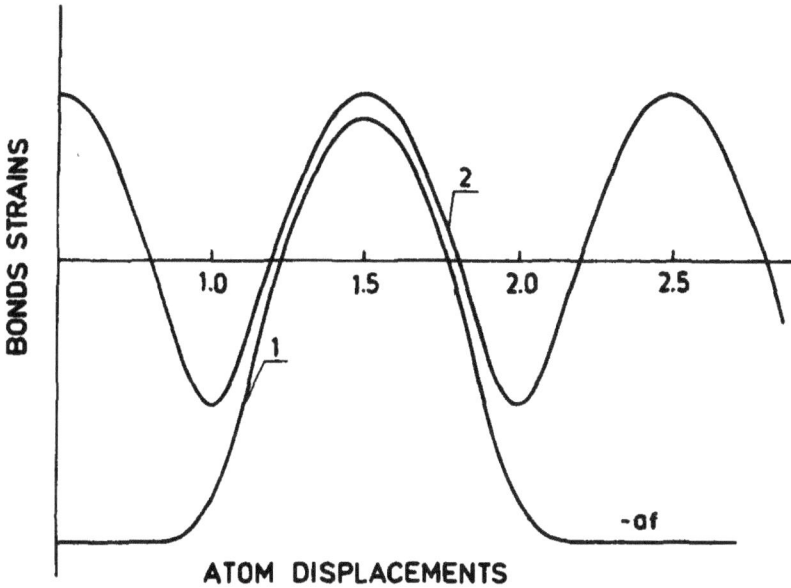

Fig. 4.13. Plot of the consecutive strains $\varepsilon(n)$ of the chemical bond versus the atom displacement $\xi(n)$ in a chain containing dislocations far apart from each other (curve 1, Eq. (4.21)) and a sequence of interacting dislocations (curve 2, Eq. (4.27)). In the first case the strains in between the dislocations reach the value $-af = -(b - a)$, i.e. the dislocations divide the interface into regions of perfect fit with the substrate. The sequence of dislocations leads to an appearance of periodical strains which change their sign.

The variation of $\varepsilon(n)$ with n is plotted in Fig. 4.13 (curve 1). As seen, far from the dislocation, $n \to \pm\infty$, the first term in the brackets goes to zero and the strains are equal to $\varepsilon(n) = -af = -(b - a)$. In other words, the bonds between the overgrowth atoms are strained to fit exactly with the spacings of the potential troughs of the substrate, and the strains are precisely equal to the natural misfit taken with a negative sign. In the core of the dislocation ($n = 0$) the strain

$$\varepsilon_c = a \left(\frac{1}{l_0} - f \right)$$

is positive as long as $1/l_0 > f$. As will be shown below, it becomes equal to zero only when the misfit reaches the so-called *limit of metastability* of the pseudomorphous state defined as $f_{ms} = 1/l_0$.

Sequence of dislocations. Making use of the substitution

$$\phi = \pi \left(\xi - \frac{1}{2} \right)$$

and after integration of (4.19) subject to the boundary condition $\xi(n = 0) = 1/2$ and $\phi = 0$ we obtain the general solution of (4.18):

$$\frac{\pi n}{k l_0} = \int_0^\phi \frac{d\varphi}{\sqrt{1 - k^2 \sin^2 \varphi}} = F(\phi, k) , \qquad (4.22)$$

where $F(\phi, k)$ is the incomplete elliptic integral of first kind [Janke, Emde and Lösch 1960], or by inversion

$$\xi(n) = \frac{1}{2} + \frac{1}{\pi} \, \text{am} \left(\frac{\pi n}{k l_0} \right) , \qquad (4.23)$$

where $\text{am}(F(\phi, k), k)$ denotes the elliptic amplitude and $k < 1$ is the modulus of the elliptic integrals. At $k = 1$ (4.23) turns into (4.20).

A graphical representation of (4.23) is given in Fig. 4.12 (curve 2). When the dislocations are far apart, $\omega \to 0$ and $k \to 1$. The nearer the dislocations are to each other, the greater ω is and the smaller than unity k becomes. Thus the modulus of the elliptic integrals determines the spacing between the dislocations, or in other words, the mean dislocation density on the one hand and the dislocation width on the other. The latter is now smaller than that of a single dislocation and is equal to $l = k l_0$.

The dislocation spacing measured in number of atoms can be easily calculated from (4.22) and reads

$$L = \frac{1}{\pi} 2 k l_0 K(k) , \qquad (4.24)$$

where $K(k) = K(\pi/2, k)$ is the complete elliptic integral of the first kind. The reciprocal of the dislocation spacing gives the mean dislocation density in the ground state:

$$f_d = \frac{\pi}{2 k l_0 K(k)} . \qquad (4.25)$$

For values of k near to unity the elliptic integral $K(k)$ can be approximated by $K(k) \cong \ln[4/(1 - k^2)^{1/2}]$ and the mean dislocation density reads

$$f_d = \frac{\pi}{2 k l_0 \ln[4/(1 - k^2)^{1/2}]} . \qquad (4.26)$$

It is immediately seen that at $k = 1$ the dislocation spacing tends to infinity and the mean dislocation density to zero. The overlayer is strained to fit exactly the periodicity of the substrate, its atomic spacing being equal to that of the latter.

In fact the mean dislocation density f_d is just the part of the natural misfit given by Eq. (4.11) which is accommodated by misfit dislocations. The mean atomic spacing $\bar{b}$ is equal to

$$\bar{b} = \frac{a}{N-1} \sum_{-N/2}^{N/2} (\xi_{n+1} - \xi_n + 1) \cong \frac{a}{L} \int_{-L/2}^{L/2} \left(\frac{d\xi}{dn} + 1 \right) dn = a + \frac{a}{L} \, ,$$

from which (4.11) follows.

The elastic strain of the consecutive springs is now

$$\varepsilon(n) = a \left(\frac{d\xi(n)}{dn} - f \right) = a \left(\frac{\sqrt{1 - k^2 \cos^2 \pi\xi(n)}}{kl_0} - f \right) , \qquad (4.27)$$

which reduces to (4.21) at $k = 1$. The strain varies periodically with the spring number, compression and expansion alternating with a period equal to the dislocation spacing (Fig. 4.13, curve 2). In the cores of the dislocations the strain $\varepsilon_c = a(1/kl_0 - f)$ is now greater than that in the core of a single dislocation. In between the dislocations, $\xi = 0, 1$, the strain

$$\varepsilon = a \left(\frac{\sqrt{1 - k^2}}{kl_0} - f \right)$$

no longer reaches the maximum strain $\varepsilon = -af$ as shown in Fig. 4.13 (curve 2). With increasing dislocation density (decreasing k) the strain in the dislocation cores increases and the one in between the dislocations decreases in absolute value, i.e. the strain varies more and more symmetrically around the zero. When the sum of the positive strains become equal to that of the negative strains, or in other words, when the areas under the $\varepsilon(n)$ curve from either side of the zero become equal, the mean atomic spacing $\bar{b} = b$. When the latter takes place, the natural misfit is completely accommodated by misfit dislocations, or in other words, by the periodical strain connected with them. When the positive and negative areas under the $\varepsilon(n)$ curve are not equal their difference gives the part of the natural misfit which is accommodated by homogeneous strain.

We can now find the energy of the system. For this purpose we have to substitute the solution (4.23) into the continuum approximation of (4.15):

$$E = \frac{1}{2}\gamma a^2 \int\limits_{-L/2}^{L/2} \left[\left(\frac{d\xi}{dn}\right) - f\right]^2 dn + \frac{1}{2}W \int\limits_{-L/2}^{L/2} (1 - \cos 2\pi\xi)\, dn \ . \quad (4.28)$$

Substituting (4.23) into the second integral of the right-hand side of (4.28) gives

$$\frac{1}{2}W \int\limits_{-L/2}^{L/2} (1 - \cos 2\pi\xi)\, dn = W l_0^2 \int\limits_{-L/2}^{L/2} \left[\left(\frac{d\xi}{dn}\right)^2 - \frac{1 - k^2}{k^2 l_0^2}\right] dn$$

and (4.28) turns into

$$E = W l_0^2 \int\limits_{-L/2}^{L/2} \left[2\left(\frac{d\xi}{dn}\right)^2 - 2f\frac{d\xi}{dn} + f^2 - \frac{1 - k^2}{k^2 l_0^2}\right] dn \ .$$

Substituting (4.19) into the above expression and carrying out the integration give the energy per atom (Fig. 4.18):

$$\mathcal{E} = \frac{E}{L} = W\left(\frac{4E(k)}{\pi k}l_0 f_{\mathrm{d}} - \frac{1 - k^2}{k^2} + l_0^2 f^2 - 2l_0^2 f f_{\mathrm{d}}\right) \ , \quad (4.29)$$

where f_{d} is given by (4.25) and

$$E(k) = \int\limits_{0}^{\pi/2} \sqrt{1 - k^2 \sin^2 \psi}\, d\psi$$

is the complete elliptic integral of second kind.

Minimization of the energy per atom with respect to the mean dislocation density f_{d} gives

$$\frac{d\mathcal{E}}{df_{\mathrm{d}}} = 2W l_0 \left(\frac{2E(k)}{\pi k} - l_0 f\right) \ , \quad (4.30)$$

where the relationships $d[E(k)/k]/dk = -K(k)/k^2$ and $d[kK(k)]/dk = E(k)/(1 - k^2)$ are used.

The condition for the lowest energy state then reads

$$f = f_{\mathrm{s}} = \frac{2}{\pi l_0}\frac{E(k)}{k} \ . \quad (4.31)$$

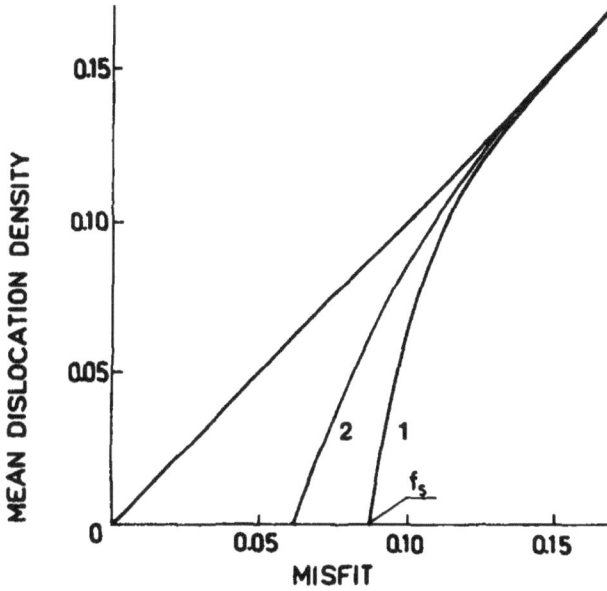

Fig. 4.14. Plot of the mean dislocation density $f_d = (\bar{b} - a)/a$ vs natural misfit $f = (b - a)/a$ in the 1D model (curve 1) and the 2D model (curve 2). The plot represents in fact the dependence of the average atomic spacing $\bar{b}$ on the bulk spacing b. At large enough values of the misfit $\bar{b} \to b$ and the misfit is accommodated completely by misfit dislocation. At misfits smaller than the stability limit $f_s, \bar{b} = a$ and the misfit is accommodated completely by homogeneous strain. The overgrowth is pseudomorphous with the substrate. The straight line gives the case $\bar{b} = b$ (after van der Merwe [1975]).

Substituting $E(k)/k$ from (4.31) into (4.29) gives the energy of the ground state:

$$\mathcal{E}_0 = W l_0^2 f^2 - W \frac{1 - k^2}{k^2} = \frac{1}{2} \gamma a^2 f^2 - W \frac{1 - k^2}{k^2}. \qquad (4.32)$$

This means that as long as $k = 1$, $\mathcal{E}_0 = 0.5 \gamma a^2 f^2$ and the pseudomorphous state is always the ground state.

Beyond the limit of stability the system in the ground state contains misfit dislocations whose density is determined by Eq. (4.25). Excluding k from (4.25) and (4.31) we find the mean dislocation density in the ground state as a function of the natural misfit. The dependence is plotted in Fig. 4.14 (curve 1). As seen the dislocation density is equal to zero up to the stability limit f_s and then sharply increases and goes asymptotically to the value of the natural misfit. In fact, this is a plot of the mean atom spacing $\bar{b}$ relative to a as a function of the natural spacing b.

Let us analyze Eq. (4.29) in more detail. At $k = 1$ (dislocations **far apart**) it turns into

$$\mathcal{E} = \frac{4}{\pi} W l_0 f_\mathrm{d} + W l_0^2 (f^2 - 2f f_\mathrm{d}) = \frac{4}{\pi} W l_0 f_\mathrm{d} \left(1 - \frac{f}{f_\mathrm{s}}\right) + \mathcal{E}(0) , \quad (4.33)$$

where $\mathcal{E}(0) = W l_0^2 f^2$ is the energy of the pseudomorphous state and f_s is the limit of the stability of the pseudomorphous state (see Eq. (4.37)).

The term

$$\mathcal{E}_\mathrm{d} = \frac{4}{\pi} W l_0 f_\mathrm{d}$$

represents the energy of a single misfit dislocation or a single soliton

$$\mathcal{E}_1 = \frac{4}{\pi} W l_0 \qquad\qquad (4.34)$$

multiplied by the density of the dislocations, f_d. In other words, $\mathcal{E}_\mathrm{d}$ is the energy of the misfit dislocations.

The second term

$$\mathcal{E}_\mathrm{hs} = W l_0^2 (f^2 - 2f f_\mathrm{d}) = \frac{1}{2}\gamma a^2 (f^2 - 2f f_\mathrm{d}) \cong \frac{1}{2}\gamma a^2 f_\mathrm{e}^2$$

is in fact the energy of the homogeneous strain. It turns out that in the case of noninteracting misfit dislocations ($k = 1$), the energy is a sum of the energy of the homogeneous strain and the energy of the misfit dislocations, i.e. $\mathcal{E} = \mathcal{E}_\mathrm{d} + \mathcal{E}_\mathrm{hs}$.

The difference of the energies at $k = 1$ and $k < 1$ represents obviously the energy of interaction of the misfit dislocations. The latter is implicitly accounted for in the k-containing terms. The energy of interaction of a pair of dislocations has been calculated to give [Villain 1980; Theodorou and Rice 1978] the following asymptotic expression:

$$\mathcal{E}_\mathrm{int} = \varkappa \mathcal{E}_\mathrm{d} \exp\left(-\frac{\pi}{f_\mathrm{d} l_0}\right) , \qquad\qquad (4.35)$$

which is valid for dislocations far apart and $\varkappa$ is a constant of the order of unity [Bak and Emery 1976; Theodorou and Rice 1978]. The exponential behavior of the interaction energy obviously reflects the dependence of k on the mean dislocation density (4.26).

The energy of the incommensurate state then becomes

$$\mathcal{E} = \mathcal{E}_1 f_\mathrm{d} \left[1 - \frac{f}{f_s} + \varkappa \exp \left(-\frac{\pi}{f_\mathrm{d} l_0} \right) \right] + \mathcal{E}(0) . \tag{4.36}$$

As shown above the pseudomorphous state is always the ground state as long as $k = 1$. When $k < 1$ the state with dislocations becomes the ground state, their density being determined by the value of k. Then from (4.31), with $k = 1$, one obtains

$$f = f_s = \frac{2}{\pi l_0} , \tag{4.37}$$

which appears as the limit of stability of the pseudomorphous state. The limit of stability of a state with a particular density of dislocations is given by Eq. (4.31). Frank and van der Merwe calculated the value of l_0 under the condition that the energies ψ_{AA}, ψ_{BB} and ψ_{AB} are equal and found $l_0 = 7.35$. Hence $f_s \cong 9\%$ in this particular case.

As the strains (and the stresses) change their signs periodically (Fig. 4.13) there are obviously springs which are unstrained. If we cut such springs the parts of the chain to the left and to the right of the cut will remain in equilibrium with their free ends. This means that when a chain with a finite length is in equilibrium its hypothetical end springs with numbers $n = -1$ and $n = N$ are unstrained. In such a case the end atoms will have a specific displacement ξ_0. It can be found by assuming the periodic variation of the strain $\varepsilon(n)$ given by (4.27) crosses zero:

$$\cos \pi \xi_0 = \left(\frac{1}{k^2} - l_0^2 f^2 \right)^{1/2} .$$

As seen ξ_0 depends on the value of the misfit. When the misfit increases the end atom climbs the slope of its respective potential trough. Obviously, there exists a critical value of the misfit at which the end atom is just on top of the hill between its respective and the neighboring potential troughs, i.e. $\xi_0 = \pm 1/2$. If the misfit is infinitesimally increased the end atom will go down in the neighboring trough. The situation is equivalent to the generation of a new misfit dislocation at the chain end. Mathematically this means that $l_0^2 f^2 > 1/k^2$ and the quantity under the square root becomes negative. On the other hand, the misfit cannot be too small as this means that the quantity under the square root will become larger than unity and the equation will have no solution. Physically this means that under some critical value of the misfit an existing dislocation should leave the chain at its free end. The above considerations are illustrated in Fig. 4.15 where the chains are shown in a folded form [Dubnova and Indenbom 1966]. We have to imagine that each atom occupies the same position but in a separate potential trough. Figure 4.15(a) demonstrates a chain which does not contain misfit dislocations. All displacements lie between $-1/2$ and $1/2$. If the natural misfit is equal to zero all atoms will lie in the bottoms of the potential troughs, i.e. $\xi(n) = 0$ (Fig. 4.15(b)). Increasing the misfit leads to a situation when the end atoms reach the tops of the respective hills, i.e. $\xi(0) = -1/2$ and $\xi(N-1) = 1/2$ (Fig. 4.15(c)). The difference of the overall lengths of the chains shown in Figs. 4.15(b) and (c) is precisely equal to a. This means that in Fig. 4.15(b) N atoms are distributed over N potential troughs whereas in Fig. 4.15(c) N atoms are distributed over $N+1$ troughs. The analogous situation for a chain containing one dislocation is shown in Figs. 4.15(d) and (e). In Fig. 4.15(d) N atoms are distributed over $N+1$ troughs while in Fig. 4.15(e), over $N+2$ troughs.

The necessary and sufficient condition for a finite chain in a certain state (dislocated or not, depending on the value of k) to exist in equilibrium with its free ends is that the upper bound $\varepsilon(\xi_0 = 1/2)$ of the periodic variation of the strain $\varepsilon(n)$ (Eq. 4.27) be positive and the lower bound $\varepsilon(\xi_0 = 0)$ be negative. This leads to

$$\frac{1}{k^2} - l_0^2 f^2 \leq 1 \quad \text{or} \quad \frac{1}{k^2} - 1 \leq l_0^2 f^2 \,,$$

$$\frac{1}{k^2} - l_0^2 f^2 \geq 0 \quad \text{or} \quad l_0^2 f^2 \leq \frac{1}{k^2} \,,$$

or in other words, to

$$\left(\frac{1}{k^2} - 1 \right)^{1/2} \leq l_0 f \leq \frac{1}{k} \,. \tag{4.38'}$$

The inequalities (4.38') determine the interval of misfit in which there exist solutions of the sine–Gordon equation for a finite chain. Outside of this interval there are no solutions and the corresponding configurations do not exist. For a chain which is pseudomorphous with the substrate ($k = 1$) this interval is

$$0 \leq f \leq \frac{1}{l_0} \,. \tag{4.38''}$$

This means that the pseudomorphous state will be stable up to $f \leq f_s$ and will exist, but not as a ground state, at misfits larger than f_s but smaller than the limit of metastability

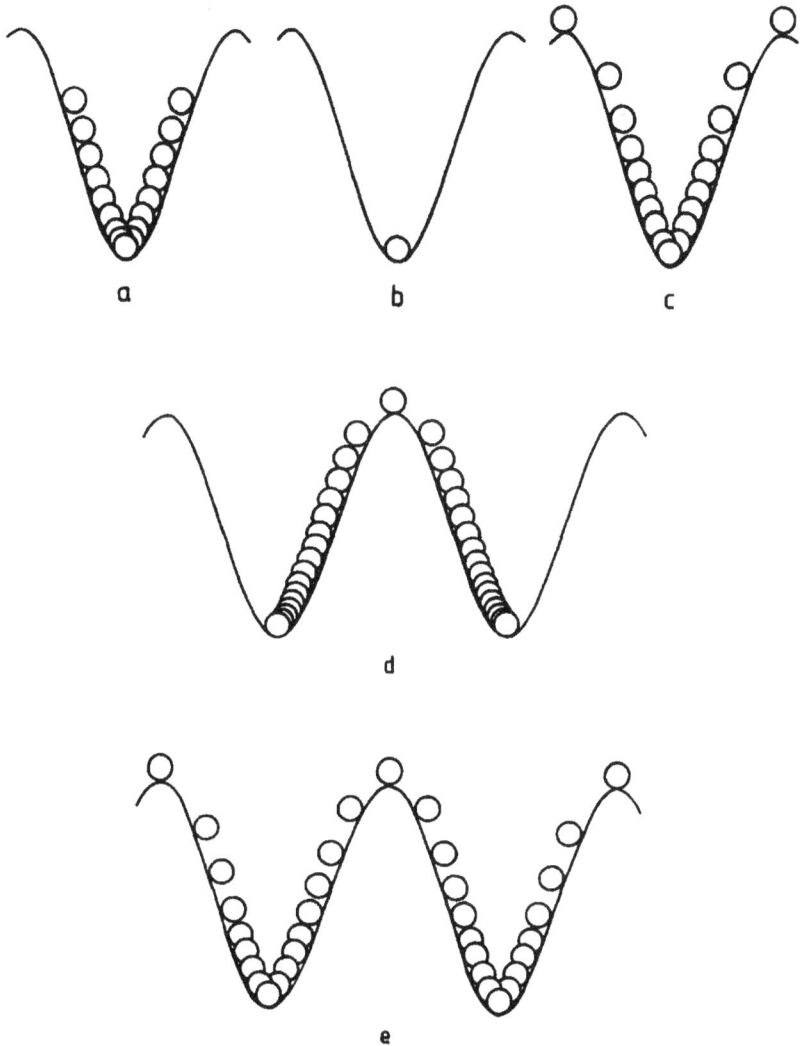

Fig. 4.15. Illustration in a folded form of chains at different values of the misfit and containing different numbers of dislocations. Imagine that each atom has the same displacement as shown in the figure but is positioned in the next potential trough. (a) A chain without a dislocation at a misfit smaller than the metastability limit f_{ms}. (b) A chain without a dislocation at $f = 0$. All the atoms are situated exactly at the bottoms of the corresponding potential troughs. (c) A chain without a dislocation at $f = f_{ms}$. The end atoms are exactly on top of the crests between their respective troughs and the next ones. (d) A chain containing one dislocation at $f = 0$. (e) A chain containing one dislocation at $f = f_{ms}$.

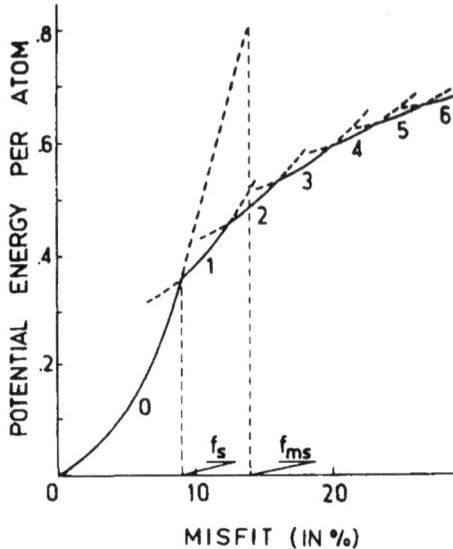

Fig. 4.16. Misfit dependence of the potential energy per atom of a finite chain. The plot consists of a sequence of parabolic segments corresponding to states with increasing number of dislocations denoted by the figure at each segment. The segments intersect each other at the corresponding stability limits given by Eq. (4.31). f_s and f_{ms} denote, respectively, the limits of stability (Eq. (4.37)) and metastability (Eq. (4.39)) of the pseudomorphous state. The solid line gives the ground state whereas the dashed lines represent the corresponding metastable states.

$$f_{ms} = \frac{1}{l_0} . \tag{4.39}$$

Summarizing, we conclude that chains containing misfit dislocations will be stable beyond a misfit determined by Eq. (4.31). The value of the stability limit depends in this case on the density of the dislocations, or in other words, on the value of k. The region of metastability of dislocated chains is now shifted to larger limits according to (4.38′) determined again by the value of k. It follows that the misfit dependence of the energy of a finite chain of atoms will consist of intersecting parabolic segments (Eq. 4.29), each segment corresponding to a particular number of misfit dislocations increased by one (Fig. 4.16). The intersections define the corresponding stability limits, the first one being given by Eq. (4.37). Each segment is confined in a misfit interval determined by (4.38′). The envelope of the parabolic segments gives the ground state of the energy of infinitely long chain.

4.2.4.4. 2D model of Frank and van der Merwe

In considering this problem we follow the analysis of van der Merwe [1970, 1973]. We consider an overlayer with rectangular symmetry and natural atom spacings b_x along the x axis and b_y along the y axis. The substrate periodic potential is given by Eq. (4.6).

We enumerate the overlayer atoms and the substrate potential troughs by n in the x direction and by m in the y direction. Then the Cartesian coordinates of an atom n, m from an arbitrarily placed origin of the coordinate system are

$$X_{nm} = a_x(n + \xi_{nm}), \qquad Y_{nm} = a_y(m + \eta_{nm}) . \tag{4.40}$$

As shown in Fig. 4.17 the linear and shear strains in the film are

$$\varepsilon_x = \frac{1}{b_x}(X_{n+1,m} - X_{nm} - b_x) ,$$

$$\varepsilon_y = \frac{1}{b_y}(Y_{n,m+1} - Y_{nm} - b_y) ,$$

$$\varepsilon_{xy} = \frac{1}{b_y}(X_{n,m+1} - X_{nm}) + \frac{1}{b_x}(Y_{n+1,m} - Y_{nm}) .$$

Making use of (4.40) gives

$$\varepsilon_x = \frac{a_x}{b_x}(\xi_{n+1,m} - \xi_{nm} - f_x) , \tag{4.41'}$$

$$\varepsilon_y = \frac{a_y}{b_y}(\eta_{n,m+1} - \eta_{nm} - f_y) , \tag{4.41''}$$

$$\varepsilon_{xy} = \frac{a_x}{b_y}(\xi_{n,m+1} - \xi_{nm}) + \frac{a_y}{b_x}(\eta_{n+1,m} - \eta_{nm}) , \tag{4.41'''}$$

where

$$f_x = \frac{b_x - a_x}{a_x} \quad \text{and} \quad f_y = \frac{b_y - a_y}{a_y} \tag{4.42}$$

are the natural misfits in the two orthogonal directions.

According to the theory of elasticity [Timoshenko 1934; van der Merwe 1973] the stresses and the energy of isotropic elastic two-dimensional continuum (rubber sheet) are given by

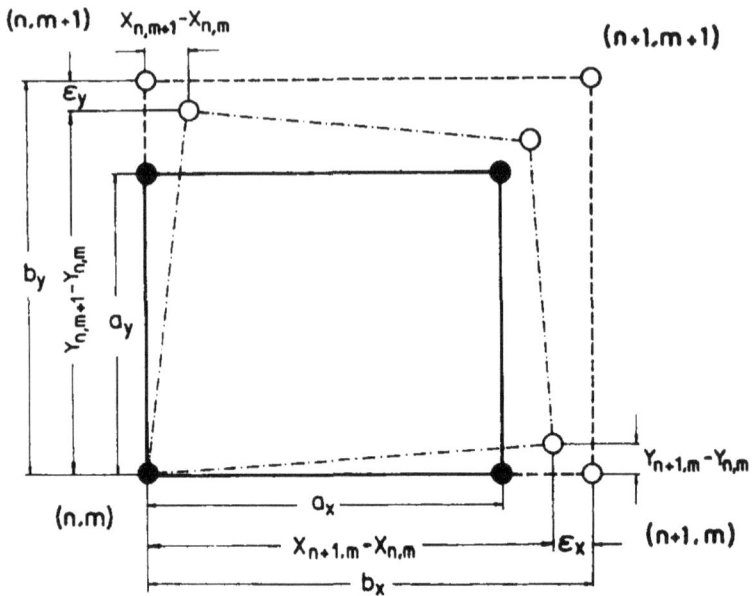

Fig. 4.17. Deformation of a rectangular unit atomic mesh of the overgrowth (open circles) from which the strains can be derived. The substrate atomic mesh is given by the filled circles (after van der Merwe [1975]).

$$T_x = \frac{2Gt}{1-\nu}(\varepsilon_x + \nu\varepsilon_y) \;,$$

$$T_y = \frac{2Gt}{1-\nu}(\varepsilon_y + \nu\varepsilon_x) \;,$$

$$T_{xy} = Gt\varepsilon_{xy} \;,$$

$$E = Gtb_xb_y\left(\frac{\varepsilon_x^2 + \varepsilon_y^2 + 2\nu\varepsilon_x\varepsilon_y}{1-\nu} + \frac{1}{2}\varepsilon_{xy}^2\right) \;, \qquad (4.43)$$

where G is the shear modulus, ν the Poisson ratio of the film material and t the thickness of the film which is equal to b_z in our particular case of monoatomic overlayer, z being the direction normal to the interface. Then $\Omega = b_xb_yb_z$ is the volume of an overlayer molecule.

Bearing in mind (4.6) the potential energy of the overlayer is

$$E = \sum_{n,m} \left\{ \frac{G\Omega}{1-\nu} \right.$$

$$\times \left[\left(\frac{a_x}{b_x} \right)^2 (\xi_{n+1,m} - \xi_{nm} - f_x)^2 + \left(\frac{a_y}{b_y} \right)^2 (\eta_{n,m+1} - \eta_{nm} - f_y)^2 \right.$$

$$\left. + 2\nu \frac{a_x a_y}{b_x b_y} (\xi_{n+1,m} - \xi_{nm} - f_x)(\eta_{n,m+1} - \eta_{nm} - f_y) \right]$$

$$\left. + \frac{1}{2} G\Omega \left(\frac{a_x}{b_y}(\xi_{n,m+1} - \xi_{nm}) + \frac{a_y}{b_x}(\eta_{n+1,m} - \eta_{nm}) \right)^2 \right\}$$

$$+ \frac{1}{2} \sum_{n,m} [W_x(1 - \cos 2\pi \xi_{nm}) + W_y(1 - \cos 2\pi \eta_{nm})] \; . \tag{4.44}$$

The conditions of equilibrium of the nmth atom now read

$$\frac{\partial E}{\partial \xi_{nm}} = \frac{\partial E}{\partial \eta_{nm}} = 0 \; .$$

Applying the first condition gives

$$(\xi_{n+1,m} - 2\xi_{nm} + \xi_{n-1,m}) + \frac{1}{2}(1 - \nu) \left(\frac{b_x}{b_y} \right)^2 (\xi_{n,m+1} - 2\xi_{nm} + \xi_{n,m-1})$$

$$+\nu \frac{a_y b_x}{a_x b_y} (\eta_{n,m+1} - \eta_{nm} - \eta_{n-1,m+1} + \eta_{n-1,m})$$

$$+\frac{1}{2}(1 - \nu)\frac{a_y b_x}{a_x b_y} (\eta_{n,m-1} - \eta_{nm} - \eta_{n+1,m-1} + \eta_{n+1,m})$$

$$= \frac{\pi}{2l_x^2} \sin 2\pi \xi_{nm} \; . \tag{4.45}$$

In the continuum limit

$$\xi_{n+1,m} - 2\xi_{nm} + \xi_{n-1,m} \cong \frac{\partial^2 \xi}{\partial n^2} \; ,$$

$$\eta_{n,m+1} - \eta_{nm} - \eta_{n-1,m+1} + \eta_{n-1,m}$$

$$= \eta_{n,m-1} - \eta_{nm} - \eta_{n+1,m-1} + \eta_{n+1,m} \cong \frac{\partial^2 \eta}{\partial n \partial m}$$

and (4.45) turns into

$$\frac{\partial^2 \xi}{\partial n^2} + \frac{1}{2}(1+\nu)\frac{a_y b_x}{a_x b_y}\frac{\partial^2 \eta}{\partial n \partial m} + \frac{1}{2}(1-\nu)\left(\frac{b_x}{b_y}\right)^2 \frac{\partial^2 \xi}{\partial m^2} = \frac{\pi}{2 l_x^2} \sin 2\pi \xi_{nm} . \quad (4.46')$$

The second equilibrium condition gives the corresponding equation for η_{nm}:

$$\frac{\partial^2 \eta}{\partial m^2} + \frac{1}{2}(1+\nu)\frac{a_x b_y}{a_y b_x}\frac{\partial^2 \xi}{\partial n \partial m} + \frac{1}{2}(1-\nu)\left(\frac{b_y}{b_x}\right)^2 \frac{\partial^2 \eta}{\partial n^2} = \frac{\pi}{2 l_y^2} \sin 2\pi \eta_{nm} ,$$

$$(4.46'')$$

where

$$l_x = \left(\frac{G\Omega a_x^2}{W_x(1-\nu)b_x^2}\right)^{1/2} , \qquad l_y = \left(\frac{G\Omega a_y^2}{W_y(1-\nu)b_y^2}\right)^{1/2} . \qquad (4.47)$$

In the case of quadratic symmetry of the contact planes the above set of equations simplifies to [Frank and van der Merwe 1949b]

$$\frac{\partial^2 \xi}{\partial n^2} + \frac{1}{2}(1+\nu)\frac{\partial^2 \eta}{\partial n \partial m} + \frac{1}{2}(1-\nu)\frac{\partial^2 \xi}{\partial m^2} = \frac{\pi}{2 l^2} \sin 2\pi \xi_{nm} , \quad (4.48')$$

$$\frac{\partial^2 \eta}{\partial m^2} + \frac{1}{2}(1+\nu)\frac{\partial^2 \xi}{\partial n \partial m} + \frac{1}{2}(1-\nu)\frac{\partial^2 \eta}{\partial n^2} = \frac{\pi}{2 l^2} \sin 2\pi \eta_{nm} , \quad (4.48'')$$

where

$$l = \left(\frac{G\Omega a^2}{W(1-\nu)b^2}\right)^{1/2} . \qquad (4.49)$$

If we are looking for solutions with edge type dislocations the mixed derivatives $\partial^2 \xi/\partial m^2$, $\partial^2 \eta/\partial n^2$, $\partial^2 \xi/\partial n \partial m$ and $\partial^2 \eta/\partial n \partial m$ vanish (in fact we neglect the shear strains $\varepsilon_{xy} = 0$) and the set of equations (4.46) turns into a set of two independent sine–Gordon equations

$$\frac{\partial^2 \xi}{\partial n^2} = \frac{\pi}{2 l_x^2} \sin 2\pi \xi_{nm} , \qquad (4.50')$$

$$\frac{\partial^2 \eta}{\partial m^2} = \frac{\pi}{2 l_y^2} \sin 2\pi \eta_{nm} . \qquad (4.50'')$$

One could consider the following limiting cases:

(i) $f_y \cong 0$, $\xi = \xi(n)$ and $\eta = $ const (corrugation of the substrate surface in one direction only). The system (4.50) reduces to

$$\frac{\partial^2 \xi}{\partial n^2} = \frac{\pi}{2 l_x^2} \sin 2\pi \xi, \qquad \frac{\pi}{2 l_y^2} \sin 2\pi \eta = 0 . \qquad (4.51)$$

The solution of the first equation is given in the previous chapter. The solution of the second one is $\eta = 0$. The result is a sequence of edge type misfit dislocation lines parallel to the y axis and a homogeneous strain $\varepsilon_y = -f_y a_y/b_y$ in the y direction. Such situation is observed in the case of epitaxial growth of tetragonal $MoSi_2$ on (100)-Si [Chen, Cheng and Lin 1986]. The epitaxial orientations and the natural misfits in the two orthogonal directions are $(100)[004]_d \parallel (100)[220]_s$, $f_x = 2.34\%$ and $f_y = 0.1\%$, and $(111)[11\bar{2}]_d \parallel (100)[2\bar{2}0]_s$, $f_x = 2.21\%$ and $f_y = 0.1\%$.

(ii) $f_x \neq f_y \neq 0$, $\xi = \xi(n)$ and $\eta = \eta(m)$. This is the general case leading to a cross grid of misfit dislocations. Particular cases are $f_x = f_y$ (quadratic symmetry of the contact planes) and $f_x \approx -f_y$. The latter is observed in some rare cases such as, for instance, in the epitaxial growth of tetragonal $MoSi_2$ on (100)-Si [Chen, Cheng and Lin 1986] where the epitaxial orientation is $(110)[004]_d \parallel (001)[220]_s$. The natural misfits in both orthogonal directions are $f_x = 2.34\%$ and $f_y = -1.69\%$. We will not treat this case here because considerations of the model with a more realistic anharmonic potential show a considerable influence of the sign of the natural misfit on the properties of the model.

Following the same procedure as before we find the solutions for $\xi(n)$ and $\eta(m)$ which have the same form as Eq. (4.22) with the only exception that l_0 is replaced by l_x and l_y. The first and the second integrals now read

$$\frac{d\xi}{dn} = \frac{\sqrt{1 - k_x^2 \cos^2 \pi\xi}}{k_x l_x} , \tag{4.52}$$

$$\frac{\pi n}{k_x l_x} = F\left[k_x, \pi\left(\xi - \frac{1}{2}\right)\right] , \tag{4.53}$$

and the corresponding solutions for $\eta(m)$ have the same form.

Substituting the solutions into the continuum approximation of (4.44) (with $\varepsilon_{xy} = 0$)

$$E = \int_{-L_x/2}^{L_x/2} dn \int_{-L_y/2}^{L_y/2} dm \left\{ W_x l_x^2 \left[2\left(\frac{d\xi}{dn}\right)^2 - 2f_x \frac{d\xi}{dn} + f_x^2 - \frac{1-k_x^2}{k_x^2 l_x^2}\right] \right.$$

$$+ W_y l_y^2 \left[2\left(\frac{d\eta}{dm}\right)^2 - 2f_y \frac{d\eta}{dm} + f_y^2 - \frac{1-k_y^2}{k_y^2 l_y^2}\right]$$

$$\left. + 2\nu \sqrt{W_x W_y} l_x l_y \left(\frac{d\xi}{dn} - f_x\right)\left(\frac{d\eta}{dm} - f_y\right) \right\} \tag{4.54}$$

gives for the energy per overlayer atom the expression

$$\mathcal{E} = \frac{E}{L_x L_y} = W_x \left(\frac{4E(k_x)}{\pi k_x} l_x f_d(x) - \frac{1 - k_x^2}{k_x^2} + l_x^2 f_x^2 - 2l_x^2 f_x f_d(x) \right)$$

$$+ W_y \left(\frac{4E(k_y)}{\pi k_y} l_y f_d(y) - \frac{1 - k_y^2}{k_y^2} + l_y^2 f_y^2 - 2l_y^2 f_y f_d(y) \right)$$

$$+ 2\nu \sqrt{W_x W_y} l_x l_y \left[f_d(x) - f_x \right] \left[f_d(y) - f_y \right] , \qquad (4.55)$$

where $f_d(x)$ and $f_d(y)$ are the mean dislocation densities, and k_x and k_y are the corresponding moduli of the elliptic integrals in the two orthogonal directions x and y, respectively.

In the case of quadratic symmetry ($f_x = f_y = f$, $f_d(x) = f_d(y) = f_d$, $k_x = k_y = k$, $W_x = W_y = W$),

$$\mathcal{E} = 2W \left(\frac{4E(k)}{\pi k} l f_d - \frac{1 - k^2}{k^2} + l^2 f^2 - 2l^2 f f_d \right) + 2\nu W l^2 (f_d - f)^2 . \quad (4.56)$$

As seen the first term in the round brackets (which is multiplied by 2 to account for the grid of misfit dislocations) in (4.56) is identical in form with Eq. (4.29). Both expressions differ only in the second term which includes explicitly the Poisson ratio. The last term in (4.55) and (4.56) contains the difference $f_d - f = f_e$ which is in fact the residual homogeneous strain and should vanish at large misfits when $f_d = f$ in the ground state.

Minimization of the energy with respect to $f_d(x)$ (or $f_d(y)$) yields the condition for the ground state:

$$f_x = \frac{2E(k_x)}{\pi k_x l_x} + \nu \left(\frac{W_y l_y^2}{W_x l_x^2} \right)^{1/2} \left[f_d(y) - f_y \right] . \qquad (4.57)$$

The corresponding expression for f_y can be easily obtained. Bearing in mind (4.47) the latter simplifies to

$$f_x = \frac{2E(k_x)}{\pi k_x l_x} + \nu \frac{a_y b_x}{b_y a_x} \left[f_d(y) - f_y \right] . \qquad (4.58)$$

The condition $k_x = 1$ gives the limit of stability of the pseudomorphous state

$$f_s(x) = \frac{2}{\pi l_x} - \nu \frac{a_y b_x}{b_y a_x} f_y , \qquad (4.59)$$

which for quadratic symmetry reduces to

$$f_s(2D) = \frac{2}{\pi l(1+\nu)} \ . \tag{4.60}$$

Note that in the 2D case the stability limit $f_s(x)$ in one of the directions depends on the misfit in the other orthogonal direction. The latter reflects the increased dimensionality of the system.

Substituting the elliptic integrals $E(k_x)$ and $E(k_y)$ from (4.58) the respective expression for f_y in (4.55) gives the energy of the ground state:

$$\mathcal{E}_{\min} = W_x l_x^2 f_x^2 + W_y l_y^2 f_y^2 - W_x \frac{1-k_x^2}{k_x^2} - W_y \frac{1-k_y^2}{k_y^2}$$

$$+ 2\nu\sqrt{W_x W_y}\, l_x l_y \left[f_x f_y - f_d(x) f_d(y) \right] \ . \tag{4.61}$$

In the case of quadratic symmetry (4.61) reduces to

$$\mathcal{E}_{\min} = 2W \left(l^2 f^2 (1+\nu) - \frac{1-k^2}{k^2} - \nu l^2 f_d^2 \right) \ ,$$

which for $k = 1$ gives the energy of the commensurate state

$$\mathcal{E}_{\min} = 2W l^2 f^2 (1+\nu) \ , \tag{4.62}$$

in which the term $1 + \nu$ containing Poisson's ratio accounts for the dimensionality of the system.

4.2.4.5. Comparison of 2D and 1D models

In order to compare numerically the 1D and 2D models we have to first find a relation between the shear modulus G and the force constant γ. For a quadratic symmetry of the contact planes this relation reads [van der Merwe 1973]

$$\gamma b = \mathbf{E} b^2 \ ,$$

where $\mathbf{E}$ is Young's modulus. The theory of elasticity [Hirth and Lothe 1968] gives the relationship between Young's and the shear moduli:

$$\mathbf{E} = 2G(1+\nu) \ .$$

Upon substitution we find

$$\gamma = 2G(1+\nu)b \ . \tag{4.63}$$

Then from (4.17) and (4.49) follows

$$l = \frac{l_0}{\sqrt{1 - \nu^2}} . \tag{4.64}$$

From (4.37), (4.60) and (4.64) we find

$$f_s(2D) = f_s(1D) \left(\frac{1 - \nu}{1 + \nu}\right)^{1/2} . \tag{4.65}$$

With a reasonable value for $\nu = 1/3$, $[(1 - \nu)/(1 + \nu)]^{1/2} = 1/\sqrt{2}$ or $f_s(2D) = f_s(1D) / \sqrt{2}$. Recalling that $f_s(1D) \cong 9\%$ with $l_0 \cong 7$, $f_s(2D) \cong 6\%$, or in other words, Poisson's effect in 2D systems leads to a considerable decrease of the stability limit.

The dependence of the mean dislocation density f_d on the natural misfit in the case of quadratic symmetry of the contact planes has the same behavior as shown in Fig. 4.14 (curve 2). The only exception is that it does not increase as steeply as the mean dislocation density $f_d(1D)$ in the 1D case.

Figure 4.18 demonstrates the dependence of the energy per atom in both the 1D and 2D models. The difference is more pronounced in the lowest energy state (curves 1 and 2). The difference of the energies when they are not in the ground state (curves 1' and 2') is due to the last term of Eq. (4.56) which vanishes at large misfits.

4.2.4.6. *Application of 1D model to thickening overlayer*

Obviously the 1D Frenkel–Kontorova model is inadequate for describing the case of thickening overlayers as three very important factors are not taken into account. The first is the rigidity of the substrate, which is believed to be valid for very thin deposits not exceeding one or two monolayers. The second is the strain gradient normal to the interface when the latter is resolved in a sequence of misfit dislocations. As long as the underlying monolayer is homogeneously strained to fit the substrate the upper one is strained to the same degree. After breaking up of the commensurability the amplitude of the periodic strain in every next monolayer is smaller than that of the previous one. The mathematical treatment of the problem is formidable [Stoop and van der Merwe 1973]. The third is Poisson's effect in a direction normal to the interface. Obviously, if, for example, $b > a$ and at least part of the natural misfit is accommodated by homogeneous strain, the overgrowth will be compressed in a direction parallel to the interface.

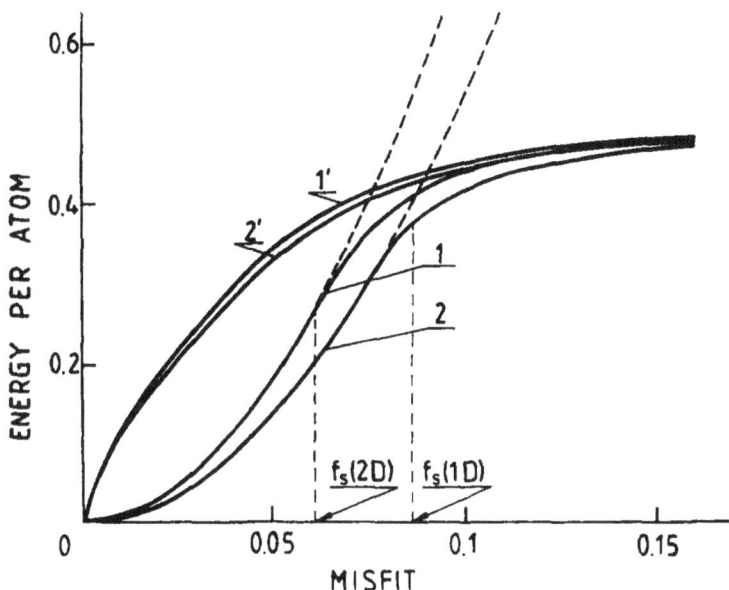

Fig. 4.18. Plots of the misfit energy per atom versus misfit in 1D model (curves 2 and 2′) and 2D model with quadratic symmetry (curves 1 and 1′). Curves 1 and 2 represent the ground state energies. Curves 1′ and 2′ represent the energies when the misfit is completely accommodated by misfit dislocations ($\bar{b} = b$) and the homogeneous strain is equal to zero (after van der Merwe [1975]).

At the same time, the overgrowth will be expanded in a direction normal to the interface.

Assuming as a first approximation that (i) the substrate is rigid, (ii) no strain gradient normal to the interface exists and (iii) the normal strain due to Poisson's effect is negligible, one can obtain qualitative results which give a good enough impression concerning the properties of a "thick" overlayer [van der Merwe *et al.* 1986].

We simulate the thick overgrowth by 1D chains of atoms "piled up" one on top of the other. The film will in general be pseudomorphous with the substrate up to some critical thickness which is a function of the natural misfit. This follows from the fact that homogeneous strain accumulates linearly with increasing film thickness and at some value of the latter the strain energy becomes greater than the energy of the misfit dislocations. Then the commensurability breaks down and misfit dislocations are introduced at the interface. The homogeneous strain is replaced by a periodic strain and the overgrowth lattice is on average relaxed. Obviously, if the misfit is small the critical thickness will be large, and vice versa. The aim of this oversimplified model is to give some indication of the misfit dependence of the equilibrium critical thickness.

The thickness t of the film is then given by

$$t = t_n = nb \, , \tag{4.66}$$

where n denotes the number of atom chains "piled" up one on top of the other.

Another oversimplified assumption concerns the force constant or "rigidity" of the overlayer

$$\gamma_n = n\gamma \, . \tag{4.67}$$

Then

$$l_n = \left(\frac{\gamma_n a^2}{2W} \right)^{1/2} = \left(\frac{n\gamma a^2}{2W} \right)^{1/2} = l_0 \sqrt{n} \, . \tag{4.68}$$

Replacing l_0 by l_n in the 1D model gives the solution of the model of thick overlayer. Thus the limit of stability becomes

$$f_s(n) = \frac{2}{\pi l_n} = \frac{2}{\pi l_0 \sqrt{n}} = \frac{f_s}{\sqrt{n}} \, . \tag{4.69}$$

The commensurability breaks down when $k = 1$ and this happens at some critical thickness

$$n = n_c = \frac{t_c}{b} \tag{4.70}$$

at a misfit

$$f = f_s(n_c) = f_s / \sqrt{n_c} \, . \tag{4.71}$$

It follows from (4.70) and (4.71) that the equilibrium critical thickness beyond which the film will no longer be pseudomorphous with the substrate is

$$t_c = b \left(\frac{f_s}{f} \right)^2 \, . \tag{4.72}$$

As follows from this oversimplified model the critical thickness for pseudomorphous growth decreases steeply with increasing lattice misfit and goes to infinity when the misfit vanishes. This behavior agrees qualitatively with the experimental evidence and, as will be shown below, appears as a better approximation for large misfits. Obviously, Eq. (4.72) cannot be compared quantitatively with experimental data but it has one important advantage. It can be used to predict qualitatively the influence of the anharmonicity of the interatomic forces on the cirtical thickness of expanded and compressed epitaxial films as will be shown in Sec. 4.2.4.7.1.

4.2.4.7. 1D model with non-Hookean interatomic forces

One of the basic restrictions of the model adopted originally by Frank and van der Merwe [1949] which makes it applicable for small lattice misfits only is the purely elastic interactions between neighboring atoms as a substitute of the real interatomic forces (see Fig. 4.7). This restriction can be relaxed by replacing the harmonic approximation with one of the more realistic pairwise potentials (4.1)–(4.3). Besides, one can use a combination of a Toda potential and a real potential such that the respective repulsive branches coincide. Then they will differ only for values of r larger than r_0, and by comparison of the results one can distinguish the purely anharmonic effects from those due to the nonconvexity of the real potentials. This is shown in Fig. 4.6 where the Toda potential is plotted with $\alpha = 2$ and $\beta = 6$ together with the generalized Morse potential (4.3) with $\mu = 4$, $\nu = 3$ and $V_0 = 1$. The repulsive branches are indistinguishable and the two potentials differ perceptibly only at $r > 1.2r_0$. Moreover, the harmonic approximations of both potentials coincide (the broken curve) so that we can refer our results to those obtained with the harmonic potential.

In principle an anharmonic potential can be constructed by joining two parabolic segments, $V(r \leq r_0) = \frac{1}{2}\gamma_1(r-r_0)^2$ and $V(r \geq r_0) = \frac{1}{2}\gamma_2(r-r_0)^2$ with $\gamma_1 > \gamma_2$. This potential does not display a finite force at large r as in the case of the Toda potential. Nevertheless, we will use the latter in our considerations and will study only the effect of the anharmonicity.

4.2.4.7.1. Effect of anharmonicity in epitaxial interfaces

Making use of the Toda potential (4.4) and the substrate periodic potential (4.8), the potential energy of the chain reads [Milchev and Markov 1984; Markov and Milchev 1984a]

$$E = \sum_{n=0}^{N-2} \left(\frac{\alpha}{\beta} \exp[-\beta a(\xi_{n+1} - \xi_n - f)] + \alpha a(\xi_{n+1} - \xi_n - f) - \frac{\alpha}{\beta} \right)$$

$$+ \frac{W}{2} \sum_{n=0}^{N-1} (1 - \cos 2\pi \xi_n) . \tag{4.73}$$

The equilibrium condition $\partial E/\partial \xi_n = 0$ leads to the set of equations

$$e^{-\beta a(\xi_1 - \xi_0 - f)} - 1 = -\frac{\pi W}{\alpha a} \sin 2\pi \xi_0 \, ,$$

$$e^{-\beta a(\xi_{n+1} - \xi_n - f)} - e^{-\beta a(\xi_n - \xi_{n-1} - f)} = \frac{\pi W}{\alpha a} \sin 2\pi \xi_n \, , \qquad (4.74)$$

$$e^{-\beta a(\xi_{N-1} - \xi_{N-2} - f)} - 1 = \frac{\pi W}{\alpha a} \sin 2\pi \xi_{N-1} \, ,$$

which turns into the harmonic set (4.16) when expanding the exponents in Taylor series up to the linear term at $\beta \to 0$.

In the continuous limit the system (4.74) turns into an anharmonic analog of the sine–Gordon equation [Milchev and Markov 1984] and an analytical solution can be found [Milchev 1986]. On the other hand, the discrete system (4.74) can be easily solved numerically and the properties of the model studied.

The disparity in the structural properties of the overgrowth with respect to the sign of the misfit is clearly demonstrated in Fig. 4.19 [Markov and Milchev 1985] where the variation of the strains of the consecutive springs $\varepsilon_n = \xi_{n+1} - \xi_n - f$ is shown. As seen the expanded dislocationless chain ($f = -10\%$) is in a much better fit with the substrate periodicity than is the compressed one ($f = 10\%$). Neglecting the deviations near the chain ends the strains of the springs in the expanded chain are exactly equal to the absolute value of the lattice misfit. In the compressed chains, however, the strains approach the lattice misfit but do not become equal to it. The latter means that an expanded overgrowth adheres more strongly to the substrate than does the compressed one.

One of the most significant results of the anharmonic model is the split of the limits of stability f_s and metastability f_{ms} with respect to the misfit sign. As shown in Fig. 4.20 [Markov and Milchev 1984b], increasing the degree of anharmonicity β results in a reduction in the values of f_s and f_{ms} for compressed chains ($b > a$) and in an increase in the absolute values of f_s and f_{ms} for expanded chains ($b < a$). The respective values for the harmonic model are given by the dashed lines. Thus the harmonic limit of stability $f_s^h = \pm 8.6\%$ splits into $+6.7\%$ and -12.2% whereas the limit of metastability $f_{ms}^h = \pm 13.6\%$ splits into $+10.2\%$ and -23.2% at some average degree of anharmonicity $\beta = 6$. Therefore a pseudomorphous overlayer can be in a state of stable (below f_s) or metastable (below f_{ms}) equilibrium up to quite different stability limits at positive and negative incompatibility with the substrate.

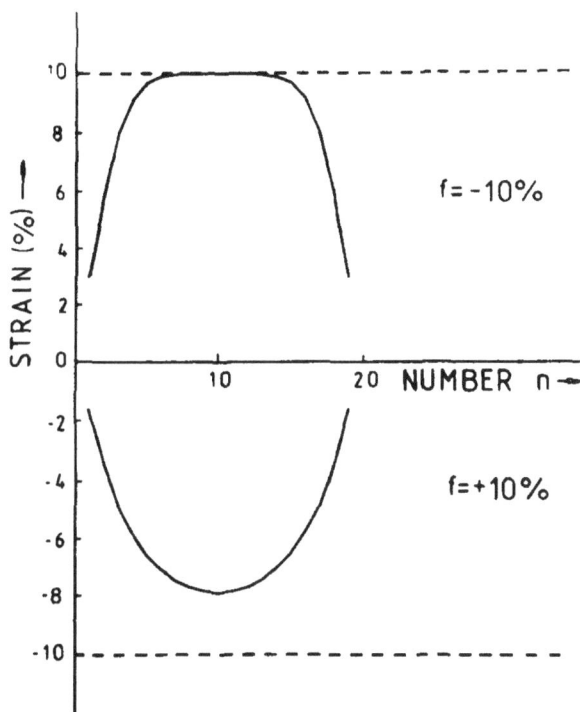

Fig. 4.19. Plot of the strains of the consecutive springs in compressed ($f = 10\%$) and expanded ($f = -10\%$) anharmonic Toda chains. (I. Markov and A. Milchev, *Surf. Sci.* **145**, 313 (1984). By permission of Elsevier Science Publishers B.V.)

Another very important conclusion which follows from the split of the critical misfits with respect to the misfit sign is connected with the critical thickness for pseudomorphous growth. As discussed above (Eq. 4.72) the latter is qualitatively proportional to the square of the limit of stability f_s. It should be expected that the critical thickness for pseudomorphous growth will be 3 to 4 times greater when the natural misfit is negative rather than positive, if all the other parameters remain unchanged. This prediction of the model seems particularly important for the epitaxial growth of semiconductor films and strained layer superlattices where the dangling bonds associated with the misfit dislocations have a deleterious effect on the properties of the corresponding heterojunctions. LPE grown $In_x Ga_{1-x} As_y P_{1-y}$ on (100) InP shows clear asymmetric behavior of the critical thickness for pseudomorphous growth with the sign of the misfit

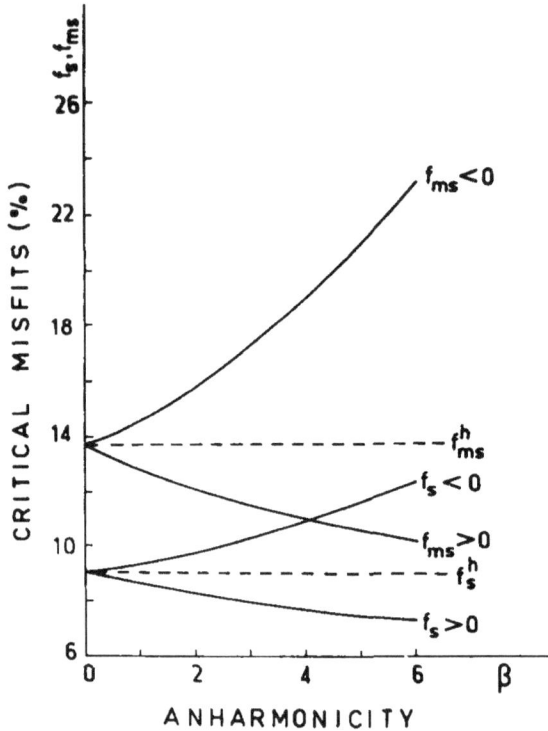

Fig. 4.20. The characteristic split of the limits of stability f_s and metastability f_{ms} of the pseudomorphous state with the sign of the natural misfit when anharmonic (Toda) interactions between the overgrowth atoms are adopted. The critical misfits are plotted versus the degree of anharmonicity β of the Toda potential. The dashed lines give the critical misfits f_s^h and f_{ms}^h in the harmonic approximation. (I. Markov and A. Milchev, *Surf. Sci.* **145**, 313 (1984). By permission of Elsevier Science Publishers B.V.)

[Krasil'nikov *et al.* 1988]. The critical thickness of the expanded epilayers is always greater than that of the compressed epilayers (Fig. 4.21). The same is observed in MBE grown $In_x Ga_{1-x}$ As on (100) InP [Franzosi *et al.* 1986].

It follows that an epitaxial film with a given thickness and different values of the misfit in different crystallographic directions x and y can be pseudomorphous with the substrate when the absolute values of the negative misfit are even larger than the values of the positive misfit at a different epitaxial orientation. An excellent example is the deposition of tetragonal and hexagonal $MoSi_2$ (t-$MoSi_2$ and h-$MoSi_2$) on the (111) and (100) faces of

h, мкм

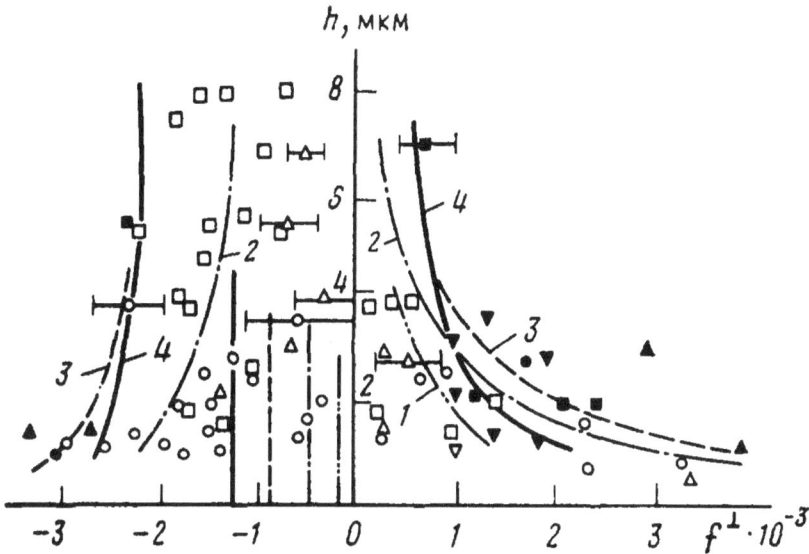

Fig. 4.21. Plot of the critical thickness for pseudomorphous growth versus the natural misfit of LPE grown $In_xGa_{1-x}As_yP_{1-y}$ on InP(001). The misfit is varied by changing the alloy composition. The asymmetrical behavior around a zero misfit is clearly seen. (V. Krasil'nikov, T. Yugova, V. Bublik, Y. Drozdov, N. Malkova, G. Shepenina, K. Hansen and A. Rezvov, *Sov. Phys. Crystallogr.* **33**, 874 (1988).)

Si, respectively [Chen, Cheng and Lin 1986; Lin and Chen 1986]. In the first case, the epitaxial orientation is $(110)[004]_d \parallel (111)[20\bar{2}]_s$, the values of the natural misfit are 2.34% and 2.21%, and the epitaxial interface is resolved in a hexagonal grid of misfit dislocations. In the second case, the epitaxial orientation is $(\bar{2}4\bar{2}3)[2\bar{1}\bar{1}2]_d \parallel (001)[2\bar{2}0]_s$, $f_x = -2.89\%$, $f_y = -1.84\%$, and the film is pseudomorphous with the substrate. Even more illustrative is the case where the epitaxial orientation is $(111)[11\bar{2}]_d \parallel (111)[20\bar{2}]_s$ and the values of the misfit are 2.21% and -2.68%. Instead of a hexagonal grid of dislocations, which is expected on the base of the harmonic model, a set of parallel dislocation lines is observed. The film is partially pseudomorphous even though the absolute value of the negative misfit is larger than the value of the positive one.

Figure 4.22 shows the characteristic split of the misfit dependence of the mean dislocation density f_d with respect to the misfit sign [Markov and Milchev 1984b]. The stepwise behavior is due to the finite length of the chains used for the computation. It must not be confused with the "devil staircase" [Aubry 1983]. As seen, f_d^- is always smaller than f_d^+ although the difference gradually decreases at large natural misfits. It can also be seen that the harmonic approximation is much closer to the positive misfit curve. What is more important, however, is that the curves, although shifted from the harmonic one, preserve their continuous character. In other words, the transition from the pseudomorphous ($f_d = 0$) to the completely dislocated ($f_d = f$) state is gradual and there is a misfit interval in which homogeneous strain and misfit dislocations coexist.

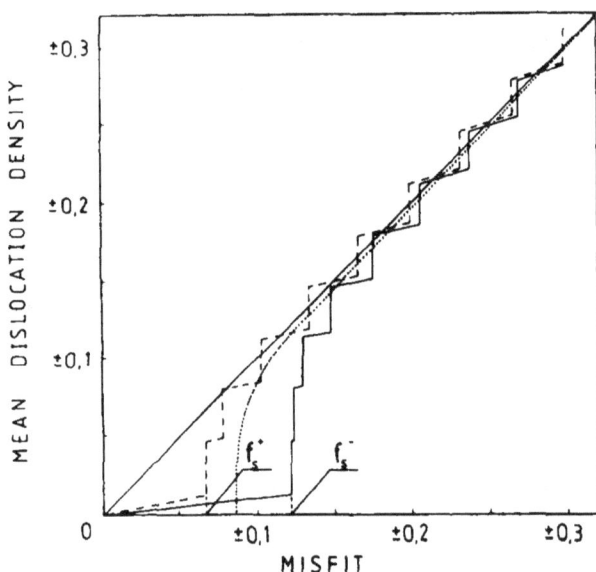

Fig. 4.22. Plot of the mean density of misfit dislocations of the ground state in the anharmonic Toda chain versus positive (dashed line) and negative (solid line) natural misfit. The curves are shown in one and the same quadrant for easier comparison. f_s^+ and f_s^- denote the corresponding limits of stability. The dotted line gives for comparison the mean dislocation density in the continuous harmonic model of Frank and van der Merwe [1949a]. (I. Markov and A. Milchev, *Surf. Sci.* **145**, 313 (1984). By permission of Elsevier Science Publishers B.V.)

The misfit dependence of the ground state energy per atom is shown in Fig. 4.23 [Markov and Milchev 1984b] for both positive (the dashed line) and negative (the solid line) misfits. The curves consist of a series of curvilinear segments as in the harmonic case (see Fig. 4.16). The segments again correspond to different numbers of misfit dislocations increasing from zero by one. It is seen that in the case of a compressed chain and particularly

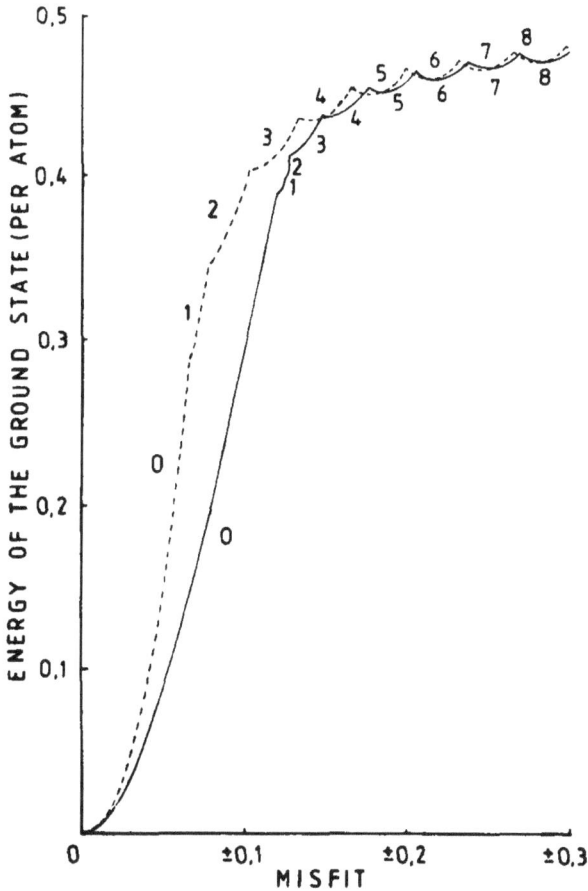

Fig. 4.23. Plot of the energy of the ground state per atom of the anharmonic Toda chain versus positive (dashed line) and negative (solid line) natural misfit. The separate curvilinear segments represent states with different numbers of misfit dislocations denoted by the figure at each segment ($N = 30$, $\alpha = 2$, $\beta = 6$). (I. Markov and A. Milchev, *Surf. Sci.* **145**, 313 (1984). By permission of Elsevier Science Publishers B.V.)

at small misfits the energy is considerably higher. At larger misfits the energy curves go closer and merge eventually. At low misfits, both positive and negative, the first sum in Eq. (4.73) (the strain energy) is dominant. At a positive misfit the steeper repulsive branch of the interatomic potential is mainly involved and accordingly the energy is higher than in the case of a negative misfit, where the strain energy is determined by the weaker attractive part of the interaction. At larger misfits, both positive and negative, the second sum in (4.73) predominates and the energy difference between the two cases gradually vanishes. It is worth noting that the harmonic curve (not shown in the figure) is again much closer to the positive misfit curve. The above result is in agreement with the finding of Murthy and Rice [1990] that the interface energy of the low positive misfit couple Cu/Ni(001) ($f = 2.56\%$) is considerably greater than that of the negative misfit couple Ni/Cu(001) ($f = -2.49\%$).

It can be concluded that the negative misfit appears to be more favorable than the positive misfit for epitaxial growth of thin films. If several epitaxial orientations are possible for a given overgrowth material on the same substrate plane, the orientation connected with a negative misfit should be favored as it is connected with a lower energy. An example for that is the orientation of Ag on (001) GaAs mentioned above [Massies and Linh 1982a,b,c]. At temperatures lower than 200°C the epitaxial orientation is $(110)[111]_{Ag} \parallel (00\bar{1})[\bar{1}10]_{GaAs}$ with $f_x = 2.23\%$ and $f_y = -3.62\%$. At higher temperatures the overgrowth is in parallel orientation $(001)[010]_{Ag} \parallel (00\bar{1})[010]_{GaAs}$ and the lattice misfit in both orthogonal directions is negative: $f_x = f_y = -3.62\%$.

4.2.4.7.2. *Influence of nonconvexity in epitaxial interfaces*

A. *Model*

The effect of anharmonicity can be more or less intuitively predicted from the asymmetry of the interatomic potential. This is not, however, the case with more real potentials where the nonconvex character leads to an existence of a maximal force between the atoms at the inflection point and to distortion of the chemical bonds when stretched out beyond the latter. That is why we will consider this case in more detail.

In order to study the effect of the nonconvexity of the real potentials we will use the generalized Morse potential (4.3). The latter has an inflection point

$$r_i = r_0 + \frac{\ln(\mu/\nu)}{\mu - \nu} \qquad (4.75)$$

beyond which the second derivative d^2V/dr^2 becomes negative and has a minimum at

$$r_m = r_0 + 2\frac{\ln(\mu/\nu)}{\mu - \nu} \ . \qquad (4.76)$$

The value of the minimum

$$\left(\frac{d^2V}{dr^2}\right)_{r=r_m} = -V_0\mu^2 \left(\frac{\nu}{\mu}\right)^{2\mu/(\mu-\nu)} \tag{4.77}$$

determines the maximum driving force for distortion to occur.

Making use of the generalized Morse potential (4.3) the potential energy of a "real" chain consisting of N atoms reads

$$E = V_0 \sum_{n=0}^{N-2} \left(\frac{\nu}{\mu-\nu}e^{-\mu a(\xi_{n+1}-\xi_n-f)} - \frac{\mu}{\mu-\nu}e^{-\nu a(\xi_{n+1}-\xi_n-f)}\right)$$

$$+ \frac{W}{2}\sum_{n=0}^{N-1}(1-\cos 2\pi\xi_n) \ . \tag{4.78}$$

Accordingly the system of equations giving the equilibrium displacements of the atoms reads

$$e^{-\mu a(\xi_1-\xi_0-f)} - e^{-\nu a(\xi_1-\xi_0-f)} = -A\sin 2\pi\xi_0 \ ,$$

$$e^{-\mu a(\xi_{n+1}-\xi_n-f)} - e^{-\nu a(\xi_{n+1}-\xi_n-f)}$$

$$-e^{-\mu a(\xi_n-\xi_{n-1}-f)} + e^{-\nu a(\xi_n-\xi_{n-1}-f)} = -A\sin 2\pi\xi_n \ , \tag{4.79}$$

$$e^{-\mu a(\xi_{N-1}-\xi_{N-2}-f)} - e^{-\nu a(\xi_{N-1}-\xi_{N-2}-f)} = A\sin 2\pi\xi_{N-1} \ ,$$

where

$$A = \frac{\pi W(\mu-\nu)}{a\mu\nu V_0} \ . \tag{4.80}$$

B. Existence of solutions

An examination of (4.79) concerning the existence of solutions is not possible in the general case as the equations are not solvable with respect to the highest variable therein. That is why we will consider the simplest case of a Morse chain ($\mu = 2\omega/b$, $\nu = \omega/b$), bearing in mind that examination of other cases (e.g., $\mu = 3$, $\nu = 1$), although more complicated, leads to the same conclusion.

The set of difference equations governing the behavior of a Morse chain can be written in terms of the strains $\varepsilon_n = \xi_{n+1} - \xi_n - f$ instead of the displacements ξ_n in the form [Markov and Trayanov 1988]

$$\xi_1 = -f_i - \frac{b}{\omega a} \ln \left(1 \pm \sqrt{1 - \frac{2\pi b W}{\omega a V_0} \sin 2\pi \xi_0} \right) ,$$

$$\varepsilon_{n+1} = -f_i - \frac{b}{\omega a}$$

$$\times \ln \left[1 \pm \sqrt{1 - 4 \left(e^{-\omega a \varepsilon_n / b} - e^{-2\omega a \varepsilon_n / b} + \frac{\pi b W}{2\omega a V_0} \sin 2\pi \xi_n \right)} \right] ,$$

$$\varepsilon_{N-1} = -f_i - \frac{b}{\omega a} \ln \left(1 \pm \sqrt{1 + \frac{2\pi b W}{\omega a V_0} \sin 2\pi \xi_{N-1}} \right) , \tag{4.81}$$

where
$$f_i = -\frac{b \ln 2}{\omega a}$$

is the misfit which corresponds exactly to the inflection point of the Morse potential (see Eq. (4.75)).

One may look for a solution of Eqs. (4.81) provided the logarithmic terms therein are well-defined analytical functions, i.e. when their arguments are non-negative. This condition is fulfilled when the discriminants D under the square roots are positive for positive signs before the square roots, or positive but smaller than unity for negative signs before the roots. When the sign before the root is positive for $D > 0$ the corresponding strain ε_{n+1} is always smaller than $-f_i$. In the other case of negative sign for $0 < D < 1$ the strain ε_{n+1} will be greater than $-f_i$. This emphasizes the fundamental role that the inflection of the real potential plays. The latter becomes clearer when distortion of the chains at negative misfits takes place. Thus when the strains of both the long and short bonds are greater than $-f_i$, the negative sign only enters the equations. When the strains of the short and long bonds are smaller and greater than $-f_i$, respectively, positive and negative signs alternate in the consecutive equations of the system (4.81).

The condition for existence of solutions $D > 0$ leads to the inequalities

$$\frac{\omega V_0}{2b} > \frac{\pi W}{a} \sin 2\pi \xi_0 ,$$

$$\frac{\omega V_0}{2b} > \frac{2\omega V_0}{b} \left(e^{-\omega a \varepsilon_n / b} - e^{-2\omega a \varepsilon_n / b} \right) + \frac{\pi W}{a} \sin 2\pi \xi_n ,$$

$$\frac{\omega V_0}{2b} > -\frac{\pi W}{a} \sin 2\pi \xi_{N-1} , \tag{4.82}$$

which means that in order to have solutions of the system (4.81) the resultant force exerted on the nth atom by the $(n-1)$th and by the substrate must be smaller than the theoretical tensile strength $\sigma_{th} = \omega V_0/2b$ of the Morse potential. If this is not the case the corresponding bond will break up and the chain will loose its integrity. This is just what happens to the most expanded bonds in the cores of the misfit dislocations at positive misfits. The dislocations core bonds are stretched out and when their strains ε_c become equal to $-f_i$, or in other words, when the force applied to the bonds becomes equal to σ_{th}, the chains break up.

Figure 4.24 represents the strain ε_c of the bonds in the cores of the dislocations as a function of the natural misfit. It is interesting to note that $\varepsilon_c = 0$ when the misfit reaches the limit of metastability of the given state just as in the harmonic model. At some critical misfit the core strain ε_c reaches $-f_i$ with infinite slope and the chain breaks up just in the dislocation core. This does not mean that bonds that are stretched out more than $-f_i$ cannot exist. As will be shown below in the case of chain distortion, bonds dilated much more than $-f_i$ can exist without rupture. The explanation is simple if one looks at Fig. 4.11. In the case of a positive misfit the dislocation represents an empty trough and the atoms on both sides of the core bond are located in such a way that the force exerted by the substrate is destructive. It is just the opposite in the case of distorted chains at negative misfits greater in absolute values than $-f_i$. The force exerted by the substrate on the atoms on both sides of the more expanded bonds is not destructive but tends to keep them together (see Fig. 4.6).

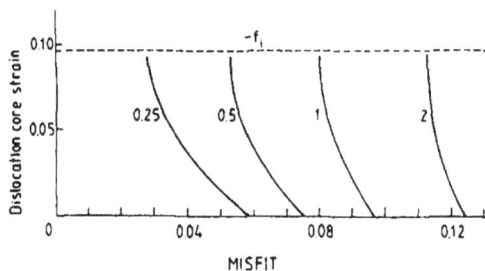

Fig. 4.24. Variation of the strain in the cores of the misfit dislocations versus positive misfit at different values of the relative substrate modulation W/V_0, given by the figure at each curve. The dashed line gives the maximum tensile strain permitted. (I. Markov and A. Trayanov, *J. Phys. C: Solid State Phys.* **21**, 2475 (1988). By permission of IOP Publishing Ltd.)

C. *Distortion of chemical bonds*

We consider first an infinite chain. The undistorted state is one in which all atoms are equally spaced at a distance equal to the substrate potential period a (Fig. 4.25(a)). A distorted chain can be dimerized so that short and long bonds alternate [Haas 1978, 1979] (Fig. 4.25(b); see also Marchand, Hood and Caillé [1988]). As mentioned above, this phenomenon is due to the fact that the average energy of one long and one short bond is smaller than the energy of a bond of intermediate length. In a dimerized chain the displacements of the consecutive atoms are equal in absolute value and opposite in sign: $\xi_{n+1} = \xi_{n-1} = -\xi_n$.

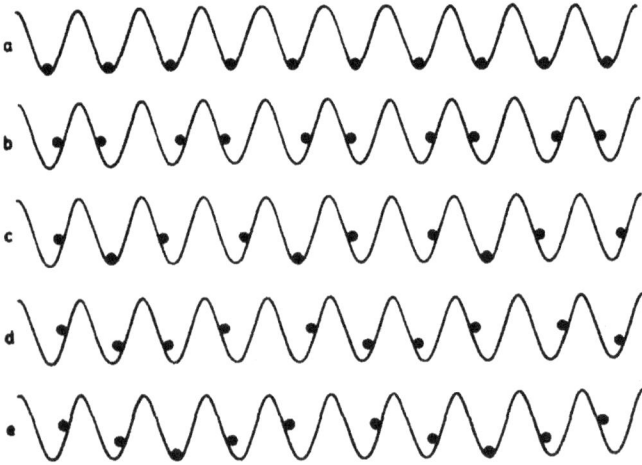

Fig. 4.25. Distortion patterns in the chain model of Frank and van der Merwe with real interactions: (a) undistorted state, (b) dimerised state, (c) trimerised state, (d) tetramerised state, (e) pentamerised state. (I. Markov and A. Trayanov, *J. Phys. C: Solid State Phys.* **21**, 2475 (1988). By permission of IOP Publishing Ltd.)

Obviously, a strong substrate–deposit interaction ($W \gg V_0$) favors the undistorted structure. A distorted structure will be tolerated when the ratio W/V_0 is small enough. Thus, applying the condition $\xi_{n+1} = -\xi_n$ to (4.79) for W/V_0 in the limit $\xi_n \to 0$ one obtains

$$\frac{W}{V_0} = \frac{2\mu\nu a^2}{(\mu - \nu)\pi^2} \left(\nu e^{\nu a f} - \mu e^{\mu a f} \right) . \tag{4.83}$$

This dependence of W/V_0 vs f outlines an area in which the dimerization is energetically favored. It is plotted in Fig. 4.26 (curve A). As seen, it starts

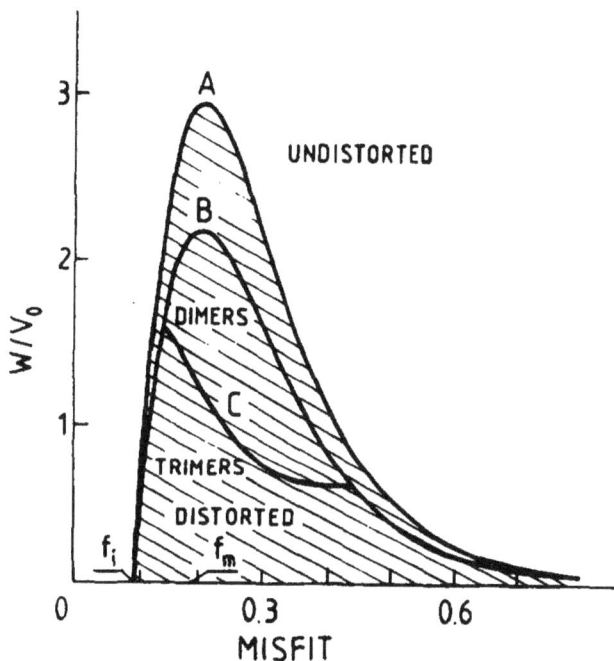

Fig. 4.26. Phase diagram W/V_0 versus misfit of existence and stability of distorted and undistorted states. Curve A outlines the area of existence and stability of dimerised state. Curve B outlines the region of existence of trimers. Curve C divides the regions of stability of dimerised and trimerised states. (I. Markov and A. Trayanov, *J. Phys. C: Solid State Phys.* **21**, 2475 (1988). By permission of IOP Publishing Ltd.)

at $f = f_i$ given by

$$f_i = \frac{r_0 - r_i}{a} = -\frac{\ln(\mu/\nu)}{a(\mu - \nu)} \tag{4.84}$$

and displays a maximum at $f = f_m$ (cf. (4.76)):

$$f_m = \frac{r_0 - r_m}{a} = -\frac{2\ln(\mu/\nu)}{a(\mu - \nu)} = 2f_i . \tag{4.85}$$

The maximum value reads

$$\frac{W_m}{V_0} = -\frac{2a^2}{\pi^2 V_0}\left(\frac{d^2 V}{dr^2}\right)_{r=r_m} = \frac{2\mu^2 a^2}{\pi^2}\left(\frac{\nu}{\mu}\right)^{2\mu/(\mu-\nu)} \tag{4.86}$$

which corresponds to the maximum driving force (4.77) for distortion to occur. Clearly dimerization cannot take place when $W > W_m$.

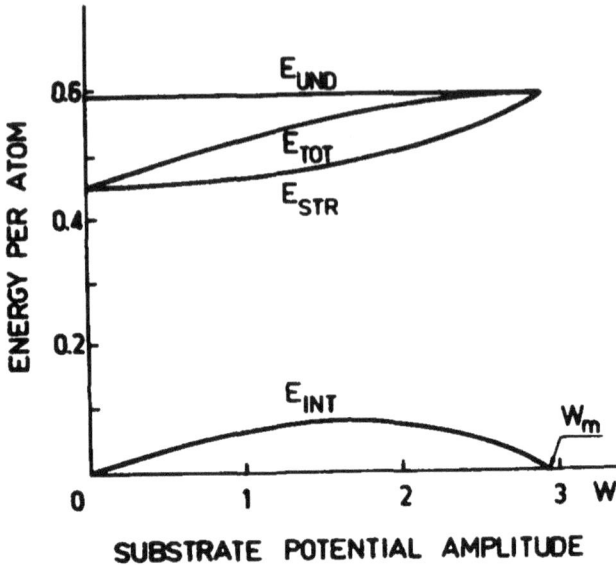

Fig. 4.27. Plot of the energy per atom of a distorted (dimerised) chain versus substrate modulation W (in units of V_0) at a constant misfit $f = f_m$. E_{str}, E_{int} and $E_{tot} = E_{str} + E_{int}$ denote the energies of the strain (the first sum in Eq. (4.78)) and the interfacial bonding (the second sum in Eq. (4.78)). E_{und} denotes the energy of the undistorted state. As seen E_{int} vanishes at the maximum substrate modulation W_m (Eq. (4.86)) (see Fig. 4.26). (I. Markov and A. Trayanov, *J. Phys. C: Solid State Phys.* **21**, 2475 (1988). By permission of IOP Publishing Ltd.)

The energy of the distorted state consists of two parts (Fig. 4.27): the strain energy E_{str} composed by the energies of the short and long bonds, and the energy of interaction with the substrate, E_{int}, due to the displacements of the atoms from the bottoms of the potential troughs. As can be seen the total energy of the distorted state, $E_{tot} = E_{str} + E_{int}$, is smaller than the energy of the undistorted state E_{und} and merges with it when the maximum value W_m is reached. Recall that at $W > W_m$ the distortion disappears.

We consider further the formation of trimers or alternation of two equally short and one long bonds (Fig. 4.25(c)). Within the trimer $\xi_{n+1} = -\xi_{n-1}$ and $\xi_n = 0$. The curve that outlines the area of existence of trimers lies under that of dimers (Fig. 4.26, curve B), its maximum value at $f = f_m$ being given by

$$\frac{W_{m3}}{V_0} = \frac{3\mu^2 a^2}{2\pi^2} \left(\frac{\nu}{\mu}\right)^{2\mu/(\mu-\nu)} . \tag{4.87}$$

Curve C in Fig. 4.26 separates the regions of stability of dimers and trimers. Below it dimers still exist but not as a ground state.

In the same way we consider tetramers, pentamers, etc. (Figs. 4.25(d) and 4.25(e)). By repeating the same procedure we find that the regions of existence of polymers with degrees of polymerization higher than 2 are included in that of dimers. It follows that curve A separates the regions of stability of distorted and undistorted states. It can be concluded that at $W > W_m$ no distortion of infinitely long chains takes place, irrespective of the value of the natural misfit. Besides, the higher the degree of polymerization the smaller the value of W at which the corresponding polymers are energetically favored. In the limit $W \to 0$, the degree of the energetically favored polymers goes to infinity, which in practice means disappearance of distortion (this is equivalent to alternation of an infinite number of short bonds and one long bond which in fact means undistorted structure).

Let us consider now a chain of finite length. It will be distorted if appropriate values of W/V_0 and f are selected. If this is not the case, the middle part of the chain will not be distorted but it turns out that the end parts of the chain will always be distorted as long as $|f| > |f_i|$, irrespective of the value of W/V_0. The latter is evidently due to the asymmetry of the atomic interactions near the free ends. As will be shown below, this edge effect leads to significant results concerning the metastability limit of the pseudomorphous state and the activation energy for introduction of dislocations at the free ends.

A two-dimensional distortion of the chemical bonds (clustering of 4 and 8 atoms) has been theoretically predicted with the help of the embedded atom method in a Ni monolayer grown on Ag(001) by Bolding and Carter [1992]. The absolute value of the negative misfit is very large, $f = -13.9\%$. The growth of the second monolayer causes a relaxation of the distortion of the first monolayer bonds, i.e. atoms of the first monolayer tend to occupy the bottoms of the potential troughs of the silver substrate. The bonds in the second monolayer become distorted but their distortion is weaker. After deposition of four monolayers the distortion of the bonds between the atoms of the first monolayer practically vanishes. Thus, the Ni atoms closest to the substrate are under the largest uniform expansive strain and the strain diminishes away from the contact plane.

Epitaxial Growth

D. *Width of misfit dislocations*

As shown above, the width of a single misfit dislocation in the harmonic model, l_0, is determined solely by the energetic parameters of the system and does not depend on the natural misfit. It is quite clear that the anharmonicity of the real potential will strongly affect the dislocation width. This follows from the fact that the dislocations have different configuration at different signs of the natural misfit (Fig. 4.11). In the harmonic limit both the repulsive and attractive branches of the interatomic potential are equally steep and the dislocation width is one and the same.

The dislocation width can be expressed as a function of the core bond strain ε_c:

$$l = \frac{1}{(d\xi/dn)_{n=0}} = \frac{a}{|\varepsilon_c + af|} \ . \tag{4.88}$$

Note that ε_c has the same sign as f. In the harmonic case, $\varepsilon_c^+ = \varepsilon_c^- = a(1/l_0 - f)$ (Eq. (4.21)) and $l = l_0$. This is not, however, the case when the interatomic potential is asymmetric. Different branches of the potential determine the values of ε_c and l. The situation becomes even more complicated when the interatomic potential is not only asymmetric, as is the Toda potential, but is nonconvex like the real potential.

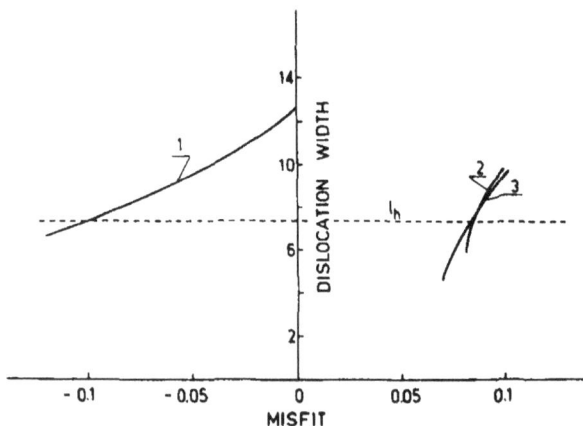

Fig. 4.28. Variation of the width of the misfit dislocations with natural misfit in a real chain: curve 1 — negative misfit; curve 2 — positive misfit; curve 3 — positive misfit with Toda potential ($\alpha = 2$, $\beta = 6$). The dashed line denoted by l_h gives the harmonic reference with $\gamma = 12$ ($N = 80, \mu = 4, \nu = 3, W/V_0 = 1$). (I. Markov and A. Trayanov, *J. Phys. C: Solid State Phys.* **21**, 2475 (1988). By permission of IOP Publishing Ltd.)

Figure 4.28 shows the dependence of the dislocation width on the misfit for long enough chains to rule out the effects of the free ends. As seen (curve 2), l increases with positive misfit in a limited interval of the latter. The right-hand limit is in fact the limit of metastability f_{ms}^+ of the configuration with particular density of dislocations as in the harmonic case. The core strain $\varepsilon_c = 0$ and $l_{max} = 1/f_{ms}^+$. The left-hand limit f_r is determined by the condition $\varepsilon_c = -f_i$ which means a rupture of the core bond. Then $l_{min} = 1/(-f_i + f_r)$. For $f < f_r$, the condition of existence (4.82) is no longer fulfilled. Thus, the interval of existence of solutions with a given dislocation density in real compressed chains is not determined by the conditions for generation and escape of dislocations at the free ends, but by those for generation and destruction of dislocations. The latter takes place by breaking up of the bonds in the cores of the dislocations. Curve 3 gives the dislocation width as obtained with the help of the Toda potential ($\alpha = 2$, $\beta = 6$, Fig. 4.6). Obviously, the tendency is the same, as the potential of Toda also displays a finite force at large r (Fig. 4.8(a)).

In the case of negative misfit (curve 1), however, nothing dramatic happens. The dislocation width again increases with the latter. The strain in the core of the dislocation is compressive and slowly increases in absolute value with increasing misfit. The condition (4.82) is always fulfilled.

E. Energy

The behavior of the energy versus misfit curve of a real chain differs qualitatively from that in the harmonic case (see Fig. 4.16), particularly at positive misfits. At small values of $W(W/V_0 < 0.5)$ the positive misfit dependence of the energy is similar to the harmonic one. At larger values of W, however, the $E(f)$ dependence consists of curvilinear segments which do not intersect each other (Fig. 4.29(a)) due to the rupture of the core bonds. This tendency becomes stronger with increasing W (Fig. 4.29(b)) and in the case of shorter chains the effect of core bond rupture is so strong that the segments do not overlap and gaps appear between them in which no solutions of the system (4.79) exist. This is seen more clearly in Fig. 4.30, where the dependence of the energy on chain length is shown at a constant value of the positive misfit. Gaps without any solution exist for short chains and disappear for longer chains. The energy shows a sawtooth behavior and the introduction of each new dislocation is connected with an abrupt energy drop, which is uncharacteristic for the harmonic model.

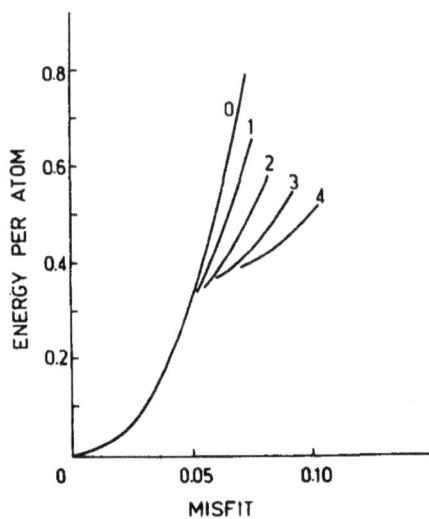

(a)

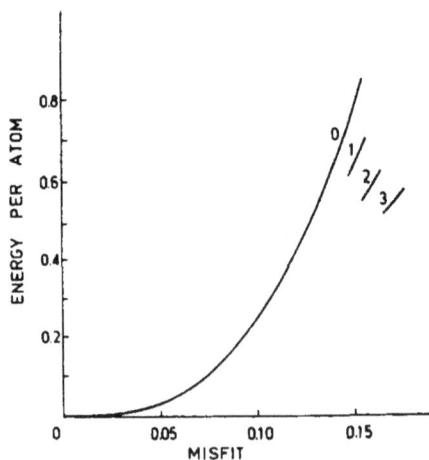

(b)

Fig. 4.29. Variation with positive misfit of the energy per atom in units of W :
(a) $W/V_0 = 0.5$; (b) $W/V_0 = 4$. The figure at each curvilinear segment denotes the
number of dislocations ($N = 80$, $\mu = 4$, $\nu = 3$). (I. Markov and A. Trayanov, *J. Phys.
C: Solid State Phys.* **21**, 2475 (1988). By permission of IOP Publishing Ltd.)

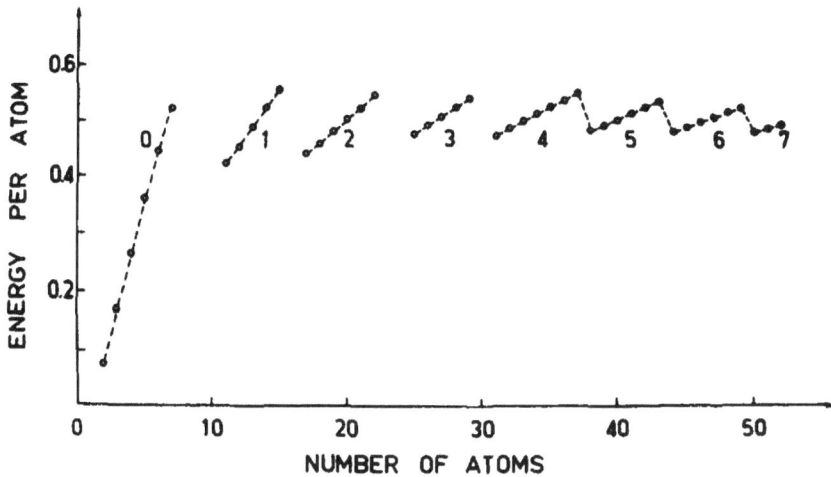

Fig. 4.30. Dependence of the energy per atom in units of W on the chain length. The number of dislocations in the ground state is denoted by the figure on each segment. The gaps without solutions at small chain lengths due to breaking of the dislocation core bonds are clearly demonstrated ($W/V_0 = 4$, $f = 0.2$, $\mu = 4$, $\nu = 3$). (I. Markov and A. Trayanov, J. Phys. C: Solid State Phys. **21**, 2475 (1988). By permission of IOP Publishing Ltd.)

The above result leads to a definite conclusion concerning the process of layer-by-layer growth of epitaxial overlays which are compressed and strongly bound to the substrate. Small monolayer islands are coherent with the substrate. After incorporating some more adatoms a dislocation is introduced at the free boundary but its core bond is stretched out more than the theoretical tensile strength of the material. The overlayer island thus breaks up into two smaller islands. This process continues until the density of such small coherent islands becomes large enough. Then they begin to coalesce with each other to produce bigger islands. The gaps shown in Fig. 4.30 disappear and the overgrowth islands can grow further by the incorporation of single adatoms. This process takes place if the misfit is larger than the stability limit f_s^+. If this is not the case, the overlayer islands grow by the incorporation of single adatoms and are pseudomorphic with the substrate until complete coverage of the latter. As the coalescence begins at a later stage of growth the monolayer film will consist of a large number of small monolayer islands. The adatom concentration on top of the small islands is insufficient to give rise to nucleation of the upper monolayer [Markov and Stoyanov 1987] and hence the formation of the latter will be delayed. Thus layer-by-layer growth and in turn the much slower damping of the RHEED intensity oscillations will be favored at positive misfit and strong bonding across the interface. Recently Becker et al. [1993] reported fragmentation upon high temperature annealing of 2D Ag islands deposited on Pt(111) in the submonolayer region. When deposited at temperatures below 500 K the silver formed large 2D islands pseudomorphous with the substrate. After annealing at higher temperature the silver islands broke down into islands consisting most probably of 7 or 12 atoms. These smaller islands were found to be nearly relaxed. When the deposition was carried out at temperatures higher than 500 K the silver film grew as small islands from the very beginning. The Ag is strongly bound to the Pt and the lattice misfit is positive and large enough (4.3%).

The form of the $E(f)$ dependence shown in Fig. 4.29, which is due to the rupture of the dislocation core bonds, leads to a new definition of the limit of stability f_s^+ of the pseudomorphous state. It is now determined by the condition of existence of a dislocation with core bond stress smaller than the theoretical tensile strength and coincides with the critical misfit f_r for rupture of the most expanded core bonds.

The case of negative misfit is quite different. The core bond strain ε_c is compressive and the force exerted by the substrate is not destructive (Fig. 4.11). It follows that expanded epilayers cannot break up in the cores of the dislocations at negative misfits. On the other hand, at misfits larger in absolute value than f_i the chains distort in between the dislocations and a rupture there is again excluded.

Figure 4.31 shows the energies of chains without and containing one dislocation. The chains are distorted at $|f| > |f_i|$. The curves intersect with each other at the limit of stability f_s^-. As seen the energies are very close particularly at misfits greater in absolute value than f_i. Obviously the contribution of the dislocation energy is small compared with the contribution of the chain distortion. At positive misfit the energy gain due to the introduction of dislocations is much greater (Fig. 4.29).

There are, however, two peculiarities at negative misfits which are uncharacteristic for the harmonic case. First, at strong enough bonding across the interface ($W/V_0 \geq W_s/V_0 = 2$, see the × sign in Fig. 4.36) the energies of the chains without and containing one dislocation do not intersect. The energy of the commensurate state goes asymptotically to the energy of the incommensurate state, being always lower than the latter. It follows that the limit of stability disappears at strong bonding and epilayers which are thin enough to fulfill the requirements of the model will be pseudomorphous with the substrate irrespective of the natural misfit.

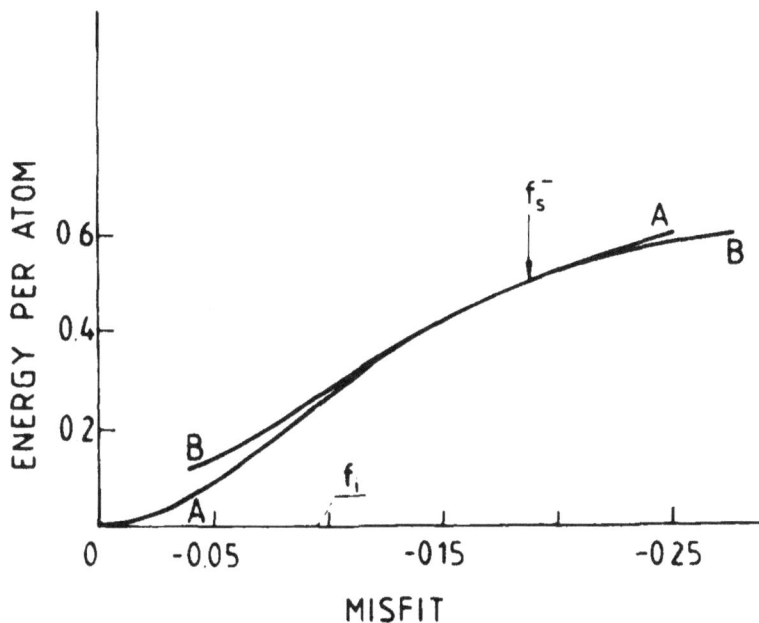

Fig. 4.31. Variation with negative misfit of the energy per atom: curve A — a chain without a dislocation; curve B — a chain containing one dislocation ($W/V_0 = 1, N = 100, \mu = 4, \nu = 3$). Note that there are no singular points at the inflection misfit f_i. (I. Markov and A. Trayanov, J. Phys. C: Solid State Phys. 21, 2475 (1988). By permission of IOP Publishing Ltd.)

The second consequence of the use of real potential in the 1D model of Frank and van der Merwe, which is absent in the original model, is that solutions of coherent configurations, although not in a ground state, exist at any value of the negative misfit. The condition is that the potential troughs be sufficiently deep ($W/V_0 \geq W_{ms}/V_0 = 0.25$). It follows that the metastability limit f_{ms}^- of the pseudomorphous state also disappears (see the × sign in Fig. 4.37). This is due to the polymerization of the free ends of the chains even under conditions (W, f) where the ground state is the undistorted state. Due to the chain end distortion the end atoms do not climb the slopes of the potential troughs with increasing misfit as shown in Figs. 4.15(c) and (e), which excludes the possibility of spontaneous introduction of dislocations at the free ends.

Another consequence of the distortion of the chain ends at $|f| > |f_i|$ is that the activation energy for introduction of a dislocation at the free ends is greater than that in compressed chains. Owing to the chain distortion the

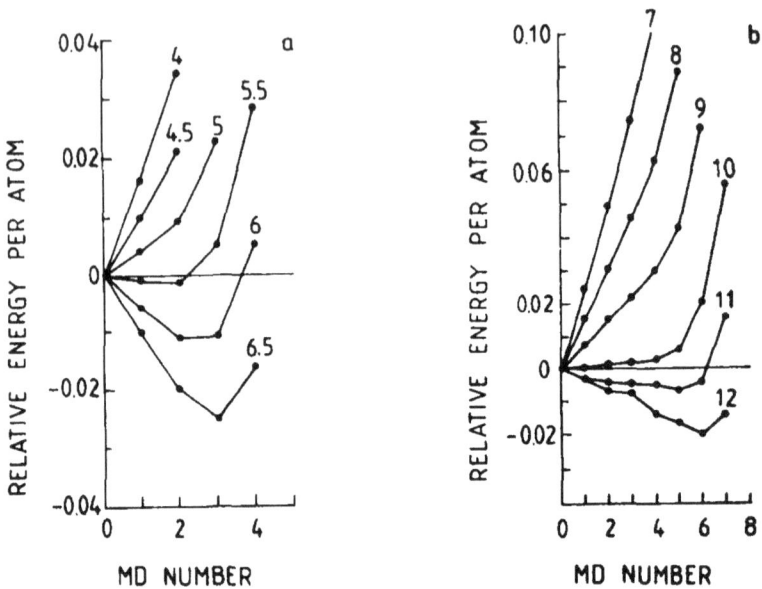

Fig. 4.32. Plot of the potential energy per atom relative to the energy of the commensurate state $\mathcal{E}(0)$ against the number of misfit dislocations in the chain for different values
of the negative misfit (in percent) given by the figures at the curves. (a) $W/V_0 = 0.25$;
(b) $W/V_0 = 0.62$. (I. Markov and A. Trayanov, *J. Phys.: Condens. Matter* **2**, 6965
(1990). By permission of IOP Publishing Ltd.)

end atoms of the chain are always near the bottoms of the potential troughs.
Hence, the introduction of a new dislocation requires overcoming a much
higher energy barrier than in the case of a positive misfit, particularly at
stronger bonding across the interface. It follows that expanded overlayers
can exist in metastable state without dislocations at higher temperatures
compared with compressed epitaxial films.

Figure 4.32 is a plot of the potential energy per atom of a real chain with
respect to the commensurate state $\mathcal{E} - \mathcal{E}(0)$, as a function of the dislocation
density f_d for different values of the negative misfit and at two different
values of the relative substrate modulation W/V_0. As in the harmonic case
one can see that the energy of the commensurate state is an additive term
to the energy of the incommensurate state. For $W/V_0 = 0.25$ (Fig. 4.32(a))
no linear dependence of energy versus number of dislocations is observed.
The reason for this is that the atoms experience the convex part of the
potential as in the harmonic case and hence the interaction between the

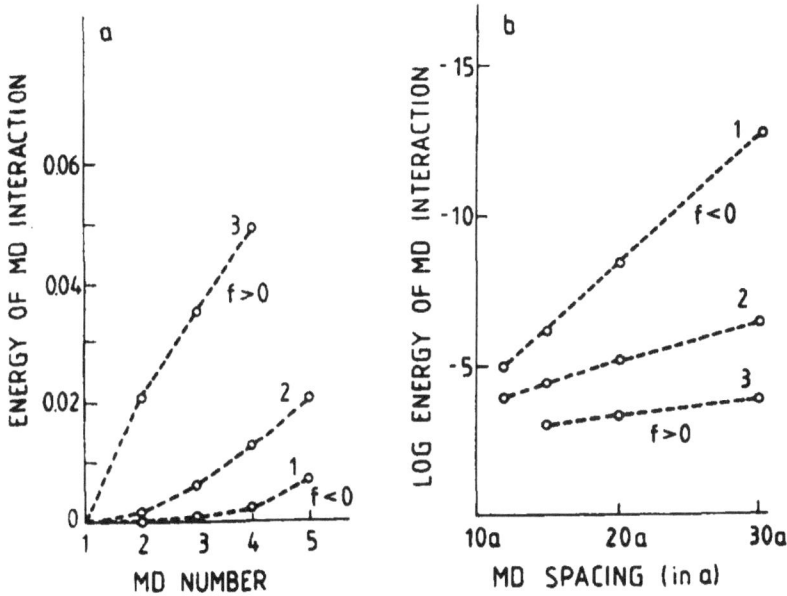

Fig. 4.33. Behavior of the pair energy of interaction of misfit dislocations: (a) dependence of the energy on the number of the dislocations for positive and negative values of the natural misfit; (b) logarithmic plot of the energy of interaction against the dislocation spacing (reciprocal of the MD density). Curves 2 in both figures represent the harmonic limit ($N = 60$, $f = \pm 7\%$, $W/V_0 = 0.5$, $\mu = 4$, $\nu = 3$). (I. Markov and A. Trayanov, *J. Phys.: Condens. Matter* **2**, 6965 (1990). By permission of IOP Publishing Ltd.)

dislocations contributes significantly to the total energy. The same $\mathcal{E}(f_d)$ dependence is observed for positive misfits irrespective of the value of the relative interfacial bonding W/V_0. However, this is not the case for a stronger interfacial bonding, $W/V_0 = 0.62$ (Fig. 4.32(b)). The energy is a linear function of the dislocation density f_d up to very high values of the latter. The atoms experience the nonconvex part of the interatomic potential, the bonds between the atoms are distorted and the interaction between the dislocations is suppressed.

This is clearly seen in Fig. 4.33 where the split of the dislocation interaction energy with respect to the misfit sign is shown. The latter is considerably larger for a positive misfit than for a negative one (Fig. 4.33(a)). The data from Fig. 4.33(a) are plotted in semilogarithmic scale in Fig. 4.33(b). As seen, for a positive misfit both in the real case

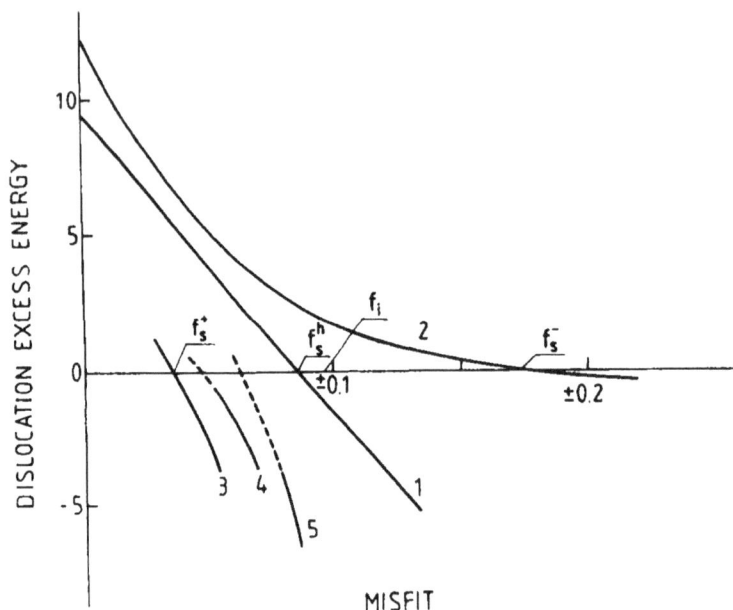

Fig. 4.34. Misfit dependence of the slopes $d\mathcal{E}/df_{\mathrm{d}}$ of the linear parts of the $\mathcal{E}(f_{\mathrm{d}})$ curves shown in Fig. 4.33. The straight line 1 illustrates the harmonic limit of Frank and van der Merwe. Curve 2 shows the negative misfit behavior with $W/V_0 = 1$. Note that there is no singular point at the inflection misfit f_{i}. Curves 3, 4 and 5 represent the positive misfit cases with $W/V_0 = 0.25$, 0.5 and 1.0, respectively ($N = 60$, $\mu = 4$, $\nu = 3$). (I. Markov and A. Trayanov, *J. Phys.: Condens. Matter* **2**, 6965 (1990). By permission of IOP Publishing Ltd.)

and in the harmonic limit, the plot deviates slightly from linearity at high dislocation densities (small dislocation spacing). In the case of negative misfit, the interaction energy depends exponentially on the dislocation density even at high values of the latter.

In the harmonic limit the derivative of the energy with respect to the mean dislocation density, $d\mathcal{E}/df_{\mathrm{d}}$, is a linear function of the misfit (see Eq. (4.30)). It has a slope equal to $2Wl_0^2$ and its intercepts with the abscissa and ordinate are equal to the stability limit f_s and the energy of a single dislocation, $4Wl_0/\pi$, respectively. The slopes of the linear parts of the curves in Fig. 4.32, i.e. the slopes $d\mathcal{E}/df_{\mathrm{d}}$, are presented in Fig. 4.34 as a function of the natural misfit. The straight line 1 shows the behavior of the harmonic approximation with $\gamma = \mu\nu V_0 = 12$ discussed above. Curve 2 represents the negative misfit dependence of $d\mathcal{E}/df_{\mathrm{d}}$ of a real chain with

$\mu = 4$, $\nu = 3$, $V_0 = 1$ and $W = 1$. Curves 3, 4 and 5 give the positive misfit dependencies of real chains with $W = 0.25$, 0.5 and 1.0, respectively. All curves except for the harmonic one show strong nonlinearity and their curvatures have opposite signs. Curves 4 and 5 do not intersect the abscissa due to the rupture of the most expanded bonds in the cores of the misfit dislocations.

As shown by Markov and Trayanov [1990] the data from the numerical solutions fit with the semiempirical expression

$$\mathcal{E} = \mathcal{E}_1^0 f_{\mathrm{d}} e^{-f/f_i} \left(1 - \frac{f}{f_{\mathrm{s}}} \right) + \mathcal{E}(0) , \qquad (4.89)$$

which is analogous to (4.33). It turns out that this expression describes all the data surprisingly well. Moreover, the harmonic limit for which $f_i = \infty$ is also formally included. Then by analogy with (4.33) we can write the following expression for the energy of a single misfit dislocation:

$$\mathcal{E}_1 = \mathcal{E}_1^0 e^{-f/f_i} , \qquad (4.90)$$

where $\mathcal{E}_1^0$ is the energy of a single dislocation at $f = 0$. Since $f_i < 0$, it follows that $\mathcal{E}_1$ is a decreasing function of the negative misfit and an increasing function of the positive misfit.

The zero energy of a single dislocation, $\mathcal{E}_1^0$, is shown in Fig. 4.35 as a function of $(\mu\nu V_0 W)^{1/2}$ ($V_0 = $ const). The straight line 1 represents the harmonic reference. Curve 3 gives the energy of a negative dislocation (two atoms in a trough, Fig. 4.11(b)) whereas curve 2 gives the energy of a positive dislocation (an empty trough, Fig. 4.11(a)). The negative dislocation energy is computed directly as the energy of the incommensurate state of a long enough chain containing only one dislocation at $f = 0$ ($\mathcal{E}(0) = 0$), while the positive dislocation energy is calculated through Eq. (4.90). The difference of the energies clearly reflects the anharmonicity of the real potential. In expanded chains (negative misfit) the atoms in the dislocation interact through the steeper repulsive branch of the potential and the zero energy $\mathcal{E}_1^0$ is greater than that in compressed chains where the weaker attractive branch operates.

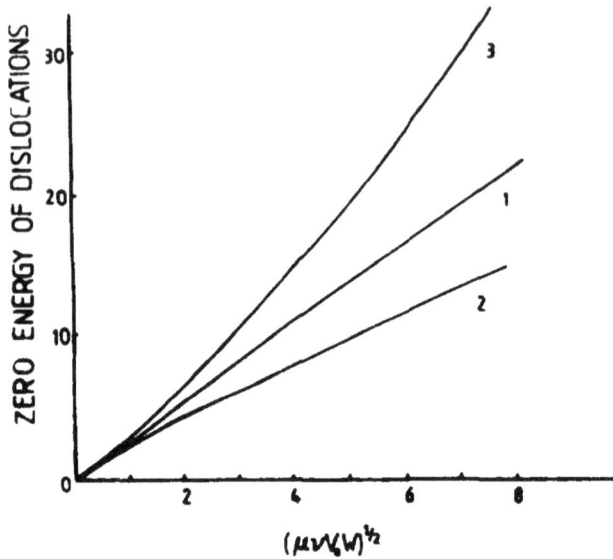

Fig. 4.35. Dependence of the energy at zero misfit of the static solitons on $(\mu\nu V_0 W)^{1/2}$. The straight line 1 presents the harmonic limit of Frank and van der Merwe. Curve 2 gives the energy of a positive misfit dislocation (an empty trough or a light wall, Fig. 4.11(a)). Curve 3 shows the energy of a negative dislocation (two atoms in a trough or a heavy wall, Fig. 4.11(b)). (I. Markov and A. Trayanov, *J. Phys.: Condens. Matter* 2, 6965 (1990). By permission of IOP Publishing Ltd.)

F. *Limits of stability*

As shown above the limits of stability and metastability split with respect to the sign of the misfit when anharmonicity is "switched on." Figure 4.36 illustrates the split of the stability limit f_s as a function of the substrate modulation W. The harmonic case is given by the straight line denoted by f_s^h. The corresponding curves for Toda chain ($\alpha = 2$, $\beta = 6$) are also given for comparison. As can be seen, the positive stability limit f_s^+ lies nearer to the reference harmonic curve than the anharmonic Toda curve. This is not surprising, bearing in mind that in the real model at positive misfit the stability limit of the pseudomorphous state is determined not by the equality of the energies of states with zero and one dislocations, but by the limit of rupture of the core bonds, f_r. The latter is shifted to greater values of the misfit. This is the reason why f_s^+ lies nearer the harmonic reference than the Toda curve. On the other hand, the negative stability limit f_s^- is shifted to greater absolute values than in the Toda case. This is

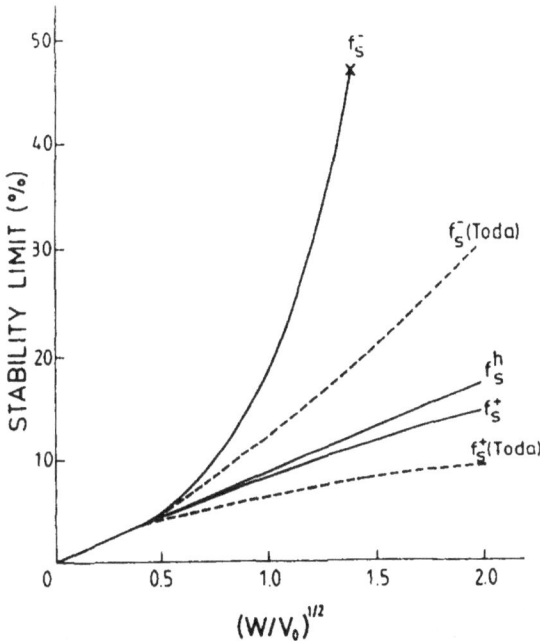

Fig. 4.36. Variation of the limits of stability, f_s^+ and f_s^-, with the square root of the relative substrate modulation $(W/V_0)^{1/2}$ ($\mu = 4$, $\nu = 3$). The straight line denoted by f_s^h denotes the harmonic limit. The limits of stability of the anharmonic Toda model ($\alpha = 2, \beta = 6$) are also included as dashed lines for comparison. The negative stability limit f_s^- terminates at $W/V_0 = 2$ (see the × sign). (I. Markov and A. Trayanov, *J. Phys. C: Solid State Phys.* **21**, 2475 (1988). By permission of IOP Publishing Ltd.)

easily understandable considering the shape of the corresponding attractive branches of the two potentials (Fig. 4.36). What is more important is that f_s^- disappears after some critical substrate–deposit bond strength $W = 2V_0$ (note the × sign at the corresponding curve). Beyond this value the pseudomorphous state is always the ground state.

Contrariwise, the positive metastability limits f_{ms}^+ for the real and Toda potentials (Fig. 4.37) overlap, which reflects the coincidence of the respective repulsive branches. However, f_{ms}^- disappears beyond some critical value of the potential amplitude $W_{ms}/V_0 = 0.25$ (note the × sign) governed by the condition $f_{ms}(W_{ms}) = f_i$. As mentioned above the chain ends are distorted and spontaneous generation of dislocations at the free ends never takes place whereas Toda limit still exists.

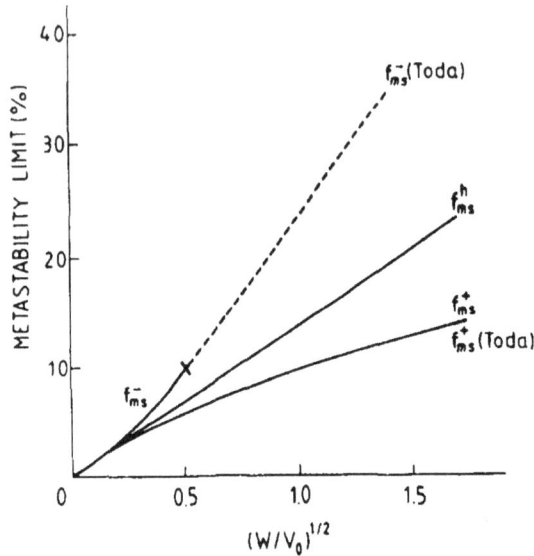

Fig. 4.37. Variation of the limits of metastability, f_{ms}^+ and f_{ms}^-, with the square root of the relative substrate modulation $(W/V_0)^{1/2}$ ($\mu = 4, \nu = 3$). The straight line denoted by f_{ms}^h gives the harmonic limit. The limits of metastability of the anharmonic Toda model ($\alpha = 2, \beta = 6$) are also included as dashed lines for comparison. The curves in the real and the Toda models coincide for positive misfit due to the coincidence of the repulsive branches of both potentials. The negative metastability limit f_{ms}^- terminates at $W/V_0 = 0.25$ (see the × sign) whereas the Toda limit still exists (the dashed line). (I. Markov and A. Trayanov, *J. Phys. C: Solid State Phys.* 21, 2475 (1988). By permission of IOP Publishing Ltd.)

G. *Mean dislocation density*

The mean dislocation density f_d in the ground state of a real chain is given in Fig. 4.38 as a function of the natural misfit f [Markov and Trayanov 1990]. The curves for positive and negative values of f are presented in one and the same quadrant for easy comparison. The smooth curve represents the continuum limit of the harmonic model. The stepwise behavior is due to the finite size of the chain ($N = 60$). The splitting of the two curves around the harmonic reference is due to the anharmonicity of the real potential. The positive misfit curve is considerably nearer to the harmonic limit than the respective curve for Toda chain. This is due to the limited interval of existence of the dislocated state as a result of the rupture of the most expanded bonds in the cores of the dislocations in compressed chains. What is more interesting is that the commensurate—

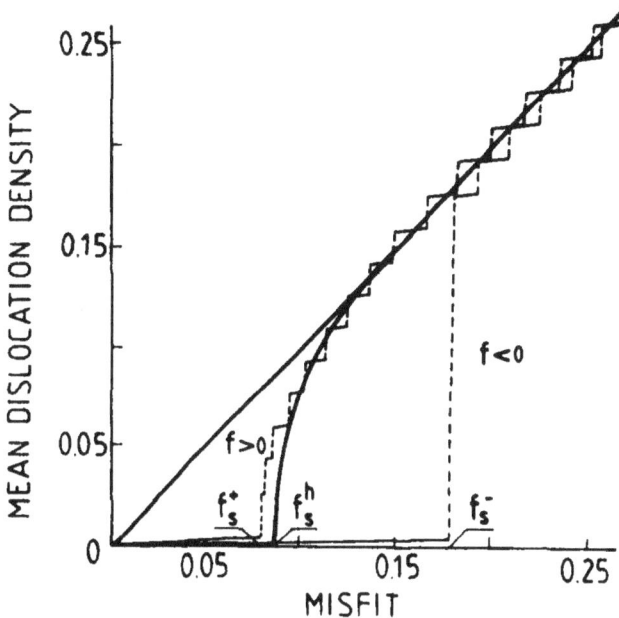

Fig. 4.38. Plot of the mean dislocation density in a real chain against the natural misfit for both positive (left) and negative (right) misfits. The harmonic continuous approximation of Frank and van der Merwe [1949b] is presented by the smooth curve for comparison. The two curves are plotted in the same quadrant for easy comparison ($\mu = 4, \nu = 3, W/V_0 = 1, N = 60$). (I. Markov and A. Trayanov, *J. Phys.: Condens. Matter* **2**, 6965 (1990). By permission of IOP Publishing Ltd.)

incommensurate (CI) transition is continuous in compressed chains but abrupt in expanded ones going by a single jump from zero to the maximum density of the dislocations.

As is well known from the harmonic model the continuous behavior of the CI transition is due to the energy of dislocation interaction. As shown above, in anharmonic chains the energy of dislocation interaction depends on the sign of the misfit — it is much smaller in expanded chains, rather than in compressed ones. This is obviously due to the fact that dislocation interaction is realized through the weaker attractive branch of the interatomic potential in expanded chains and through the steeper repulsive branch in compressed ones. This explains the abrupt behavior of the CI transition in expanded chains.

H. Effect of anharmonicity and nonconvexity in epitaxial growth

It is thus evident that in addition to its anharmonicity the fundamental characteristic which distinguishes the nonconvex interatomic potential from the harmonic approximation is not the finite energy of dissociation of two neighboring atoms but its inflection. It leads to two effects. First, there is the rupture of the bonds in the dislocation cores in compressed overlayers, and, second, there is the distortion of the chemical bonds in expanded films. In summary, the replacement of the harmonic interactions by more realistic interatomic forces in the Frank–van der Merwe model results in the following more important conclusions concerning the growth of thin epitaxial films:

(i) Compressed epilayers can crack along the dislocation lines.

(ii) The limits of stability and metastability of the pseudomorphous state are much greater in absolute value in expanded rather than in compressed epilayers.

(iii) Thin expanded pseudomorphous films should be stable beyond some critical interfacial bonding W_s, irrespective of the absolute value of the natural misfit, and should always exist in metastable state beyond some critical interfacial bond strength $W_{ms} \ll W_s$.

(iv) The activation barrier for introduction of dislocations at the free ends is higher in expanded rather than in compressed films, and therefore the expanded films can withstand higher temperatures in pseudomorphous state than compressed films.

(v) The equilibrium critical thickness for pseudomorphous growth should be much greater for expanded rather than compressed films (see Eq. (4.72)).

(vi) The mean dislocation density should be smaller in expanded rather than in compressed epilayers for one and the same film thickness.

(vii) The natural misfit in expanded epilayers is entirely accommodated either by homogeneous strain or by misfit dislocations without intermediate state.

One of the most important consequences of the nonconvexity of the real interactions from a technological viewpoint concerns crack formation. Π-shaped cracks have been observed in compressed Ge films deposited on Si [Tkhorik and Khazan 1983]. Cracks have also been found in compressed garnet films grown on garnet substrates [Miller and Caruso 1974]. In this case the cracks were observed in slightly, rather than strongly, compressed samples, in agreement with the predictions of the model. Olsen, Abrahams and Zamerowski [1974] observed unidirectional cracks in both expanded and compressed epilayers of $In_x Ga_{1-x} P$ deposited on (100) GaAs. They found that stretched layers cracked at smaller misfits than compressed layers, in contradiction with the prediction of the above model. Cracks in expanded $In_xGa_{1-x}As$ and $In_xAl_{1-x}As$ layers grown by MBE on (100)InP and in expanded $In_xGa_{1-x}As_yP_{1-y}$ layers grown by LPE on (100)InP were observed and studied by Franzosi *et al.* [1988]. These authors found that the cracks propagate deeply into the InP substrate. The same phenomenon has been established also for misfit dislocations in MBE-grown $In_xGa_{1-x}As/(100)InP$ single heterostructures irrespective of the sign of the natural misfit [Franzosi *et al.* 1985]. The misfit dislocations are "squeezed" into the substrate due to the stress in the overgrowth. This clearly shows the connection between the misfit dislocations and the formation of cracks. Obviously, the cracking phenomenon should be studied in more detail both experimentally and theoretically. Experimental observations concerning cracking of epitaxial films of different materials have been summarized by Tkhorik and Khazan [1983].

4.2.5. van der Merwe model of thick overgrowth

The case of thick overgrowth is considered in a more or less similar way. By "thick" we mean mathematically infinite. In this case (Fig. 4.39) the two crystal halves A and B with quadratic symmetry of the contact planes and atomic spacings a and b, respectively, are considered as elastic continua with shear moduli G_a and G_b and Poisson's ratios ν_a and ν_b, respectively [van der Merwe 1963a]. An important feature of the system consisting of two semi-infinite crystals is that the homogeneous strain is equal to zero and the natural misfit is accommodated entirely by misfit dislocations. As shown in Sec. 4.2.4.4 the energy of a cross grid of misfit dislocations of an edge type to a first approximation represents a sum of two arrays of dislocations parallel to the two orthogonal directions. In other words, the energies of the two arrays are additive and we can consider them independently assuming a misfit in one direction only.

Here we allow both crystals to be elastically strained. In order to describe the displacements at either side of the interface, we introduce a reference lattice C with a parameter c [van der Merwe 1950, 1973] along the x axis:

$$Pb = (P+1)a = \left(P + \frac{1}{2}\right)c \ ,$$

where $P = a/(b-a)$ is an integer.

Assuming for definiteness that $b > a$ we can imagine that the two lattices A and B are generated from C by contraction of A and expansion of B. Then the reference lattice spacing reads

$$\frac{2}{c} = \frac{1}{a} + \frac{1}{b}$$

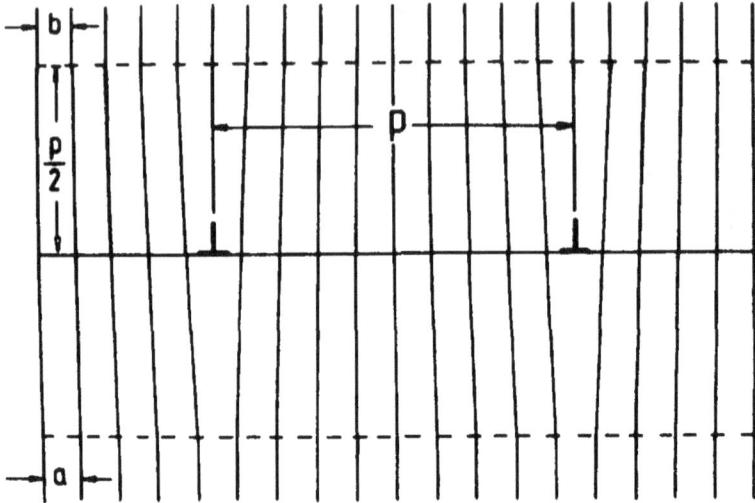

Fig. 4.39. Model of epitaxial interface between two semi-infinite crystals resolved in a sequence of misfit dislocations spaced at an average distance p. The dashed lines located at a distance $p/2$ from the contact plane show the boundary beyond which the periodic strains originating from the dislocations practically vanish (after van der Merwe [1950]).

or

$$c = \frac{ab}{\frac{1}{2}(a+b)} \ . \tag{4.91}$$

The vernier of misfit or the dislocation spacing p is given by

$$p = \left(P + \frac{1}{2}\right) c = \frac{ab}{(b-a)} \ . \tag{4.92}$$

Bearing in mind that both lattices A and B are strained it is reasonable to define the misfit as

$$f = \frac{c}{p} = \frac{b-a}{\frac{1}{2}(a+b)} \tag{4.93}$$

rather than as $f = 1/P = (b-a)/a$ as in the rigid substrate model.

Assume now that the atoms of A and B are located at points of the reference lattice and there is an atom of B exactly opposite to any atom of A. Then we allow the atoms of A and B to occupy their natural positions in their respective lattices. The atoms of A and B will be displaced with respect to the respective positions of the reference lattice C. The displacement between the atoms with respect to each other will be

$$U = \frac{c}{2} + \frac{c}{p}x \, ,$$

where the first term $c/2$ locates the origin of the x axis at a dislocation line while the second term cx/p accounts for the linear increase of the atom spacing due to the vernier of misfit. If we now allow elastic displacements $u_a(x)$ and $u_b(x)$ of the corresponding atoms of the A and B lattices, then the displacement of a B atom with respect to the corresponding A atom will be

$$U = \frac{1}{2}c + \frac{c}{p}x + u_b(x) - u_a(x) \, . \tag{4.94}$$

As in the previous case (Eq. (4.8)), each half-crystal exerts a periodic potential on the atoms of the other half in the form

$$V = \frac{G_i c^2}{4\pi^2 d} \left[1 - \cos\left(2\pi \frac{U}{c} \right) \right] \, , \tag{4.95}$$

where G_i is the shear modulus at the interface and $d \cong c$ is the separation of the atoms of the adjoining crystal planes.

The mathematical treatment of the problem, although involving much greater difficulties, leads to expressions for U similar to (4.20) and (4.23). Thus for dislocations far apart ($b \to a$, $p \to \infty$) and, assuming for simplicity, $\nu_a = \nu_b = \nu$ and $G_a = G_b = G_i = G$, the solution reads

$$\frac{U}{c} = \frac{1}{2} + \frac{1}{\pi} \arctan\left(\frac{x}{x_0} \right) \, , \tag{4.96}$$

where $x_0 = c/2(1 - \nu)$.

The general solution reads [van der Merwe 1975]

$$\frac{U}{c} = \frac{1}{2} + \frac{1}{\pi} \arctan\left[\left(\sqrt{1 + \lambda^{-2}} + \lambda^{-1} \right) \tan\left(\pi\frac{x}{p} \right) \right] \, , \tag{4.97}$$

where

$$\lambda = 2\pi \frac{G'}{G_i} \frac{c}{p} \tag{4.98}$$

and

$$\frac{1}{G'} = \frac{1 - \nu_a}{G_a} + \frac{1 - \nu_b}{G_b} \, . \tag{4.99}$$

It is immediately seen that (4.96) can be easily obtained in the limit $p \to \infty$ with $G_a = G_b = G_i$. Equations (4.96) and (4.97) show a behavior similar to that plotted in Fig. 4.12, curves 1 and 2, respectively. In other words, as in the case of a monolayer overgrowth the interface is resolved in a sequence of misfit dislocations.

The energy of the misfit dislocations (the homogeneous strain is absent) is naturally divided into two parts. The first is the energy of interaction between the atoms of the two crystal halves:

$$E_i = \frac{1}{p} \int\limits_{-p/2}^{p/2} \frac{G_i c^2}{4\pi^2 d} \left[1 - \cos\left(2\pi \frac{U}{c} \right) \right] dx$$

$$= \frac{G_i c^2}{4\pi^2 d} \left(1 + \lambda - \sqrt{1 + \lambda^2} \right) , \qquad (4.100)$$

which is obtained by substitution of (4.97) into the integral of (4.100) and carrying out the integration.

The second is the energy of the periodic elastic strain which is distributed in the two crystal halves A and B. For the average strain energy per atom stored in that part of the crystal B which extends from the interface to a distance h from the latter, one obtains

$$E_e^b(h) = -\frac{(1 - \nu_b)c^2 G' G_i}{4\pi^2 G_b d}$$

$$\times \lambda \left[\ln \left(\frac{1 - A^2}{1 - A^2 e^{-2H}} \right) + \frac{H A^2 e^{-2H}(H - 1 + A^2 e^{-2H})}{(1 - \nu_b)(1 - A^2 e^{-2H})^2} \right] ,$$

$$(4.101)$$

where

$$H = 2\pi \frac{h}{p}$$

and

$$A = \sqrt{1 + \lambda^2} - \lambda .$$

The limit $H \to \infty$ gives the total strain energy per atom in the crystal **B**:

$$E_e^b = -\frac{(1 - \nu_b)c^2 G' G_i}{4\pi^2 G_b d} \ln(1 - A^2)$$

$$= -\frac{(1 - \nu_b)c^2 G' G_i}{4\pi^2 G_b d} \lambda \ln \left(2\lambda\sqrt{1 + \lambda^2} - 2\lambda^2 \right) . \qquad (4.102)$$

The analogous expression can be written down for the strain energy per atom stored in the crystal A. The ratio of the strain energies gives the distribution of the strain in the two crystal halves:

$$\frac{E_e^b}{E_e^a} = \frac{(1 - \nu_b)G_a}{(1 - \nu_a)G_b} \ .$$

As seen the "stiffer" the crystal the less strained it is and vice versa.

Bearing in mind (4.99) the total strain energy per atom in the bicrystal connected with the misfit dislocations is then

$$E_e = E_e^b + E_e^a = -\frac{G_i c^2}{4\pi^2 d} \lambda \ln\left(2\lambda\sqrt{1 + \lambda^2} - 2\lambda^2\right) \ . \tag{4.103}$$

Summing up (4.100) and (4.103) gives the total energy per atom of the misfit dislocations:

$$E_d = \frac{G_i c^2}{4\pi^2 d}\left[1 + \lambda - \sqrt{1 + \lambda^2} - \lambda \ln\left(2\lambda\sqrt{1 + \lambda^2} - 2\lambda^2\right)\right] \ . \tag{4.104}$$

The energies E_e, E_i and E_d are plotted as functions of the misfit c/p in Fig. 4.40. As can be seen the strain energy E_e is greater than the energy of interaction E_i for small misfits. At larger misfits E_e gradually diminishes and the total energy approaches a constant value:

$$E_d^\infty = \frac{G_i c^2}{4\pi^2 d} \ , \tag{4.105}$$

which can be taken as a measure of the interfacial bonding, analogous to W in the monolayer model. This should be the energy if the two lattices A and B were assumed rigid. Then the relative displacements of the atoms with respect to each other would be given by $U/c = 1/2 + x/p$.

One of the most important results of this analysis, which is closely connected with the mechanism of growth of thin films, concerns the distribution of strain energy with the distance from the interface. Making use of (4.101) and (4.102) gives for the fraction of average strain energy per atom stored beyond a distance h from the interface

$$\Delta E_e^b(h) = \frac{E_e^b - E_e^b(h)}{E_e^b}$$

$$= \left(\ln\left(1 - A^2 e^{-2H}\right) - \frac{HA^2 e^{-2H}(H - 1 + A^2 e^{-2H})}{(1 - \nu_b)(1 - A^2 e^{-2H})^2}\right)$$

$$\times \left[\ln(1 - A^2)\right]^{-1} \ . \tag{4.106}$$

The latter is plotted in Fig. 4.41. As seen it decreases rapidly with the distance from the interface and practically vanishes at a distance equal to $p/2$. It follows that, first, we can qualify a deposit as thick when it is thicker than one half of the dislocation spacing, and, second, beyond this

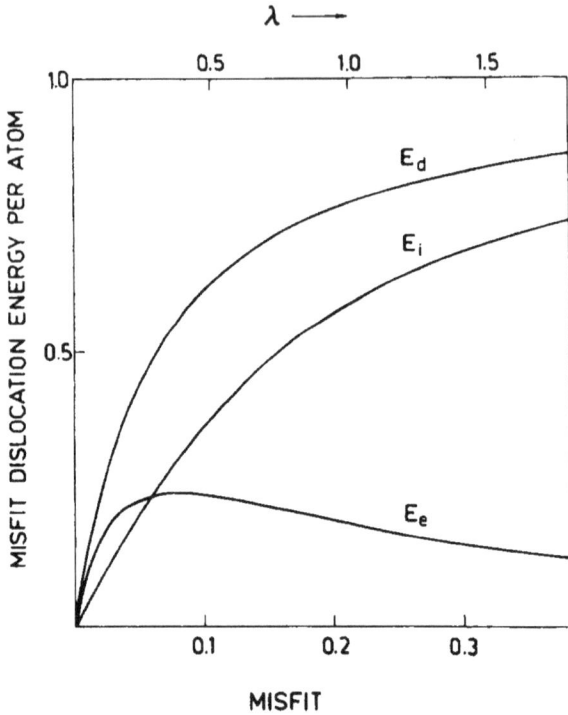

Fig. 4.40. Misfit dependence of the strain energy E_e, the interaction energy E_i and the total dislocation energy $E_d = E_e + E_i$ in units of $G_i c^2/4\pi^2 d$. The upper axis shows the variation of the parameter λ (after van der Merwe [1950]).

thickness the atoms of the deposit will not "feel" the presence of the foreign substrate.

The gradual decrease of the fraction of the elastic energy $\Delta E_e^b(h)$, which is stored in the deposit crystal beyond a distance h from the interface, is illustrated schematically in Fig. 4.42. The amplitude of the periodic variation of the bond strains is greatest for the first monolayer. Expansion and compression alternate periodically with a period equal to the dislocation spacing. The amplitude gradually decreases in every next monolayer. Beyond a thickness $h \cong p/2$, the amplitude becomes practically equal to zero and the atoms become equidistant. However, if the deposited film is thinner than $p/2$ a periodic variation of the bond lengths on the surface of the film should be detected. Such a variation is really observed on the surface of overlayers of Fe(110) on W(110) [Gradmann and Waller 1982] and of Cu on Pd(100) [Asonen et al. 1985]. In the former case the thickness of the film samples varied between 2 and 9 monolayers and $p/2 = 7d_{110}$ ($a_{Fe} = 2.866$ Å, $a_W = 3.165$ Å).

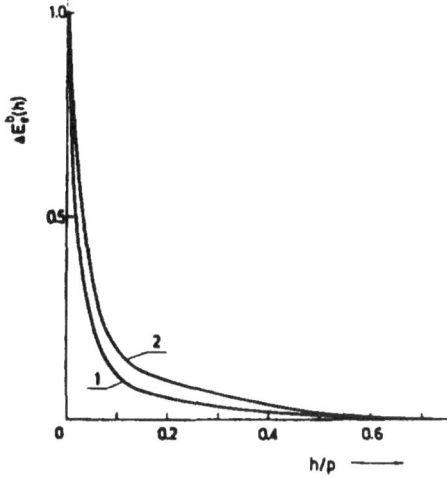

Fig. 4.41. Plots of the fraction $\Delta E_e^b(h)$ of the strain energy stored in the crystal B beyond a distance h from the contact plane as a function of h/p. Curves 1 and 2 correspond to misfits $c/p = 2\%$ and 20%, respectively ($G_a = G_b = G_i$, $\nu_a = \nu_b = \nu = 0.3$) (after van der Merwe [1975]).

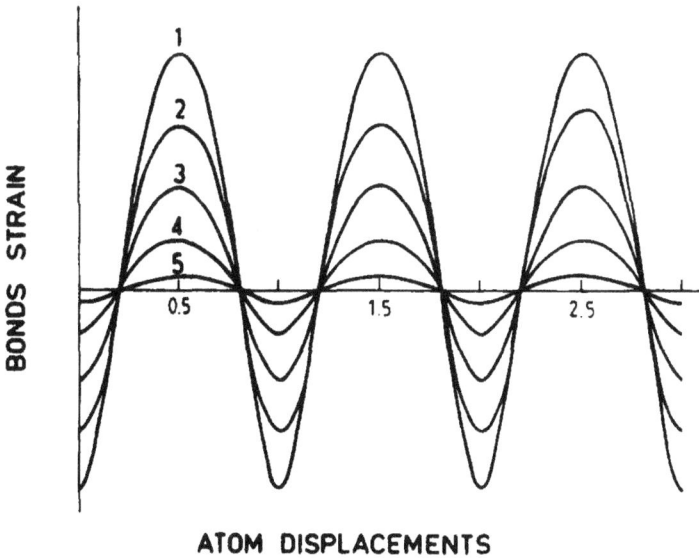

ATOM DISPLACEMENTS

Fig. 4.42. Schematic plot of the strains of consecutive bonds against the atom displacements as in Fig. 4.13 (curve 2) demonstrating the gradual decrease of the amplitudes of the periodic strains with distance from the contact plane. The figure at each curve denotes the number of the corresponding monolayer counted from the contact plane.

Let us illustrate the formulae we have just derived. In a system of low misfit, e.g., Ag on Au(001), $a_{Ag} = 2.8894$ Å and $a_{Au} = 2.8841$ Å, $c/p = 0.00184$, $c = d = 2.8867$ Å, $G_{Ag} = 3.38 \times 10^{11}$ dyne/cm^2 and $G_{Au} = 3.1 \times 10^{11}$ dyne/cm^2, $\nu_{Ag} = 0.354$ and $\nu_{Au} = 0.412$ (see Huntington [1958] and also Hirth and Lothe [1968], Appendix), $G' = 2.63 \times 10^{11}$ dyne/cm^2 and $G_i \cong (G_{Ag}G_{Au})^{1/2} = 3.24 \times 10^{11}$ dyne/cm^2. Then $G_i c^2/4\pi^2 d = 237$ erg/cm^2, $\lambda = 0.0094$ and $E_d \cong 11$ erg/cm^2. In a system with larger misfit, e.g., Ag on Cu(001), with $G_{Cu} = 5.46 \times 10^{11}$ dyne/cm^2, $\nu_{Cu} = 0.324$ and $a = 2.556$ Å, $c/p = 0.1223$, $\lambda = 0.568$, $G_i c^2/4\pi^2 d = 294$ erg/cm^2 and $E_d = 192$ erg/cm^2. It is doubtful whether the above theory can be used in the case of semiconductor materials with covalent bonds which are rather brittle and unflexible. Nevertheless, for the case of a Ge deposit on (001) Si, with $G_{Ge} = 5.64 \times 10^{11}$ dyne/cm^2, $G_{Si} = 6.42 \times 10^{11}$ dyne/cm^2, $\nu_{Ge} = 0.2$, $\nu_{Si} = 0.215$, $a_{0Ge} = 5.6575$ Å and $a_{0Si} = 5.4307$ Å, $c/p = 0.041$, $G_i \cong 6 \times 10^{11}$ dyne/cm^2, $G' = 3.78 \times 10^{11}$ dyne/cm^2 and $E_d \cong 300$ erg/cm^2. Comparing these results with the values of the specific surface energies which are usually of the order of 1×10^3 erg/cm^2, we can conclude that the theory of van der Merwe predicts reasonable values for the energy of the misfit dislocations at the interface between semi-infinite crystals.

It is worth noting, however, that a comparison of the energy of the misfit dislocations with surface energies is in principle incorrect. The energy of the misfit dislocations is often erroneously identified with the energy of the interface. As will be shown below the energy of the misfit dislocations is only a part of the energy of the interface which is due to the lattice misfit. The interfacial energy is composed of two parts. The first one is due to the difference in nature and strength of the chemical bonds in the absence of misfit. The second part, which we have just derived, is due to the lattice misfit. A good example is the energy of the interface between In$_x$Ga$_{1-x}$As and InP(001). When $x = 0.43$ the lattice misfit and the misfit energy are equal to zero. At the same time the interfacial energy is not equal to zero due to the difference of the strengths of the chemical bonds in the two materials.

4.2.6. Thickening overgrowth

As shown above a thin film consisting of one to several monolayers can be either commensurate or incommensurate with the substrate. In other words, the natural misfit can be accommodated either by homogeneous strain or by misfit dislocations or by both in the general case. In contrast, the natural misfit between an infinitely thick deposit and the substrate is accommodated entirely by misfit dislocations. Thus in overgrowth with finite thickness the homogeneous strain can eliminate partially or entirely the misfit dislocations. The purpose of this chapter is to study what will be the equilibrium (lowest energy) structure of the interface during the process of thickening of the deposit.

As mentioned above, owing to the interaction across the interface, the lattice parameter b of the overgrowth tends to take the value of the substrate lattice parameter a. On the other hand, owing to the cohesive forces between the overgrowth atoms, they tend to keep their natural spacing b. As a result, the overgrowth atoms will be spaced at some average spacing $\bar{b}$ such that $a < \bar{b} < b$.

We can define now the natural misfit as [Matthews 1975]

$$f = \frac{a - b}{b} \tag{4.107}$$

and it will be accommodated partly by homogeneous strain

$$f_e = \frac{\bar{b} - b}{b} \tag{4.108}$$

and partly by misfit dislocations

$$f_d = \frac{a - \bar{b}}{\bar{b}} , \tag{4.109}$$

so that for small misfits

$$f_e + f_d \cong \frac{\bar{b} - b}{b} + \frac{a - \bar{b}}{\bar{b}}\frac{\bar{b}}{b} = f . \tag{4.110}$$

The misfit dislocations will be spaced in the general case at a distance

$$\bar{p} = \frac{a}{f_d} = \frac{a\bar{b}}{a - \bar{b}} \tag{4.111}$$

and will have a Burger's vector

$$\bar{c} = \frac{a\bar{b}}{\frac{1}{2}(a + \bar{b})} . \tag{4.112}$$

Minimum energy considerations of epitaxial bicrystal systems [van der Merwe 1963b; Jesser and Kuhlmann-Wilsdorf 1967] have shown that initially the deposit grows pseudomorphically with the substrate up to a critical thickness t_c (see Eq. (4.72)). The natural misfit will be entirely accommodated by a homogeneous strain so that the average atomic spacing $\bar{b} = a$. Then $f_e = f$ and $f_d = 0$. The interface is not resolved into a sequence of misfit dislocations as their spacing $\bar{p}$ tends to infinity. In the case of square atomic meshes of the adjoining crystal planes the energy of the interface due to the lattice misfit will be equal to the energy of the homogeneous strain E_{hs} given by

$$E_{hs} = 2G_b t f^2 \frac{1 + \nu_b}{1 - \nu_b} . \tag{4.113}$$

Beyond the critical thickness, misfit dislocations are introduced at the interface so that initially homogeneous strain and misfit dislocations coexist. The misfit dislocation energy is then given by Eq. (4.104) but with p and c replaced by $\bar{p}$ and $\bar{c}$, respectively. With the thickening of the film the homogeneous strain gradually vanishes, and in sufficiently thick films $(t > p/2)$ the natural misfit is totally accommodated by misfit dislocations. Then $\bar{b} = b$, $f_e = 0$, $f_d = f$ and the dislocations are separated by a distance p. In other words, the other extreme case is reached.

In order to illustrate the minimum energy considerations, we plot the homogeneous strain energy E_{hs} for films of various thicknesses and the misfit dislocation energy E_d against the misfit as is done in Fig. 4.43. As we see, the homogeneous strain energy curves intersect the misfit dislocation energy curves at some critical values of the misfit f_n varying with the film thickness $n = t/b$ measured in number of monolayers. It follows that if $f > f_1$, E_{hs} is greater than E_d even for a monolayer film, and it will resolve into a sequence of misfit dislocations rather than be homogeneously strained. If $f_2 < f < f_1$, the first monolayer will be pseudomorphous with the substrate, but when a second monolayer is deposited on top of the first one, misfit dislocations will be introduced at the interface to relieve the homogeneous strain if the necessary thermal activation exists. The smaller the natural misfit, the thicker the film can grow under homogeneous strain. When the misfit is very small, the film can grow under homogeneous strain to a considerable thickness as in the growth of Ge on GaAs [Matthews, Mader and Light 1970]. For this reason superlattices such as $Al_x Ga_{1-x} As/GaAs$ can be grown without misfit dislocations at the interface. If the temperature is low enough so that misfit dislocations are not introduced during the growth, then when the film thickness exceeds the critical value, the bicrystal system will be in a metastable state and any pumping of energy will lead to nucleation of dislocations and hence to deterioration of the performance of any device made in this way.

Equating the energy $2E_d$ of a square grid of two perpendicular and noninteracting arrays of misfit dislocations from (4.104) and the energy E_{hs} of the homogeneous strain from (4.113) gives ($\nu_a = \nu_b = \nu$)

$$\frac{t_c}{c} = \frac{G_i}{8\pi^2 G_b} \frac{1-\nu}{1+\nu} \frac{f(\lambda)}{f^2} , \qquad (4.114)$$

where

$$f(\lambda) = 1 + \lambda - \sqrt{1 + \lambda^2} - \lambda \ln\left(2\lambda\sqrt{1 + \lambda^2} - 2\lambda^2\right) . \qquad (4.115)$$

For small misfits where second order terms can be neglected ($\lambda^2 \ll 1$), (4.115) reduces to

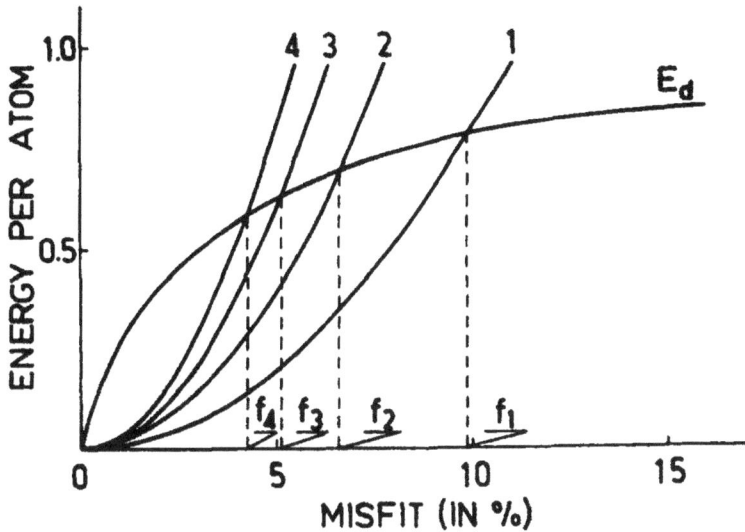

Fig. 4.43. Plot of the dislocation energy E_d and the energy of the homogeneous strain E_{hs} versus misfit of overlayers with different thicknesses shown by the figures which denote the number of monolayers. The points of intersection determine the critical misfits f_n ($n = 1, 2, \ldots$) which decrease with increasing film thickness given in number n of monolayers. (I. Markov and S. Stoyanov, *Contemp. Phys.* **28**, 267 (1987). By permission of Taylor & Francis Ltd.)

$$f(\lambda) = \lambda(1 - \ln 2\lambda) = \lambda(\ln e - \ln 2\lambda) = \lambda \ln \left(\frac{e}{2\lambda} \right) ,$$

which appears as a very good approximation [Kasper and Herzog 1977].

Accounting for (4.98) and (4.99), Eq. (4.114) turns into [van der Merwe 1973]

$$\frac{t_c}{c} = \frac{1}{4\pi(1 + \nu)(1 + G_b/G_a)} \frac{\mathfrak{C} - \ln f}{f} , \qquad (4.116)$$

where

$$\mathfrak{C} = \ln \left(\frac{(1 - \nu)(1 + G_b/G_a)G_i e}{4\pi G_b} \right) . \qquad (4.117)$$

Equation (4.116) is plotted in Fig. 4.44 (curve 1) with $G_a = G_b = G_i$. As seen, the critical thickness goes to infinity with vanishing misfit. Moreover, it decreases with increasing ratio G_b/G_a. In other words, the "stiffer" the deposit and the "softer" the substrate crystals, the lower the critical

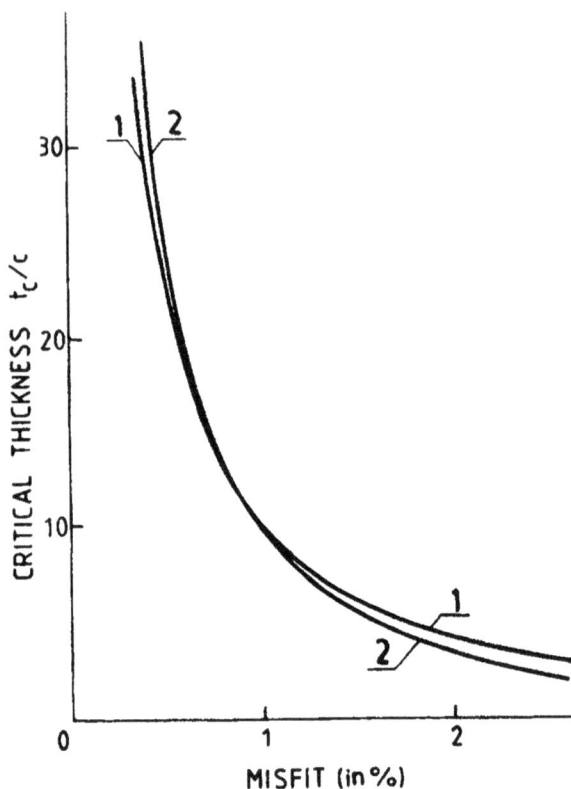

Fig. 4.44. Misfit dependence of the equilibrium critical thickness for pseudomorphous growth in units of the Burgers vector c according to the theory of van der Merwe [1973] (curve 1) (Eq. (4.116) with $G_a = G_b = G_i$, $\nu_a = \nu_b = \nu = 0.3$) and the Volterra approach of Matthews [Matthews and Blakeslee 1974] (curve 2) (Eq. (4.124) with $G_s = G_d$, $\nu = 0.3$).

thickness will be, and vice versa. A similar analysis for finding t_c for crystals with diamond lattice has been performed by Kasper and Herzog [1977].

Let us evaluate the critical thickness for some real systems. In the case of deposition of Ag on Au with the materials constants given above, Eq. (4.114) predicts $t_c/c = 168$ or $t_c \cong 485$ Å. In the case of deposition of Ag on Cu, $t_c/c = 0.7$ or dislocations will be introduced after the deposition of the first monolayer. The theory predicts a value of $t_c/c = 5.7$ for the system Ge/Si. This means that a film consisting of 5 monolayers should be pseudomorphous with the substrate, but after the deposition of the sixth monolayer the interface should resolve in a cross grid of misfit dislocations.

An example of interest is the deposition of alloys where one could change the value of the natural misfit by varying the alloy composition. One of the most studied systems is $Ge_x Si_{1-x}/Si$ [Kasper and Herzog 1977; Kohama, Fukuda and Seki 1988; Bean *et al.* 1984; Bean 1985]. The natural misfit as a function of composition x is given by $f = 0.041x$. Thus at $x = 0.5$, $f = 0.021$ and the experimentally found critical thickness is 100 Å or about 25 monolayers (Bean 1985). Calculations based on Eq. (4.114) give the value $t_c/c = 9$, i.e. approximately three times smaller than the experimentally found value.

The above disagreement can be explained bearing in mind that Eq. (4.114) gives the equilibrium critical thickness for pseudomorphous growth. In other words, this is the thickness beyond which misfit dislocations become energetically favored. However, in real experiments an energetic barrier for nucleation of dislocations should be overcome. It follows that the real critical thickness should be greater than that given above. The nucleation of the misfit dislocations has been taken into account in a series of papers (Marée *et al.* 1987; Van de Leur *et al.* 1988; Fukuda, Kohama and Ohmachi 1990; Kamat and Hirth 1990) and the interested reader is referred to them.

Recalling (4.65) we conclude that the approximate expression for $t_c(f)$, given by Eq. (4.72), is in fact determined by the intersection of the homogeneous strain energy (Eq. (4.113)) with the maximum energy of the misfit dislocations, E_d^∞, given by Eq. (4.105). Obviously, it overestimates the critical thickness.

4.2.7. *The Volterra approach*

Volterra [1907] considered the elastic properties of a hollow cylinder with inner and outer diameters r_0 and R, respectively, cut parallel to the cylinder axis (Fig. 4.45(a)). When a force parallel to the cylinder axis is applied along the cut a screw dislocation is formed (Fig. 4.45(b)). When the force applied is normal to the cylinder axis and the one edge of the cut is displaced normally with respect to the other, an edge dislocation results (Fig. 4.45(c)). The displacement along the cut of one half of the cylinder with respect to the other is just equal to the Burgers vector of the dislocation.

The strain energy of the edge dislocation per unit length is given by [Hirth and Lothe 1968]

$$E_s = \frac{Gb^2}{4\pi(1 - \nu)} \ln\left(\frac{R}{r_0}\right) ,$$

a   b   c

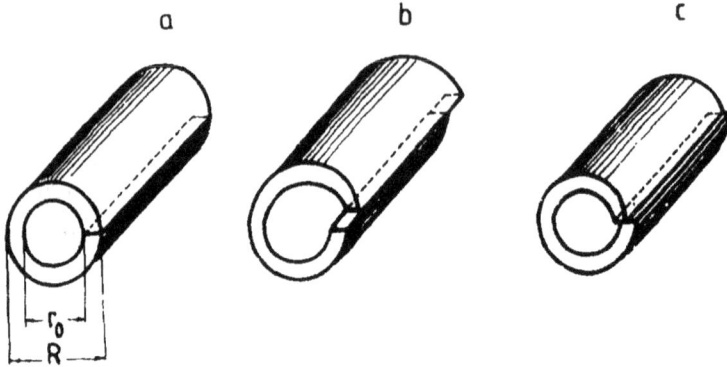

Fig. 4.45. Volterra models of (b) screw and (c) edge dislocations based on the consideration of a hollow cylinder (a) with outer diameter R and inner diameter r_0 cut parallel to the cylinder axis (after Hirth and Lothe [1968]).

where G is the shear modulus of the crystal, ν is the Poisson ratio and b is the magnitude of Burger's vector. The total energy is obtained by the addition of the core energy. The latter represents a fraction of Gb^2 and its contribution is formally accounted for by assuming $r_0 = b/\alpha$, where α is a constant varying from 1 to 2 for metals and to 4 for nonmetals. Then the total energy of an edge dislocation reads

$$E_d = \frac{Gb^2}{4\pi(1-\nu)}\left(\ln\frac{R}{b} + \ln\alpha\right) . \qquad (4.118)$$

For the energy of a misfit dislocation between two misfitting crystals with shear moduli G_s and G_d and a Poisson ratio ν, Matthews [1975] used a similar expression

$$E_d = \frac{G_s G_d c^2}{2\pi(G_s + G_d)(1-\nu)}\left[\ln\left(\frac{R}{c}\right) + 1\right] , \qquad (4.119)$$

which turns into (4.118) with $G_d = G_s = G$ and $\alpha = e$. R denotes the distance to the outermost boundary of the strain field of the dislocation, and c, which is given by Eq. (4.91), is the magnitude of Burger's vector.

If the film thickness t is smaller than the dislocation spacing $p/2$, R can be approximated by the film thickness t (Fig. 4.46(a)) and the energy reads

$$E_d = \frac{G_s G_d c^2}{2\pi(G_s + G_d)(1-\nu)}\left[\ln\left(\frac{t}{c}\right) + 1\right] . \qquad (4.120)$$

a

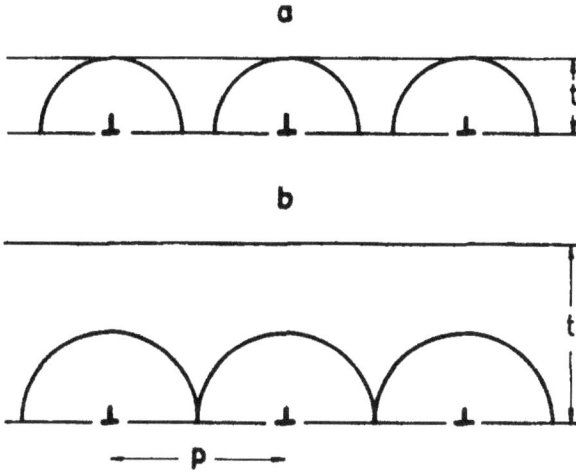

b

Fig. 4.46. For the determination of the outermost boundary R of the strain field created by the dislocation. (a) $t < p/2$, $R = t$; (b) $t > p/2$, $R = p/2$ (t: film thickness, p: dislocation spacing).

At the other extreme, $t > p/2$, R is approximated by $p/2$ (Fig. 4.46(b)). This choice of R becomes immediately understandable if one recollects the fact that the strain field of the dislocations practically vanishes beyond $p/2$ (see Fig. 4.41). Then the energy reads

$$E_{\rm d} = \frac{G_{\rm s}G_{\rm d}c^2}{2\pi(G_{\rm s} + G_{\rm d})(1 - \nu)}\left[\ln\left(\frac{p}{2c}\right) + 1\right] . \qquad (4.121)$$

In the case of quadratic interfacial symmetry ($f_x = f_y = f$) and assuming the natural misfit is accommodated partly by misfit dislocations and partly by homogeneous strain, the energy $E_{\rm m}$ due to the lattice misfit is a sum of the energies of two noninteracting arrays of misfit dislocations each with a density $f_{\rm d} = f - f_{\rm e}$ and the energy of the homogeneous strain $E_{\rm hs}$ given by Eq. (4.113). Neglecting the energy of interaction between neighboring dislocations in the array one writes

$$E_{\rm m} = 2(f - f_{\rm e})\frac{G_{\rm s}G_{\rm d}\bar{c}}{2\pi(G_{\rm s} + G_{\rm d})(1 - \nu)}\left[\ln\left(\frac{\bar{R}}{\bar{c}}\right) + 1\right] + 2G_{\rm d}tf_{\rm e}^2\frac{1 + \nu}{1 - \nu} ,$$

where $\bar{R} = t$ or $\bar{p}/2$ when t is smaller or greater than $\bar{p}/2$.

The condition $dE_m/df_e = 0$ yields the homogeneous strain f_e^* which minimizes the energy. In the first case ($t < \bar{p}/2$ and $\bar{R} = t$) one obtains for f_e^*

$$f_e^* = \frac{G_s \bar{c}}{4\pi(G_s + G_d)(1+\nu)t} \left[\ln\left(\frac{t}{\bar{c}}\right) + 1 \right] . \qquad (4.122)$$

In the second limiting case ($\bar{R} = \bar{p}/2$ and $\bar{c}/\bar{p} = f_d = f - f_e$) the minimization gives

$$f_e^* = -\frac{G_s \bar{c}}{4\pi(G_s + G_d)(1+\nu)t} \ln\left[2(f - f_e^*)\right] . \qquad (4.123)$$

The critical thickness t_c is determined by the condition $f_e = f$. Making use of Eq. (4.122) (the equilibrium thickness is usually smaller than $p/2$) gives

$$\frac{t_c}{\bar{c}} = \frac{G_s}{4\pi(G_s + G_d)(1+\nu)f} \left[\ln\left(\frac{t_c}{\bar{c}}\right) + 1 \right] . \qquad (4.124)$$

Equation (4.124) gives values for the critical thickness very close to that predicted by the theory of van der Merwe (see Fig. 4.44, curve 2). One should, however, note the uncertainty in the determination of the core energy of the dislocations.

Thus the main advantage of the Volterra approach consists in the simplicity of the expression for the dislocation energy which allows the easy treatment of more difficult problems such as the equilibrium structure of the interface between tile-shaped 3D islands and the substrate [Matthews, Jackson and Chambers 1975c], in multilayers [Matthews and Blakeslee 1974], imperfect dislocations with Burger's vectors inclined with respect to the interface, generation of misfit dislocations, etc. [Matthews 1975a, 1975b]. The interested reader is referred to the review paper of Matthews [1975b] and the references therein.

As mentioned in the previous section a considerable discrepancy is established when experimental data are compared with the theoretical expressions for the critical thickness for pseudomorphous growth. On the other hand, the problem of the critical thickness for pseudomorphous growth turned out to be very important from a technological point of view. The reasons are well described in the review paper of Hu [1991]. First, the misfit dislocations deteriorate the performance of the heterostructure devices due to the increased leakage current. On the other hand, the misfit dislocations are often generated by dislocations which are inherited from the substrate and end at the film surface (threading dislocations). The diffusion of the dopant is usually enhanced along the threading dislocation line and the latter forms the so-called "transistor pipe" connecting the emitter and the collector. Second, in uniformly strained epilayers the interatomic spacing differs from that in the unstrained (relaxed) ones, thus changing the width of the forbidden energy zone [Land *et al.* 1985]. It is thus obvious that the homogeneous strain can serve in addition to the alloy composition as a parameter for further tailoring of the heterostructure properties.

What both approaches of van der Merwe [1963b] and Matthews [1975b] to the problem of the critical thickness have in common is that the latter is inversely proportional to the lattice misfit. Recently People and Bean [1985, 1986] derived an expression for the critical thickness by comparing the energies of the homogeneous strain and the areal energy of a single screw dislocation. They found that the critical thickness is inversely proportional to the square of the lattice misfit and has an absolute value which is of one order of magnitude greater than that predicted by van der Merwe and Matthews. As their result is still under discussion (see Hu [1991]) we will not reproduce it here. The reader who is interested in the present day state of the problem of the critical thickness for pseudomorphous growth is referred to the excellent review paper of Hu [1991].

4.3. Mechanism of Growth of Thin Epitaxial Films

As mentioned in the introduction of this chapter the chemical potentials of the substrate and deposit crystals differ owing to the difference in the nature and strength of the chemical bonds on the one hand, and the lattices and lattice parameters on the other. Then, the chemical potential of the overgrowth will differ from that of the infinitely large crystal due to the difference of bonding across the interface (see Eq. (1.59)). The atoms of the deposit can be bound more loosely or more tightly to the substrate atoms than to the atoms of the same crystal. As a result the chemical potential of the first layers of the deposit will be higher or lower than the chemical potential of the infinitely large deposit crystal.

Let us consider this case in more detail beginning from the first monolayer of the deposit. Its chemical potential is equal to the work of separation of an atom from a half-crystal position taken with a negative sign (Eq. (1.59)). Assuming additivity of the bond energies the work to separate an atom from a kink position consists of two parts: the work to disrupt the lateral bonds and the work to disrupt the bonds with the substrate atoms. However, the lateral bonds are the same as in the bulk deposit crystal. Then the difference in the chemical potentials is due to the difference of bonding with the substrate. When the atoms of the deposit are more loosely bound to the foreign substrate than to the same crystal the equilibrium vapor pressure of the deposit will be higher than the equilibrium vapor pressure of the large deposit crystal and vice versa. In turn the chemical potential of the first monolayer will be higher or lower than the chemical potential of the infinitely large deposit crystal. The atoms of the second monolayer will "feel" more weakly the presence of the foreign substrate and hence their chemical potential will be closer to that of the large deposit crystal. In other words, the chemical potential of the deposit will vary from monolayer to monolayer due to the interaction with the substrate. On the other hand, the overgrowth can be pseudomorphous with the substrate or the interface can be resolved in a grid of misfit dislocations. Then the film can be homogeneously or periodically strained and the strain energy per atom should change additionally the chemical potential of the film. The homogeneous strain does not change from monolayer to monolayer of the overgrowth while the periodic strain due to the misfit dislocations attenuates with the distance from the interface (see Figs. 4.41 and 4.42). Hence the chemical potential again varies from monolayer to monolayer due to the elastic strains in addition to the bonding with the substrate [Venables 1979; Grabow and Gilmer 1988; Stoyanov 1986]. It is namely this dependence of the chemical potential of the overgrowth on its thickness which constitutes the main difference of the epitaxial growth from the usual crystal growth and which leads to the appearance of the three well-known mechanisms of epitaxial growth: (i) *Volmer–Weber* mechanism or island growth (Fig. 4.47(a)), (ii) *Frank–van der Merwe* mechanism or layer-by-layer growth (Fig. 4.47(b)) and (iii) *Stranski–Krastanov* mechanism or layer-by-layer growth followed by formation of 3D islands (Fig. 4.47(c)).

This classification has been given for the first time by Bauer [1958]. The historical reason why the island growth was named after Volmer and Weber was that Volmer was the first to develop the theory of the rate of 3D nucleation on a foreign substrate and together with Weber [Volmer and Weber 1926] interpreted the experimental data of Frankenheim [1836], which were the first laboratory experiments on epitaxial growth, in the light of his theory.

Frank and van der Merwe simulated the overgrowth by an infinite chain of atoms, thus asssuming implicitly that the overgrowth covers completely

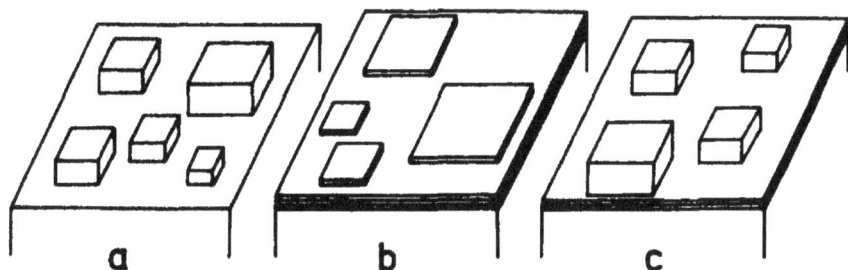

Fig. 4.47. Schematic representation of the three possible mechanisms of growth of thin epitaxial films according to the classification of Bauer [1958]: (a) Volmer–Veber mechanism or island growth, (b) Frank–van der Merwe mechanism or layer-by-layer growth, (c) Stranski–Krastanov mechanism or layer-by-layer growth followed by 3D islands. (I. Markov and S. Stoyanov, *Contemp. Phys.* **28**, 267 (1987). By permission of Taylor & Francis Ltd.)

the substrate. In other words, they assumed that the overgrowth follows the layer-by-layer pattern of growth and they did not specify at that time that different values of the energetic parameter l_0 should lead in fact to different modes of growth. This was the reason why Bauer named the layer-by-layer growth after their names.

The first paper devoted solely to the problem of the mode of epitaxial growth of thin films was written by Stranski and Kuleliev [1929]. The considerations, which were naturally based upon the concept of the work of separation of a building unit from a half-crystal position, were further developed by Stranski and Krastanov [1938]. We will outline their model in more detail as it lies in the base of our further considerations of the mode of growth.

Thus, Stranski and Kuleliev studied the stability of the first, second, third, etc., monolayers of a monovalent ionic crystal K^+A^- with a sodium chloride lattice on the surface of a bivalent ionic crystal $K^{2+}A^{2-}$ with the same lattice, assuming that no alloying takes place and that both crystals have equal lattice parameters (Fig. 4.48). As a measure of stability they accepted the equilibrium vapor pressure of each monolayer. According to Eq. (1.58) it is a function of the corresponding work of separation from the half-crystal position. As can be judged from Fig. 4.48 the ions of the first layer of K^+A^- are attracted by the underlying doubly charged ions more strongly than by the surface of their own monovalent crystal; the lateral interactions remain the same. Hence, the work of separation from the half-crystal position, $\varphi_{1/2}^{(1)}$, of the first monolayer is larger than that of

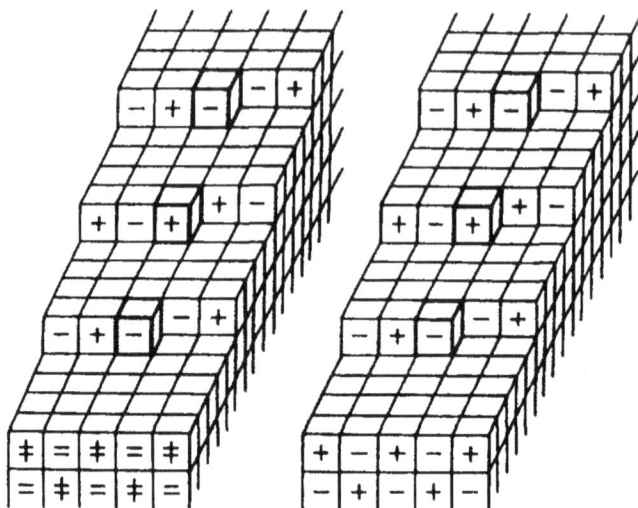

Fig. 4.48. Illustration of the model of Stranski and Kuleliev [1929] for the growth of monovalent ionic crystal K^+A^- on the surface of an isomorphous bivalent ionic crystal $K^{2+}A^{2-}$ (left), compared with the growth of the same monovalent ionic crystal K^+A^- (right). The symbols on the cube faces of the left-hand crystal represent doubly charged positive and negative ions. The stability of the consecutive monolayers is determined by the work of separation of the ions from the corresponding half-crystal positions denoted by the bold cubes. (I. Markov and S. Stoyanov, *Contemp. Phys.* **28**, 267 (1987). By permission of Taylor & Francis Ltd.)

the bulk K^+A^- crystal, $\varphi_{1/2}^{(0)}$, and, correspondingly, the equilibrium vapor pressure P_1 will be lower than that of the bulk K^+A^- crystal, P_∞. It follows that one monolayer of K^+A^- can be adsorbed on the surface of $K^{2+}A^{2-}$ at any vapor pressure higher than P_1 and lower than P_∞, in other words, at undersaturation with respect to the bulk K^+A^- crystal. The ions of the second layer are attracted by the ions of the first layer as if they are on the same crystal K^+A^- but are repulsed by the doubly charged ions of $K^{2+}A^{2-}$ more strongly than if the substrate were monovalent. Thus $\varphi_{1/2}^{(2)} < \varphi_{1/2}^{(0)}$ and $P_2 > P_\infty$. Hence, a supersaturation is required in the system in order to deposit the second monolayer. Stranski and Kuleliev concluded that every odd or even overlayer will have an equilibrium vapor pressure that is, respectively, less than or greater than P_∞; in other words the equilibrium vapor pressures of the consecutive monolayers should oscillate around the equilibrium vapor pressure P_∞. The dependence of the chemical potential of the consecutive monolayers according to the considerations of Stranski and Kuleliev is shown in Fig. 4.49. As can be seen, after a few layers of

K^+A^-, the energetic influence of the substrate vanishes ($P_n \cong P\infty$) and the film will continue to grow as if on the same crystal.

Ten years later Stranski and Krastanov [1938] extended the considerations of the same model by calculating the Gibbs free energies of formation of 2D nuclei of the first, second, third, etc., monolayers, as well as two and four monolayers thick 2D nuclei. It turned out that, after the complete coverage of the substrate crystal $K^{2+}A^{2-}$ by an adlayer of K^+A^- at undersaturation (for reasons given above), the work of formation of 2D nuclei of the second monolayer is significantly greater than that of 2D nuclei that are two monolayers thick. The reason is that the chemical potential of a bilayer deposited on the first monolayer is lower than that of a single monolayer (see Fig. 4.49). Simple considerations show that the work of separation of a whole K^+A^- molecule from a doubly high half-crystal position (Fig. 4.50) is equal to the arithmetic average of the works of separation of single ions from the kink positions of the second and third monolayers, i.e. $\varphi_{1/2}$ (K^+A^-) $= (\varphi_{1/2}^{(2)} + \varphi_{1/2}^{(3)})/2$. Then the chemical potential of the bilayer will be precisely equal to the arithmetic average of the chemical potentials of the second and third monolayers, i.e. $\mu(2 + 3) = (\mu_2 + \mu_3)/2$. Note that when writing the chemical potential of the bilayer we do not account for the chemical potential of the first monolayer. The reason is that it is more strongly bound to the substrate and is completely built up. On this account, it does not take part in the process of exchange of atoms with the vapor phase.

Thus Stranski and Krastanov predicted the possibility — admittedly for a very particular system — of the formation of nuclei with many layer thickness on the first stable adlayer (or adlayers) of the overgrowth, a mechanism of growth well known today and bearing their names. As will be shown below, the physical reason for such a mode of growth could be different and could include the lattice misfit as well. It is worth noting, however, that Stranski and Krastanov considered the many layer thick nuclei as 2D nuclei. They also considered the formation of 3D nuclei on top of the first monolayer and found that it is slightly greater than that of bilayer 2D nucleus. In our further considerations we will consider the bilayer island as a three-dimensional island.

In order to establish the factors affecting the mode of growth we will consider briefly some experimental examples, dividing them into several groups according to the different nature of the chemical bonds: metals on insulators, metals on metals, metals on semiconductors and semiconductors on semiconductors [Markov and Stoyanov 1987].

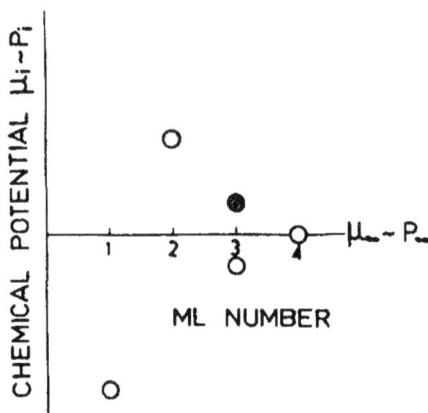

Fig. 4.49. Dependence of the chemical potential μ_i (or the equilibrium vapor pressures P_i) of the consecutive uppermost monolayers on their number as follows from the model of Stranski and Kuleliev [1929] of the growth of a monovalent ionic crystal K^+A^- on the surface of a bivalent ionic crystal $K^{2+}A^{2-}$ with the same lattice parameter (open circles). As seen, the chemical potential oscillates around the chemical potential of the bulk crystal K^+A^-. The filled circle gives the chemical potential of a bilayer formed on top of the completely built first monolayer (after Stranski and Krastanov [1938]). It is exactly equal to the arithmetic average of the chemical potentials of the separate second and third monolayers. (I. Markov and S. Stoyanov, *Contemp. Phys.* **28**, 267 (1987). By permission of Taylor & Francis Ltd.)

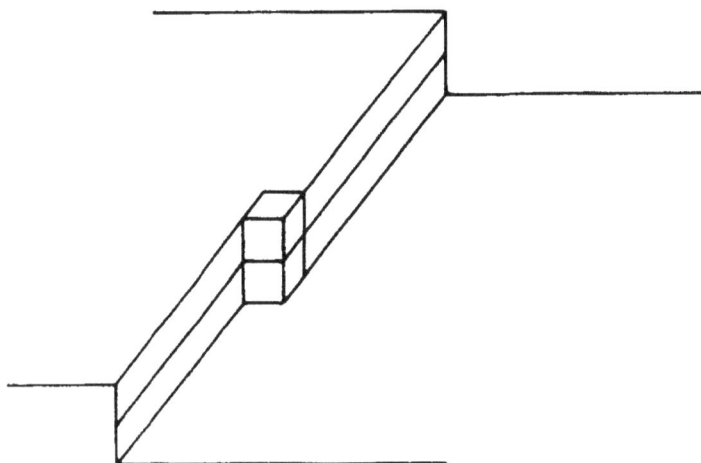

Fig. 4.50. Half-crystal position with a bilayer height (after Stranski and Krastanov [1938]).

A typical example of island growth is the deposition of metals on MgO, mica, molybdenite (MoS_2) and on alkali halide crystals such as NaCl, KCl, KBr, etc. (for a review, see Pashley [1956, 1965, 1970]; Grünbaum [1974]; Kern, LeLay and Metois [1979]). Metals such as Ag, Au, Cu, Fe, Pd, Ni, Co, etc. on these insulators comprise systems which are characterized by adhesion forces between the metal atoms and the insulator substrates that are considerably weaker than the cohesion forces within the metals themselves. An exception from the rule is the deposition of Ag on (100) MgO at extremely low temperatures. Lord and Prutton [1974] established with the help of LEED and AES that at $-200°C$ Ag grows layer-by-layer on (100) MgO crystal cleaved in vacuum chamber. At room temperature, however, Ag grows on MgO as separate islands.

In the case of deposition of metals on metals there is a greatest variety of modes of growth owing to the broad range of the ratio of adhesion to cohesion forces and of the lattice misfit as well (for a review, see Vook [1982, 1984]; Bauer [1982]). An interesting example is the deposition of Cu on the (111) plane of Ag [Horng and Vook 1974]. When less than one-third of a monolayer of Cu is deposited on the (111) plane of Ag at 210°C, the RHEED pattern consists of diffraction streaks originated from the Ag substrate plus bright spots belonging to the Cu deposit. At room temperature the RHEED pattern of the Ag substrate plus one third of a Cu monolayer exhibits intense Ag streaks with fainter streaks for Cu overgrowth, which is indicative for a very smooth surface. It follows that at high substrate temperatures Cu begins to grow as isolated particles or islands from the very beginning of the process, while at low temperatures the Cu streaks originate from a deposit with a layer structure. When more than two Cu monolayers are deposited at room temperature the material in excess of two monolayers forms 3D islands. Thus the Volmer–Weber growth mode is established at high temperatures and the Stranski–Krastanov mechanism takes place at low temperatures. Besides, the thinner the overgrowth the larger its average lattice parameter. For thicker deposits it has its bulk value. In other words, thin Cu overlayers are at least partly homogeneously strained. It should be noted that the enthalpy of evaporation of Cu (80500 cal/g.atom) is greater than that of Ag (67900 cal/g.atom), which means that the cohesion forces are expected to be stronger than the adhesion forces. Besides, the misfit is quite large in absolute value, $f = (b - a)/a = -11.5\%$.

Face-centered cubic metals grow on the most densely packed planes (110) and (100) faces of body-centered cubic metals W and Mo, usually following the layer-by-layer mechanism at low temperatures and the

Stranski–Krastanov mechanism at high temperatures [Bauer 1982]. It should be noted that the enthalpies of evaporation of Mo and W (142000 cal/g.atom and 191000 cal/g.atom, respectively) are much greater than those of fcc metals deposited which vary from 67900 cal/g.atom for Ag to 94000 cal/g.atom for Pd [Emsley 1991]. Hence adhesion forces much stronger than the cohesion forces should be expected.

A typical example is the case of deposition of Cu on (110) and (100) W [Bauer *et al.* 1974]. LEED and AES measurements have shown that at room temperature the growth follows the layer-by-layer pattern. When the substrate plane is (110) the first Cu monolayer has 1.41×10^{15} atoms cm^{-2}, which is exactly the surface density of the W atoms in the adjoining W(110) plane. Hence, the first Cu monolayer is pseudomorphous with the substrate. The second monolayer is expanded in the [100] direction and not strained in the [111] direction, thus having 1.6×10^{15} atoms cm^{-2}, which is somewhere between the surface densities of the W(110) and Cu(111) planes. If a film thicker than two monolayers is annealed at a higher temperature all material in excess of two monolayers agglomerates into isolated 3D islands, which means that the first two monolayers are stable. In the case of the (100) substrate plane, the first two monolayers have 1×10^{15} atoms cm^{-2} each, i.e. they have the same density as the W(100) plane and hence are pseudomorphous with the substrate. Even at room temperature the slope of the AES signal versus film thickness plot is very small after the second break point, which is an indication that 3D islands are formed on top of the second monolayer. This is unlike the (110) plane where 3D islands are formed on top of the second monolayer at temperatures higher than 700 K. Thus layer-by-layer growth is observed at low temperatures and Stranski–Krastanov growth at higher temperatures. The critical temperature for transition from one to the other mechanism is considerably higher for the more densely packed (110) than for the (100) substrate plane. Similar results are also obtained in the case of deposition of Ag and Au on both low index planes of W [Bauer *et al.* 1977].

When evaporating Fe on Cu(111) in the temperature range 20–400°C Gradmann, Kümmerle and Tillmanns [1976] (see also Gradmann and Till-manns [1977]) established that the LEED patterns of the Fe films cannot be distinguished from the pattern of the pure Cu substrate. Note that at temperatures below 912°C the thermodynamically stable phase of iron is α-Fe with a bcc lattice; above this temperature it is γ-Fe with an fcc lattice. Hence, the thinnest Fe films (< 20 Å) consist of the thermodynamically unstable γ-Fe and grow pseudomorphous with the Cu substrate. Moreover,

AES measurements showed that at low temperatures and/or high enough atom arrival rates, the films grow in the layer-by-layer mode, while at high temperatures and/or low atom arrival rates, island growth is observed from the very beginning of the deposition. Thus, the mode of growth depends on the supersaturation $\Delta\mu$, independently of whether the latter was changed by the temperature or by the atom arrival rate. Note that the enthalpy of evaporation of γ-Fe (96000 cal/g.atom) is higher than that of Cu (80500 cal/g.atom). The atom spacings of α-Fe and γ-Fe are 2.4823 Å and 2.578 Å, respectively, and the corresponding values of the misfit with the copper substrate are 2.88% and 0.85%.

The deposition of metals on semiconductors follows more or less the same pattern as that of metals on metals systems. Stranski–Krastanov or Volmer–Weber growth at high temperatures and layer-by-layer growth at low temperatures are usually observed. The interpretation of the observations is much more complicated in view of the different nature of the chemical bonds in the substrate and deposit crystals.

In the case of deposition of Ag on the As($00\bar{1}$) face of GaAs described above [Massies and Linh, 1982a–c] AES measurements have shown that above 100°C, 3D islands are formed on the GaAs substrate (Volmer–Weber mode), while at lower temperatures the growth is very close to the layer-by-layer pattern.

Gold and silver films on Si(111) and Si(100) are the most often studied systems ([Hanbücken and Neddermeyer 1982; Hanbücken, Futamoto and Venables 1984a, b; Hanbücken and LeLay 1986]; for a review see LeLay [1983]). A characteristic feature of the Au/Si system is the alloying which takes place even at room temperature. Silver has a much weaker tendency to make alloys with Si, and Ag adlayers on Si(111) are believed to form sharp interfaces with the Si substrate. It has been found that at room temperature the Ag films grow either via the Stranski–Krastanov mode with very flat islands [Venables, Derrien and Janssen 1980] or via the layer-by-layer mode [LeLay *et al.* 1981; Bolmont *et al.* 1981]. In a comparative study of Ag on the (111) and (100) planes of Si by AES and SEM, Hanbücken, Futamoto and Venables [1984b] found that at high temperatures (400–500°C) the 3D Ag islands are flatter on Si(111) (relative height 0.01–0.04) compared to those on Si(100) (relative height 0.3–0.6). An increase of the relative height of the 3D islands with temperature has been observed during deposition of Bi on Si(100) where a transition from layer-by-layer to Stranski–Krastanov growth at a critical temperature of about 280 K has been found [Fan, Ignatiev and Wu 1990].

The epitaxial growth of one semiconductor material on the single crystal surface of another material is closely connected with device fabrication. Here both the substrate and the deposit materials are characterized by their directional covalent bonds. The growth of elementary semiconductors Si and Ge on top of each other as well as of $Ge_x Si_{1-x}$ alloys on the (100) and (111) faces of Si is most often studied.

Considering the covalent bonds as "brittle" and "inflexible," the lattice misfit between Ge and Si (4.1%) can be accepted as very large. Narusawa, Gibson and Hiraki [1981a] (see also Narusawa and Gibson [1981b, 1982]) interpreted their data of Rutherford Backscattering Spectroscopy (RBS) that Ge films deposited at 350°C grow pseudomorphically with the substrate up to three monolayers, after which 3D islands are formed on top. Alloying does not occur and the interface is abrupt. These conclusions were later confirmed by LEED and AES experiments of Shoji *et al.* [1983] and by Asai, Ueba and Tatsuyama [1985] for both (111) and (100) faces of Si. These authors reported that at room temperature layer-by-layer growth takes place until six monolayers are formed, further deposition giving rise to amorphous Ge. Above 350°C the growth mode follows the Stranski–Krastanov pattern in which 3D islands are formed on top of three monolayers of Ge. Marée *et al.* [1987] found with the help of RBS and RHEED that 3D Ge islands were formed on four monolayers (two bilayers) which remained stable on the Si(111) substrate at temperatures up to 800°C. The results obtained by deposition at room temperature with further annealing at higher temperatures were practically the same as the ones obtained by deposition at elevated temperatures. The room temperature deposited films were continuous and smooth. The high temperature films showed drastic changes which were attributed to island formation. SEM micrograph of a Ge film, 50 monolayer thick, deposited at room temperature and annealed at 500°C for 3 min, showed the presence of 3D islands with a density approximately 1×10^9 cm^{-2}. Thus layer-by-layer growth has been established up to 500°C and Stranski–Krastanov growth beyond this temperature. Similar investigations of the deposition of Si on Ge(111) [Marée *et al.* 1987] showed that layer-by-layer growth takes place at temperatures up to 600°C which was then replaced by growth of isolated islands with approximate density 4×10^7 cm^{-2} directly on the Ge substrate (Volmer–Weber growth).

A marked dependence of the mechanism of growth on the value of the natural misfit has been observed in the case of deposition of $Ge_x Si_{1-x}$ alloys on Si(100). Kasper, Herzog and Kibbel [1975] established that when the Ge content x in the alloy exceeded 0.2 so that the lattice mismatch is larger than 0.82%, the growth proceeds by formation of 3D islands. 2D growth takes place when the Ge content is less than 0.2. The lower the Ge content the thicker the alloy film can be grown by successive monolayers pseudomorphous with the substrate. These findings were later confirmed by Bean *et al.* [1984], who found, however, that pseudomorphous 2D growth can take place for alloys with Ge content up to 0.5 and thicknesses as great as 0.25 μm.

The examples given above do not exhaust all the cases described in the literature. Deposition of noble gases is another interesting example. For instance, in the case of deposition of Xe on Si(111) at 25 K layerlike growth [Bartha and Henzler 1985] has been established, the film being pseudomorphous with the substrate. At elevated temperatures (36 K) clear island growth has been observed. The same tendency has been found in quite a different system, namely, the growth of tungsten oxide on tungsten [Lepage, Mezin and Palmier 1984]. In the low temperature regime (< 600°C) the oxide seems to grow layer by layer, while at high temperatures (> 700°C) the deposit is formed by discontinuous islands.

We can summarize the main tendencies in the mode of growth which are more or less applicable to systems involving any kind of chemical bonds as follows [Markov and Stoyanov 1987]:

1. When the interfacial bonding is weaker than the bonding in the deposit itself, the formation and growth of isolated 3D islands rather than monolayers are favored.

2. High substrate temperatures favor the growth of 3D islands either directly on the substrate (Volmer–Weber mechanism) or on one or several stable adlayers of the deposit (Stranski–Krastanov mechanism). In addition, the higher the temperature, the greater is the relative height of the 3D crystallites.

3. Higher deposition rates favor layerlike growth. The higher the deposition rate, the more layerlike the growth pattern, and vice versa.

4. The lattice misfit plays a prominent role in determining the mode of growth. The larger the misfit, the greater is the tendency towards island-like growth, and vice versa.

5. The crystallographic orientation of the substrate also affects the mechanism of growth. The more densely packed the substrate plane, the greater the tendency towards layerlike growth in comparison with the less densely packed planes. In particular, this tendency is expressed by the

fact that the more densely packed the substrate planes are, the flatter the 3D crystallites will grow.

4.3.1. *Relation of Dupré for misfitting crystals*

As discussed above, the substrate and deposit crystals differ not only in their lattices and lattice parameters but also in the nature and strength of the chemical bonds. In the case of zero misfit the quantity which gives properly the catalytic potency of the substrate, or in other words, the energetic influence of the substrate on the film growth, is the specific adhesion energy which is determined by Eq. (1.28). To account for the lattice misfit we perform the same imaginary process as described in Chap. 1 (Fig. 1.6).

We assume that two infinitely large crystals A and B have different lattice parameters a and b [van der Merwe 1979]. We cleave both crystals reversibly and isothermally, and produce two surfaces of A and two surfaces of B. We then uniformly strain the halves of one crystal, say B, to match exactly the lattice parameter of the other (A), and put the halves of A and B in contact as before. Assuming that the lateral homogeneous strain does not affect the bonding across the interface, we gain an energy $-2U_{AB}$. After that we allow the bicrystal system to relax completely so that misfit dislocations are introduced at the interface. The energy of homogeneous strain is regained completely, but the energy E_d of a cross grid of misfit dislocations is introduced. The energy balance now reads

$$2U_i = U_{AA} + U_{BB} - 2U_{AB} + 2E_d \ .$$

Then

$$\sigma_i^* = \sigma_A + \sigma_B - \beta + \mathcal{E}_d = \sigma_i + \mathcal{E}_d \qquad (4.125)$$

or

$$\sigma_i^* = \sigma_A + \sigma_B - \beta^* \ , \qquad (4.126)$$

where $\mathcal{E}_d = E_d/\Sigma$ is the misfit dislocation energy per unit area and the asterisks indicate that the quantities refer to misfitting crystals. The specific adhesion energy is now given by the difference

$$\beta^* = \beta - \mathcal{E}_d \ . \qquad (4.127)$$

As we see, the dislocation energy appears as a decrement to the binding between both crystals. This is not surprising if we recollect that when

$E_d = 0$ the atoms are not displaced from their positions at the bottoms of the potential troughs provided by the other crystal.

On the other hand, the dislocation energy appears as an increment to the specific interfacial energy which is due to the lattice misfit. The remaining part σ_i is due to the different nature and strength of the chemical bonds and does not depend on the misfit. Thus the interfacial energy $\sigma_i^* = \sigma_i + \mathcal{E}_d$ consists of two parts: a chemical part σ_i and a misfit part $\mathcal{E}_d$.

We can repeat this process assuming now that the crystal B is not infinitely thick, but thinner than the double critical thickness $2t_c$ for pseudomorphous growth (Fig. 4.51). We again cleave both crystals reversibly and isothermally, strain uniformly the halves of B to match exactly the lattice parameter of A, and put the halves of A and B in contact as before. Carrying out this operation we strain the free surfaces of B and change the specific surface energy. Drechsler and Nicholas [1967] found that the change of the specific surface energy does not exceed several percent of the absolute value. We assume that this change is much smaller than the work done to strain the crystals and neglect it. Then we allow the bicrystal systems to relax. Misfit dislocations will not be introduced at the interface as the half-thickness of B is smaller than the equilibrium critical thickness for pseudomorphous growth and the pseudomorphous film is stable. The energy balance will read

$$\sigma_i^* = \sigma_A + \sigma_B - \beta + \mathcal{E}_e(f) = \sigma_i + \mathcal{E}_e(f) , \qquad (4.128)$$

where $\mathcal{E}_e$ is the strain energy per unit area of the interface stored in crystal B. The latter is a parabolic function of the lattice misfit f.

Finally, in the general case (the crystal B is again thin but is thicker than the double critical thickness $2t_c$ for pseudomorphous growth) part of the homogeneous strain is regained and misfit dislocations are introduced at the interface but their density is partially reduced owing to the residual strain. Then [Markov 1988]

$$\sigma_i^* = \sigma_A + \sigma_B - \beta + \mathcal{E}_e(f_e) + \mathcal{E}_d(f - f_e)$$

$$= \sigma_i + \mathcal{E}_e(f_e) + \mathcal{E}_d(f - f_e) \qquad (4.129)$$

and

$$\beta^* = \beta - \mathcal{E}_e(f_e) - \mathcal{E}_d(f - f_e) , \qquad (4.130)$$

where the strain energy $\mathcal{E}_e(f_e)$ and the dislocation energy $\mathcal{E}_d(f - f_e)$ depend on the homogeneous strain f_e and the mean dislocation density $f_d = f - f_e$, respectively.

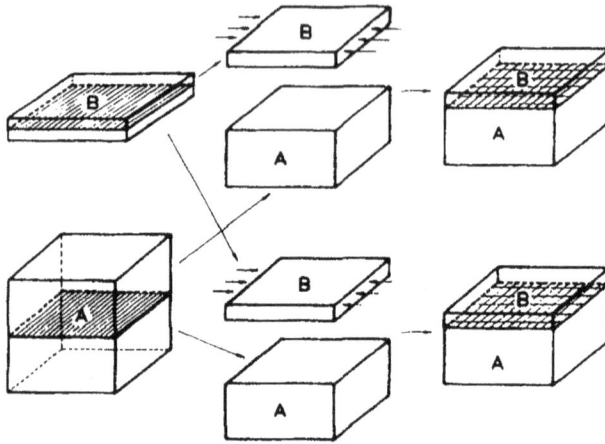

Fig. 4.51. For the derivation of the relation of Dupré in the case of thin films on a semi-infinite substrate.

4.3.2. *Thickness dependence of chemical potential*

As discussed in Chap. 1 the chemical potentials of the first deposited layers differ from the chemical potential of the bulk deposit crystal, μ_∞. First, the interaction with the substrate differs from that with the same crystal, and second, the lattice misfit leads to the appearance of homogeneous strain and/or misfit dislocations. We know that elastically strained crystals have higher chemical potentials and hence the homogeneous strain energy and the average value of the periodical strain energy due to the misfit dislocations contribute to the chemical potential of the film. On the other hand, the atom displacements affect the interaction across the interface and again lead to increase of the chemical potentials of the first few layers deposited on the foreign substrate.

The chemical potential and hence the equilibrium vapor pressure of a semi-infinite monolayer on a foreign substrate is a function of the separation work from half-crystal position $\varphi'_{1/2}$ where the prime reflects the influence of the foreign substrate. An atom in a half-crystal position is again connected with half atomic row, half crystal plane and the underlying half crystal block, but now the latter is of a different material and the corresponding energy of desorption, φ'_a, differs from the energy of desorption, φ_a, from the surface of the same crystal. Thus we can write the following general expression for the chemical potentials of the atoms in the first, second, third, etc. monolayers:

$$\mu(n) = \mu_\infty + a^2(\sigma + \sigma_i^* - \sigma_s) , \qquad (4.131)$$

where $\Delta\sigma = \sigma + \sigma_i^* - \sigma_s$ is the change of the surface energy connected with the deposition and is exactly equal to the difference of bonding between like and unlike substrates (see Chap. 1). The chemical potential of each monolayer is just equal to the chemical potential of the bulk crystal plus the difference of the bondings per atom with the like and unlike crystals. In fact Eq. (4.131) is equivalent to Eq. (1.59) with the exception that it accounts for the lattice misfit. In this respect it appears as a generalization of Eq. (1.59).

Substituting σ_i^* from (4.129) into (4.131) and replacing σ and β by φ_a and φ_a', respectively, give [Markov 1988]

$$\mu(n) = \mu_\infty + [\varphi_a - \varphi_a'(n) + \varepsilon_d(n) + \varepsilon_e(n)] , \qquad (4.132)$$

where $\varepsilon_d(n) = a^2\mathcal{E}_d(n)$ and $\varepsilon_e(n) = a^2\mathcal{E}_e(n)$ are now the energies per atom of the misfit dislocations and the homogeneous strain.

The term in the square brackets accounts for the strength of the interfacial bonding related to the strength of bonding with the same crystal. It includes the energy per atom, ε_d, of a cross grid of misfit dislocations as shown in the previous section. The last term in the brackets is the contribution of the homogeneous strain energy per atom, ε_e, to the chemical potential.

It is clear that when the substrate is of the same material as the deposit, $\varphi_a = \varphi_a'$, $f_d = f_e = 0$, $\varepsilon_d = \varepsilon_e = 0$ and $\mu(n) = \mu_\infty$. What follows is that the difference between the crystal growth and epitaxial growth is of purely thermodynamic nature. We could even define the different kinds of epitaxial growth on the base of Eq. (4.132). Thus if the main contribution to $\mu(n)$ comes from the difference in bonding, $\varphi_a - \varphi_a'(n)$, or in other words, from the nature of the chemical bonds in both partners, we have the case of heteroepitaxy. When the main contribution to $\mu(n)$ comes from the misfit energy, the bonding in both crystals remaining essentially the same, we consider that as homoepitaxial growth. Finally, when $\mu(n) = \mu_\infty$ we cannot speak of epitaxy at all.

φ_a' can be either greater or smaller than φ_a, and hence the term in the brackets can be either positive or negative and $\mu(n)$ can be either greater or smaller than μ_∞. In order to follow the $\mu(n)$ dependence we should consider the thickness dependence of the quantities involved in Eq. (4.132).

The adhesion energy per atom, φ_a', accounts only for the atomic interaction across the interface in the absence of misfit. For short range interactions it changes rapidly with film thickness, going to φ_a from above

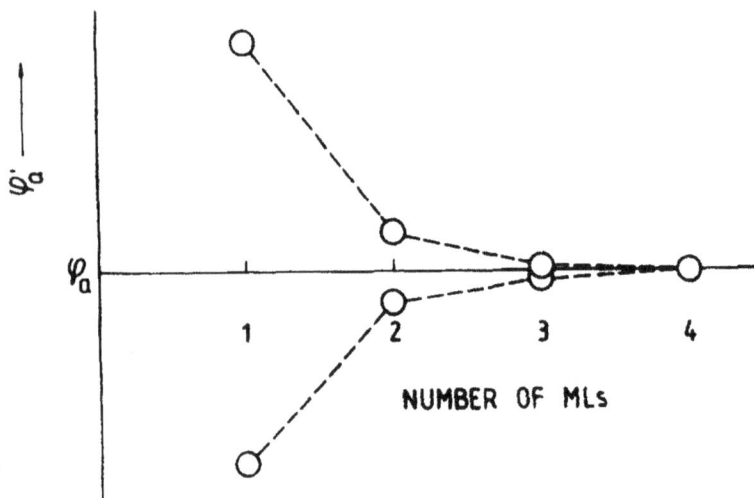

Fig. 4.52. Dependence of the energy of desorption of an atom from unlike substrate on the distance from the interface measured in number of monolayers. $\varphi'_a > \varphi_a$ corresponds to complete wetting and vice versa.

or from below (Fig. 4.52). The energetic influence of the substrate on the atoms of the second monolayer will be very weak and can be neglected except for some extreme cases.

As shown in the previous section, the thickness up to which ε_d and ε_e contribute to the chemical potential depends in a complicated way on the lattice misfit. The periodical strain due to misfit dislocations attenuates rapidly with the distance from the interface and practically vanishes beyond a distance equal to half of the misfit dislocation spacing, $\bar{p}/2$ (Fig. 4.41). The latter is inversely proportional to that part of the misfit f_d which is accommodated by misfit dislocations. Hence, $\varepsilon_d = 0$ for $t > \bar{p}/2$.

The homogeneous strain energy ε_e is a parabolic function of the homogeneous strain f_e and a linear function of the thickness t. When $t < t_c$, $\varepsilon_e = \varepsilon_e(f)$. Above the critical thickness for pseudomorphous growth, t_c, f_e and ε_e rapidly decrease and can be neglected. Hence, we can simplify our considerations assuming $\varepsilon_d = 0$ and $\varepsilon_e = \varepsilon_e(f)$ at $t < t_c$ and $\varepsilon_e = 0$ and $\varepsilon_d = \varepsilon_d(f)$ at $t > t_c$.

We consider now some typical cases. We assume first that $\varphi'_a > \varphi_a$, i.e. the adhesive forces are stronger than the cohesive ones. If the misfit is small enough so that $\varepsilon_e \ll \mu_\infty$ and t_c is large, then

$$\mu(1) = \mu_\infty + \varphi_a - \varphi'_a + \varepsilon_e < \mu_\infty$$

and

$$\mu(n) = \mu_\infty + \varepsilon_e \cong \mu_\infty, \qquad 2 \le n \le t_c/b \;.$$

If the energetic influence of the substrate is felt not only by the first monolayer but also by the atoms of the second monolayer, though much more weakly, one observes the behavior shown in Fig. 4.53 by solid circles.

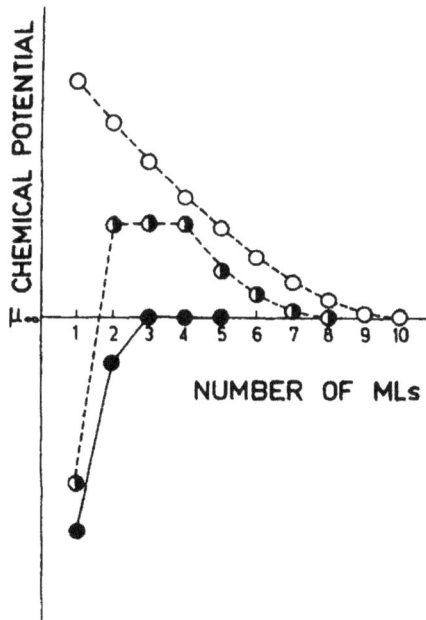

Fig. 4.53. Schematic presentation of the dependence of the chemical potential of the uppermost monolayer on film thickness in ultrathin films. Filled circles: Frank–van der Merwe mechanism or layer-by-layer growth; semifilled circles: Stranski–Krastanov mechanism or layer-by-layer growth followed by islands; empty circles: Volmer–Weber mechanism or island growth (after Stoyanov [1986]).

Let us assume now that $\varphi'_a > \varphi_a$ as before but the misfit is large, say, $f_4 < f < f_5$ (see Fig. 4.43) so that the energy of the homogeneous strain is not negligible compared with μ_∞. Then

$$\mu(1) = \mu_\infty + \varphi_a - \varphi'_a + \varepsilon_e < \mu \;,$$

$$\mu(n) = \mu_\infty + \varepsilon_e > \mu_\infty, \qquad 2 \le n \le 4 \;,$$

$$\mu(n) = \mu_\infty + \varepsilon_d(t) > \mu_\infty, \qquad n \ge 5 \;.$$

The latter $\mu(n)$ dependence is shown in Fig. 4.53 by the semifilled circles. The gradual decrease of $\mu(n)$ beyond the fourth monolayer reflects the decrease of the mean energy per atom of the periodical elastic strain due to the misfit dislocations with film thickness (see Fig. 4.42).

We assume further that $\varphi'_a < \varphi_a$. If the misfit is small enough so that t_c is large, then

$$\mu(1) = \mu_\infty + \varphi_a - \varphi'_a + \varepsilon_e(f) > \mu_\infty \ ,$$

$$\mu(n) = \mu_\infty + \varepsilon_e > \mu_\infty, \qquad 2 \le n \le t_c/b \ .$$

Beyond t_c, ε_e vanishes and misfit dislocations are introduced at the interface so that ε_e is replaced by ε_d.

If the misfit is large and in the extreme case larger than f_1, $\varepsilon_e = 0$ from the very beginning,

$$\mu(1) = \mu_\infty + \varphi_a - \varphi'_a + \varepsilon_d > \mu_\infty \ ,$$

$$\mu(n) = \mu_\infty + \varepsilon_d(n) > \mu_\infty, \qquad 2 \le n \le p/2b \ ,$$

$$\mu(n) = \mu_\infty, \qquad n > p/2b \ .$$

Let us consider the last case in more detail. The $\mu(n)$ dependence is decreasing, thus reflecting the decrease of the periodical strain with film thickness (see Fig. 4.41). Then every next monolayer will have a chemical potential smaller than that of the previous one and will start to form before the completion of the latter. In such a case the formation of islands thicker than one monolayer is thermodynamically favored and their equilibrium with the vapor phase will be realized through a half-crystal position of a multilayer height (Fig. 4.50). Then the chemical potential of the film consisting of several atomic monolayers, $\mu(1 \to n)$, will be given by the mean value of the chemical potentials of the constituent monolayers:

$$\mu(1 \to n) = \frac{1}{n} \sum \mu(n)$$

(Fig. 4.53, open circles), as discussed above.

4.3.3. *Thermodynamic criterion for modes of growth*

It follows that depending on the interrelation between the adhesive and cohesive forces on the one hand, and the value of the misfit, resulting in an interplay between the energies of the misfit dislocations and homogeneous strain, on the other, three different types of thickness dependence of the chemical potential can be distinguished:

1. $d\mu/dn < 0$ when $\varphi'_a < \varphi_a$ at any value of the misfit,
2. $d\mu/dn > 0$ when $\varphi'_a > \varphi_a$ at small misfits,
3. $d\mu/dn \lessgtr 0$ when $\varphi'_a > \varphi_a$ at large misfits.

Obviously when $d\mu/dn$ is positive the completion of the first monolayer before the start of the second one, of the second before the start of the third, etc., is thermodynamically favored — layer-by-layer growth is expected. In the opposite case the formation of a second monolayer before the completion of the first one is thermodynamically favored and the formation of 3D islands should take place. It follows that the above inequalities define the *thermodynamic criterion* for the mechanism of growth of thin epitaxial films [Stoyanov 1986; Grabow and Gilmer 1988]:

1. VOLMER–WEBER growth when $d\mu/dn < 0$,
2. FRANK–van der MERWE growth when $d\mu/dn > 0$,
3. STRANSKI–KRASTANOV growth when $d\mu/dn \lessgtr 0$.

Once the chemical potential acquires its bulk value μ_∞ the epilayer will grow further by the simultaneous growth of several monolayers [Borovinski and Tzindergozen 1968; Gilmer 1980a; Chernov 1984].

It is worth noting that the above thermodynamic criterion is in fact equivalent to that given by Bauer [1958] in terms of the specific surface energies. The latter becomes clear if one looks at Eq. (4.131).

4.3.4. *Kinetics of growth of thin epitaxial films*

We are now in a position to study the growth of thin epitaxial films bearing in mind the above thickness dependence of the chemical potential. The latter predicts the equilibrium morphology of the deposit whereas the deposition process is usually carried out under conditions far from equilibrium. So we have to study the question of how the substrate temperature and rate of deposition affect the mechanism of growth.

We consider the case of complete condensation when all atoms arriving at the crystal surface join sites of growth before re-evaporation. As in the case of growth of a defectless atomically smooth crystal face, the atoms from the vapor phase strike the substrate and, after a period of thermal accommodation, randomly walk to give rise of 2D nuclei. The 2D nuclei grow further by the attachment of adatoms diffusing to their edges on the substrate surface and on their exposed surface as well. An adatom population is formed on top of them (Fig. 4.54(a)) whose concentration $n_s(r)$ can be found upon solving the master equation (3.125) subject to the boundary conditions $n_s(r = \rho) = n_{s1}^e$ and $(dn_s/dr)_{r=0} = 0$. The solution reads (see Eqs. (3.127) and (3.135))

$$n_s(r) = n_{s1}^e + \frac{\mathcal{R}}{4D_s}(\rho^2 - r^2) , \qquad (4.133)$$

where $\mathcal{R} = P(2\pi mkT)^{-1/2}$ cm^{-2}sec^{-1} is the atom arrival rate. The quantity n_{s1}^e is the concentration of adatoms in equilibrium with the island edges given by

$$n_{s1}^e = n_{se} \exp\left(\frac{\mu(1) - \mu_\infty}{kT}\right) , \qquad (4.134)$$

where n_{se} is the adatom concentration on the surface of the same bulk deposit crystal given by Eq. (3.18).

Equation (4.133) shows a parabolic dependence of the adatom concentration on the distance from the island center (Fig. 4.55(a)) which displays a maximum

$$n_{s,max} = n_{s1}^e + \frac{\mathcal{R}}{4D_s}\rho^2$$

just over the island center. The increase in ρ leads to values of $n_{s,max}$ high enough to give rise to nuclei on top of the islands (Fig. 4.54(b)). Thus nuclei of the second monolayer appear before the completion of the first one. Once such nuclei are formed they grow initially at the expense of the atoms diffusing to their edges on the terrace between the edges of the upper and lower islands. The adatom concentration on the terrace (Fig. 4.55(b)) is given by Eq. (3.136) subject to the boundary conditions $n_s(\rho_1) = n_{s1}^e$ and $n_s(\rho_2) = n_{s2}^e$, where

$$n_{s2}^e = n_{se} \exp\left(\frac{\mu(2) - \mu_\infty}{kT}\right) \qquad (4.135)$$

is the adatom concentration which is in equilibrium with the edges of the second monolayer islands [Markov and Stoyanov 1987].

The solution reads (c.f. Eq. (3.136'))

$$n_s(r) = n_{s1}^e + \frac{\mathcal{R}}{4D_s}\left(\rho_1^2 - r^2\right) - \left(\Delta n_s + \frac{\mathcal{R}}{4D_s}\left(\rho_1^2 - \rho_2^2\right)\right) \frac{\ln\left(r/\rho_1\right)}{\ln\left(\rho_2/\rho_1\right)} , \qquad (4.136)$$

where $\Delta n_s = n_{s1}^e - n_{s2}^e$.

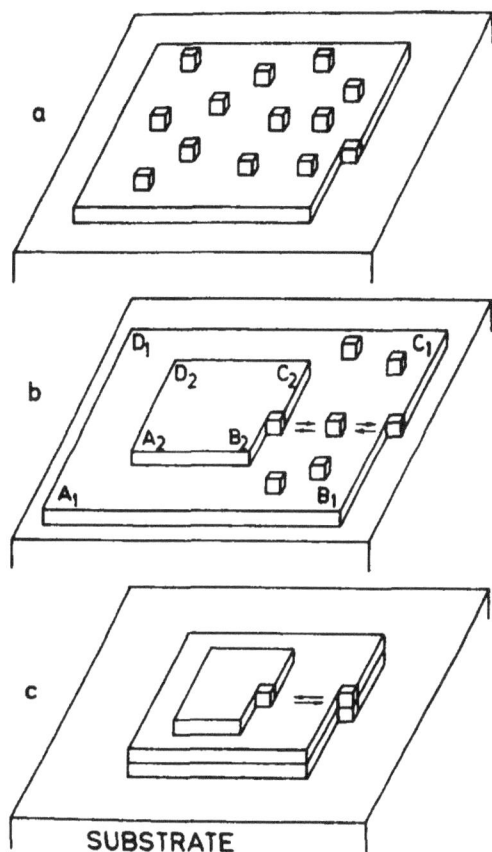

Fig. 4.54. Subsequent stages of film growth and atom exchange between the kinks and the dilute adlayer. (a) The concentration of atoms adsorbed on top of the first monolayer island increases with island size, which leads to nucleation of 2D islands of the second monolayer. (b) Surface transport from the edges $A_1B_1C_1D_1$ to the edges $A_2B_2C_2D_2$ takes place when $\mu(2) < \mu(1)$. (c) Surface transport transforms the layer configuration into a crystal of two-monolayer height, which grows further by nucleation of islands of the third monolayer. (I. Markov and S. Stoyanov, *Contemp. Phys.* **28**, 267 (1987). By permission of Taylor & Francis Ltd.)

Suppose now that $\mu(1) > \mu(2)$, i.e. $d\mu/dn < 0$. The adatom population on top of the first monolayer islands is supersaturated with respect to the bulk deposit crystal. This favors nucleation on top of the first monolayer islands and the thermodynamic driving force for nucleation to occur should be greater than $\Delta\mu = \mu(1) - \mu_\infty$. On the other hand, $n_{s1}^e > n_{s2}^e$ and *surface*

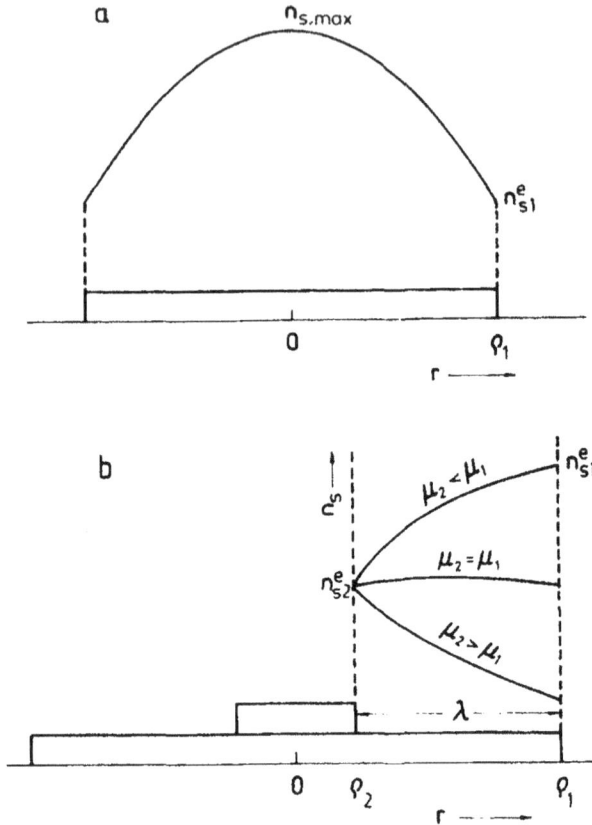

Fig. 4.55. Profile of the adatom concentration on (a) the surface of a monolayer island and (b) the terrace formed between the edges of the lower and upper monolayer islands with radii ρ_1 and ρ_2, respectively. n_{s1}^e and n_{s2}^e are the adatom concentrations in equilibrium with the corresponding island edges.

transport from the edges of the lower island to the edges of the upper island will take place whose driving force is given by $\Delta n_s/\lambda = (n_{s1}^e - n_{s2}^e)/\lambda$, where λ is the distance between the edges (Fig. 4.55(b)). Thus the upper islands will grow at the expense of the lower islands and after some time will catch up with the lower islands to produce islands with a double height (Fig. 4.54(c)). Hence, at temperatures high enough to facilitate the surface transport island growth will be observed. However, if the temperature is low the surface transport from edge to edge will be hindered so that the first monolayer islands will grow laterally to coalesce and cover completely the substrate before a significant growth on top of them takes place. Layerlike growth will occur for kinetic reasons. However, such films grown at low temperature are metastable. Upon heating they will break up and agglomerate into 3D islands. Note that the growth will not follow the true layer-by-layer mechanism (complete coverage of the substrate by one monolayer before the next one to nucleate) as the thermodynamics favor island growth.

In the opposite case, $\mu(1) < \mu(2)$ $(d\mu/dn > 0)$ (Fig. 4.55(b)), the islands of the second monolayer will have a chemical potential higher than that of the lower islands and surface transport of atoms will occur from the edges of the upper islands to the edges of the lower islands. As a result the upper islands will decay. Thus, layer-by-layer growth will be observed irrespective of the temperature.

Finally, when $d\mu/dn$ changes its sign with film thickness the first monolayers will grow layer by layer for reasons given above. Once a particular thickness is reached (the so-called Stranski–Krastanov thickness) such that the corresponding chemical potential is higher than μ_∞, 3D islands will form and grow at high temperatures. Surface transport from the edges of more elastically strained islands to the edges of islands less strained or not strained at all will take place. As a result the Stranski–Krastanov mechanism will be observed. At low temperatures the growth will proceed further by successive formation of monolayers. Again if such low temperature films are annealed at higher temperatures the material in excess of the first stable monolayers (for which $d\mu(n)/dn > 0$) will break up and agglomerate into 3D islands.

It is important to note once more that a true layer-by-layer growth takes place only when the chemical potential is an increasing function of the film thickness, i.e. $d\mu/dn > 0$. At low temperatures and $d\mu/dn < 0$, the film growth will proceed by simultaneous growth of several monolayers as shown in Chap. 3.

Thus, we have to expect a change of the mechanism of growth from layerlike to either Volmer–Weber or Stranski–Krastanov with increasing temperature. The necessary condition for a transition to occur is $d\mu/dn < 0$. In other words, the thermodynamics should favor island growth either on the substrate surface or on several stable monolayers of the deposit. Our next task is to find the critical temperature for transition to occur.

We consider the case where $\mu_1 > \mu_2 > \mu_3 \ldots$ so that island growth is expected under near-to-equilibrium conditions. The same is valid in

the case of Stranski–Krastanov growth after completion of the first stable
adlayers. As discussed above, 2D nuclei of the second, third, etc. monolayers are formed on top of the first monolayer islands which results in the
formation of flat pyramids of growth as shown in Fig. 3.26. A very nice
picture of such pyramids of growth of Cu on Ru(0001) can be seen in the
paper of Pötschke *et al.* [1991]. As the chemical potential is a decreasing
function of the monolayer number n the surface transport will be directed
from the lower to the upper steps. We make use of the solutions of the
diffusion equation (3.125) subject to the boundary conditions $n_s(\rho_1) = n_{s1}^e$,
$n_s(\rho_2) = n_{s2}^e$, $n_s(\rho_3) = n_{s3}^e$, etc. assuming rapid exchange of atoms between
the steps and the dilute adlayers on the terraces (diffusion regime). We
then obtain solutions for $n_s(r)$ on every terrace (Eq. (4.136)) and, following
the same procedure as in Chap. 3, we calculate the rates of advance of the
circular steps, $v_n = d\rho_n/dt$ [Stoyanov and Markov 1982].

Thus for the first monolayer island we have

$$\frac{d\rho_1}{dt} = \frac{\mathcal{R}}{2\pi N_0 N_s \rho_1} \left(1 - \frac{\pi \rho_1^2 N_s (1 - \rho_2^2/\rho_1^2)}{\ln(\rho_1^2/\rho_2^2)} \right) - \frac{2\Delta n_s^e D_s}{N_0 \rho_1 \ln(\rho_1^2/\rho_2^2)} \, ,$$

where N_s is the density of the growth pyramids formed by successive
nucleation per unit area of the substrate and $\Delta n_s^e = n_{s1}^e - n_{s2}^e$.

In this expression, the first term on the right-hand side which contains
$\mathcal{R}$ is always positive as the surface coverage $\pi \rho_1^2 N_s$ is smaller than unity
before the coalescence and $\rho_1 > \rho_2$. The second term which contains the
equilibrium concentration difference Δn_s^e is also positive as $n_{s1}^e > n_{s2}^e$. It
follows that v_1 can be either positive or negative, depending on the values
of the deposition rate $\mathcal{R}$ and the difference Δn_s^e which is a steep function of
the temperature. In the extreme case of an absence of deposition (annealing
at $\mathcal{R} = 0$), the first term in the right-hand side is equal to zero and $v_1 < 0$,
thus reflecting the process of detachment and transport of atoms from the
lower monolayer island edge to the edge of the upper island. The same
process takes place during deposition, but at a higher temperature, when
the negative term overcompensates the positive one. If this occurs before
the coalescence of the first monolayer islands, say, at a surface coverage
$\Theta_1 = \pi \rho_1^2 N_s < 0.5$, island growth has to be expected. On decreasing
the temperature, Δn_s^e decreases, and under a given temperature the term
containing Δn_s^e has a negligible contribution to v_1. The rate v_1 is then just
the same as in the case of deposition on its own substrate, i.e. in the case
of a growth of a bulk crystal when $\mu(n) = \mu_\infty$.

We have to solve now a set of differential equations for the rate of advance of the steps. The latter can be written in terms of surface coverages $\Theta_n = \pi \rho_n^2 N_s$ $(n = 1, 2, 3, \ldots)$ as a function of a dimensionless time $\theta = \mathcal{R}t/N_0$, which is in fact the number of monolayers deposited, in the form

$$\frac{d\Theta_1}{d\theta} = 1 - \frac{M_1 + \Theta_1 - \Theta_2}{\ln(\Theta_1/\Theta_2)} \ ,$$

$$\frac{d\Theta_n}{d\theta} = \frac{M_{n-1} + \Theta_{n-1} - \Theta_n}{\ln(\Theta_{n-1}/\Theta_n)} - \frac{M_n + \Theta_n - \Theta_{n+1}}{\ln(\Theta_n/\Theta_{n+1})} \ , \qquad (4.137)$$

$$\frac{d\Theta_N}{d\theta} = \frac{M_{N-1} + \Theta_{N-1} - \Theta_N}{\ln(\Theta_{N-1}/\Theta_N)} \ ,$$

where the subscript N denotes the uppermost monolayer and the parameters

$$M_n = \frac{4\pi D_s N_s (n_{s,n}^e - n_{s,n+1}^e)}{\mathcal{R}} \qquad (4.138)$$

include all the materials quantities and the differences of the adatom concentrations, or in other words, the differences of the chemical potentials (see Eqs. (4.134) and (4.135)).

4.3.5. Critical temperature for transition from 2D to 3D growth

A numerical analysis of (4.137) shows that the solutions for Θ_n and hence the time evolution of the shape of the growth pyramids are very sensitive to the values of M_n. The latter are strongly increasing functions of temperature and are inversely proportional to the atom arrival rate $\mathcal{R}$. When the chemical potentials are independent of the layer number, or in other words, $n_{sn}^e = n_{se}$, $M_n = 0$. In this case there is no directed surface transport between the steps and the growth pyramids preserve their shape. This means that the epitaxial film will grow as the bulk crystal face following the 2D nucleation mechanism with simultaneous growth of several layers. It is immediately seen that the system (4.137) turns into (3.144) for a pyramid of growth consisting of 2D islands with $M_n = 0$.

Let us consider the simplest case of the bilayer pyramid shown in Fig. 4.54 or Fig. 4.55(b), assuming in addition that $n_{s2}^e = n_{se}$, i.e. $\mu(2) = \mu_\infty$. This is a sufficient condition for islands growth to take place as bilayer islands will be formed as a result of it. As discussed in Chap. 3, the bilayer steps propagate more slowly than single steps and the double step will be caught up with by the upper step. Trilayer island will be formed, etc. The numerical solution of the system (4.137) in the case of a bilayer pyramid is shown in Fig. 4.56. At $M_1 = 0.25$, Θ_1 initially increases, displays a maximum $\Theta_1 = 0.5$ and then decreases. The latter means that at some stage of growth the rate of advance of the first monolayer island, $d\rho_1/dt$, becomes negative, or in other words, the first island decays and the atoms feed the second island. Then the edges of the latter catch up with the edges of the former and an island with double height is produced (Fig. 4.54(c)). The double step advances more slowly than the single step, and after some time the third step catches up the double step, thus producing an island with triple height. As a result, island growth takes place, the *kinetic criterion* for it being

$$\frac{4\pi D_s N_s (n_{s1}^e - n_{se})}{\mathcal{R}} \geq 0.25 . \tag{4.139}$$

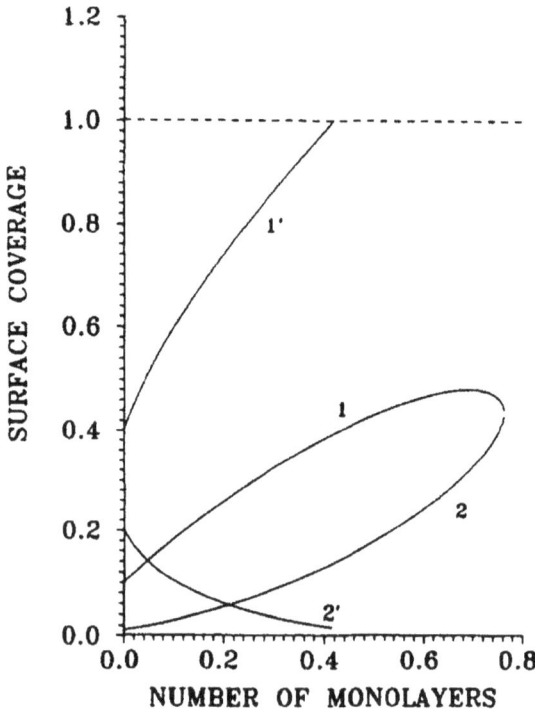

Fig. 4.56. Dependence of the surface coverages of the first (curves 1 and 1′) and second (curves 2 and 2′) monolayers on the number of monolayers deposited. Curves 1 and 2: $M_1 = 0.25$ ($\mu(1) > \mu(2)$); curves 1′ and 2′: $M = -1.5$ ($\mu(1) < \mu(2)$).

The physical meaning of this criterion becomes transparent if we write it in the form

$$\frac{4\pi D_s (n_{s1}^e - n_{se})}{\mathcal{R}/N_s} \geq 0.25 .$$

The numerator represents the total diffusion flux from the edge of the lower island to that of the upper island resulting from the difference of the equilibrium adatom concentrations. The denominator is equal to the number of atoms joining one pyramid per unit time. Therefore, the criterion simply states that in order for island growth to take place the diffusion flux from edge to edge should be equal to or larger than 25% of the total number of atoms joining the pyramid. An increase of the deposition rate $\mathcal{R}$ leads to an increase of the overall growth rate of the pyramid without affecting the diffusion flux which is responsible for the transformation of the pyramid to a 3D island. The result is a transition to layer-by-layer growth. An increase of temperature has the opposite effect: it results in a faster surface transport, which in turn facilitates the 2D to 3D transformation.

It is interesting to see what will happen when M_1 has a negative value, i.e. when $\mu(1) < \mu_\infty$. As seen in Fig. 4.56 the surface coverage of the second monolayer, Θ_2, decreases, thus reflecting the fact that the surface transport is directed from the upper to the lower islands edges. The lower islands grow at the expence of the upper ones and cover completely the substrate. True layer-by-layer growth results.

Making use of Eqs. (4.134), (3.18) and (3.20) for the transition temperature T_t from (4.136), one obtains

$$T_t = \frac{(\varphi_{1/2} - \varphi_a) - [\mu(1) - \mu_\infty] + \varphi_{sd}}{k \ln(16\pi\nu N_s/\mathcal{R})} . \tag{4.140}$$

Substituting the difference of the chemical potentials from Eq. (4.132) into Eq. (4.140) for the transition from layerlike growth to island growth, one obtains

$$T_t = \frac{[(\varphi_{1/2} - \varphi_a) - (\varphi_a - \varphi_a')] - \varepsilon_d + \varphi_{sd}}{k \ln(16\pi\nu N_s/\mathcal{R})} . \tag{4.141}$$

Bearing in mind that the lateral bonds of an atom in a kink position remain practically unchanged, $\varphi_{1/2} - \varphi_a \cong \varphi_{1/2}' - \varphi_a'$, and the energy difference in the square brackets is just equal to the energy $\varphi_{1/2}' - \varphi_a$ for the transfer of an atom from a kink position in the step of the first monolayer island to the dilute adlayer on top of it.

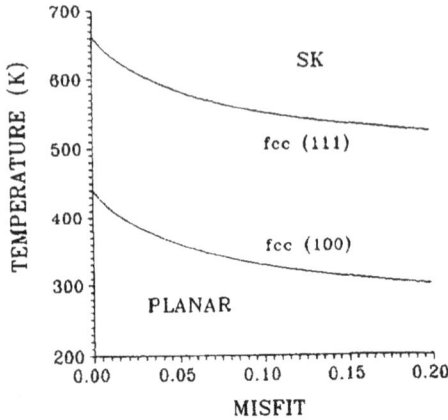

Fig. 4.57. Plot of the temperature (in K) for transition from a planar (layerlike) growth to Stranski–Krastanov growth as a function of the natural misfit at two different orientations (111) and (100) of fcc substrate crystal. An average enthalpy of evaporation, $\Delta H_{ev} \equiv \varphi_{1/2} = 70$ kcal mol^{-1}, has been taken. The misfit dislocation energy $\mathcal{E}_d$ is computed using the theory of van der Merwe (Eq. (4.104)) with an average shear modulus $G = 5 \times 10^{11}$ dyne/cm^2 and Poisson's ratio $\nu = 0.3$.

Equation (4.141) is valid for transition from layer-by-layer to island growth where the contribution of the interatomic forces across the interface to the chemical potential is greatest. In the case of transition from layer-by-layer to Stranski–Krastanov growth (the latter taking place after formation of one or two stable adlayers) $\varphi_a - \varphi_a'$ can be neglected. Assuming that the stable monolayers are pseudomorphous with the substrate and the 3D islands are relaxed, Eq. (4.140) simplifies to

$$T_t = \frac{(\varphi_{1/2} - \varphi_a) - \mathcal{E}_d + \varphi_{sd}}{k \ln(16\pi\nu N_s/\mathcal{R})} . \tag{4.142}$$

Then the surface transport will occur from the edges of more elastically strained monolayer islands to the edges of less strained, or unstrained at all, monolayer islands.

It should be noted that whereas $\varphi_{1/2}$ is characteristic for the bulk material, φ_a and φ_a' depend on the crystallographic orientation of the substrate. It follows that the critical temperatures will be higher for the (111) face of an fcc crystal ($\varphi_a = 3\psi$, $\varphi_a' = 3\psi'$) than for the (100) face ($\varphi_a = 4\psi$, $\varphi_a' = 4\psi'$). This is indeed what is observed in a series of experiments (for a review, see Markov and Stoyanov [1987]; Markov [1983]).

The temperature for transition from layer-by-layer to Stranski-Krastanov growth (Eq. (4.142)) is plotted against the natural misfit in Fig. 4.57. The misfit dislocation energy per atom $\mathcal{E}_d$ is calculated as a function of the misfit by using the theory of van der Merwe (Eq. (4.104)) with an average shear modulus $G = 5 \times 10^{11}$ dyne/cm^2 and Poisson's ratio $\nu = 0.3$. An average enthalpy of evaporation, $\Delta H_{ev} \equiv \varphi_{1/2} = 70$ kcal mol^{-1}, has been taken. As seen the temperature decreases by no more than 140–150 degrees when the misfit increases from zero to 0.20, and a transition from planar to Stranski–Krastanov growth can be expected with increasing temperature or misfit. Besides, the transition temperature for the more closed packed surface is more than 200 degrees higher in comparison with the less closed packed surface. A transition from planar to Stranski–Krastanov growth can be expected even at room temperature for the latter. As will be shown below this result is in fairly good agreement with experimental observations.

Thus the morphology of the growing epitaxial film is a result of the kinetics of deposition and can be quite different from the morphology required by the thermodynamic criterion. As seen the criterion (4.140) accounts for both the kinetics and the thermodynamics of the growth of thin epilayers.

Let us go back to some of the experimental data mentioned at the beginning of this chapter. We consider first the transition from layer-by-layer to Stranski–Krastanov growth in the case of deposition of Cu on W(110) and W(100) [Bauer et al. 1974]. The upper stable adlayer has 1.6×10^{15} atoms/cm^2, and assuming an approximate (111) structure of the adlayer the average atom spacing is 2.686 Å. Bearing in mind that the first neighbor distance of Cu is 2.556 Å, the natural misfit between the copper crystallites and the underlying strained copper adlayer is $f \cong 0.05$. $G_a = G_b = G_i = G = 5.46 \times 10^{11}$ dyne/cm^2, $\nu_a = \nu_b = \nu = 0.324$, $G' = G/2(1 - \nu)$ and $\lambda = \pi f/(1 - \nu) = 0.23$. Then using the theory of van der Merwe (Eq. (4.104)) we find $\mathcal{E}_d = 310$ erg/cm^2. Then $\varepsilon_d = a^2 \mathcal{E}_d \cong 2 \times 10^{-20}$ J/atom = 0.126 eV/atom. The enthalpy of evaporation of Cu is $\Delta H_e = 72800$ cal/mole and $\varphi_{1/2} - \varphi_a = 36400$ cal/mole = 2.5×10^{-19} J/atom = 1.58 eV/atom. With $\nu \cong 3 \times 10^{13}$ sec^{-1}, $N_s \cong 1 \times 10^{10}$ cm^{-2}, $\mathcal{R} = 5.9 \times 10^{12}$ cm^{-2}sec^{-1}, and neglecting φ_{sd}, we find $T_t = 600$ K, in good agreement with the experimentally found value 700 K. The same calculations performed for the case of deposition of Cu on W(100) give $f = 0.21$, $\lambda = 0.976$, $\mathcal{E}_d = 600$ erg/cm^2 and $\varepsilon_d = 0.245$ eV/atom. Bearing in mind that $\varphi_a = 4\psi$, $\varphi_{1/2} - \varphi_a = 26830$ cal/mole = 1.053 eV/atom,

T_t = 330 K. As mentioned above, 3D islands have been experimentally observed to form on the stable adlayers even at room temperature.

As seen, Eq. (4.142) predicts quantitatively the transition from layer to Stranski–Krastanov mode of growth and gives correctly the dependence of the transition temperature on the crystallographic orientation of the substrate. As for transition from layer to island growth we need reliable values for the adhesion energy φ'_a. It follows that Eq. (4.141) will operate satisfactorily when the enthalpies of evaporation and the bond strengths of both materials do not differ too much provided the nature of the chemical bonds is one and the same. We can assume then that the adhesion energy lies between the cohesion energies of the two materials. We will consider as an example the deposition of Ge on Si(111) and Si on Ge(111) [Marée, Barbour and van der Veen 1987].

A transition from layer to Stranski–Krastanov growth is observed when Ge is deposited on Si(111), the transition temperature being 500°C. Making use of Eq. (4.142) and following the procedure outlined above we find $\mathcal{E}_d$ = 400 erg/cm^2 and ε_d = 0.4 eV/atom, $\varphi_{1/2} - \varphi_a$ = 44750 cal/mole = 1.94 eV/atom. With N_s = 1 × 10^9 cm^{-2} and $\mathcal{R}$ = 0.1 ML per sec = 7.2 × 10^{13} cm^{-2}sec^{-1}, T_t = 480°C, in excellent agreement with the experimental observation.

The transition from layer-by-layer to island growth in the deposition of Si on Ge(111) is more difficult to handle. We assume first that the shear modulus at the interface, G_i, has a value in between the shear moduli of Si and Ge, 6.41 × 10^{11} dyne/cm^2 and 5.46 × 10^{11} dyne/cm^2, respectively. We accept the average value $G_i = (G_{Ge}G_{Si})^{1/2}$ = 5.9 × 10^{11} dyne/cm^2. Then $\mathcal{E}_d$ = 600 erg/cm^2 and ε_d = 0.6 eV/atom. The same assumption is made for evaluating φ'_a. From φ_a(Si) = ΔH_e(Si)/2 = 54450 cal/mole and φ_a(Ge) = ΔH_e(Ge)/2 = 44750 cal/mole, we find $\varphi'_a \cong [\varphi_a(\text{Si})\varphi_a(\text{Ge})]^{1/2}$ = 49360 cal/mole. Then $\varphi_{1/2} - \varphi_a$ = 54450 cal/mole = 2.362 eV/atom, $\varphi_a - \varphi'_a$ = 5090 cal/mole = 0.221 eV/atom. With $\mathcal{R}$ = 7.2 × 10^{13} cm^{-2}sec^{-1} and N_s = 4 × 10^7 cm^{-2} and neglecting again φ_{sd} for the transition temperature one finds T_t = 590°C. It is worth noting that the approximation concerning φ'_a is quite reasonable. If we approximate φ'_a by either φ_a (Si) or φ_a (Ge), then T_t = 720°C or 480°C, respectively. We can conclude that the real value of φ'_a lies, indeed, between φ_a (Si) and φ_a (Ge).

We have to bear in mind, however, that the theoretically predicted values of T_t are underestimated as the activation energies for surface diffusion have been neglected. In the case of diffusion of Si atoms on Si(111), the value 1.3 eV has been reported [Farrow 1974; Kasper 1982]. In addition

Sakamoto, Miki and Sakamoto [1990] found that the surface diffusion on a vicinal Si(111) surface is anisotropic. Another uncertainty in calculating the transition temperatures comes from the application of the theory of van der Merwe [1973] to the calculations of the misfit dislocation energy in the case of materials with covalent bonds which are considered as brittle and inflexible. Additional uncertainty comes from using the nearest neighbor model for the calculation of the desorption energies. It is thus surprising that irrespective of all the approximations made in the quantities involved in Eqs. (4.141) and (4.142), the latter are in good semiquantitative agreement with the experimental data.

Equations (4.141) and (4.142) explain readily the transition from layer-by-layer to Stranski–Krastanov or island growth with increasing misfit. As an example we will consider the deposition of $Si_{1-x}Ge_x$ on Si [Kasper, Herzog and Kibbel 1975]. When the composition x varies from 0.15 to 0.25 the natural misfit varies from 0.006 to 0.01, the parameter λ varies from 0.024 to 0.04 and ε_d varies from 0.12 to 0.175 eV. The critical temperature for 2D–3D transition decreases by about 30 degrees which is sufficient to change the mode of growth when the deposition is carried out at a constant temperature.

4.3.6. Cross hatch patterns

The theoretical model described above gives an explanation of the appearance of the so-called "cross hatch patterns" (see Franzosi *et al.* [1986] and the references therein). The latter represents an array of parallel lines or a grid of two arrays of mutually perpendicular lines on the surface of the growing epilayer where the latter is thicker than the remaining part of the film. Detailed investigation of the phenomenon in the case of growth of $In_xGa_{1-x}As$ on InP(100) [Franzosi *et al.* 1986] showed that each cross hatch line corresponds to a dislocation line. Thus the cross hatch pattern appears only when the interface is resolved in a cross grid of misfit dislocations although one-to-one correspondence between the hatch lines and the dislocation lines has never been found. Besides, cross hatch patterns have been observed on the surface of the films under both tensile and compressive stress. Cross hatch patterns have never been observed on the surface of pseudomorphous films. Bearing in mind the thermodynamical analysis of the equilibrium morphology of growing epilayers given in this chapter it is easy to assume that the parts of the film which are just over the dislocation lines are elastically relaxed (if $t < p/2$), whereas the film remains under misfit stress in between the lines. It follows that the chemical potential of the film over the dislocation lines is lower in comparison with that in the regions between the dislocation lines (Fig. 4.58) and the variation of μ is just given by the energy of the homogeneous strain ε_e. Then a surface transport of adatoms from regions with enhanced chemical potential to regions with lower chemical potential (denoted by the arrows in Fig. 4.58) takes place just as in the case shown in Fig. 4.54(b). The parts over the dislocation lines grow thicker than the remaining parts of the film and a cross hatch pattern results.

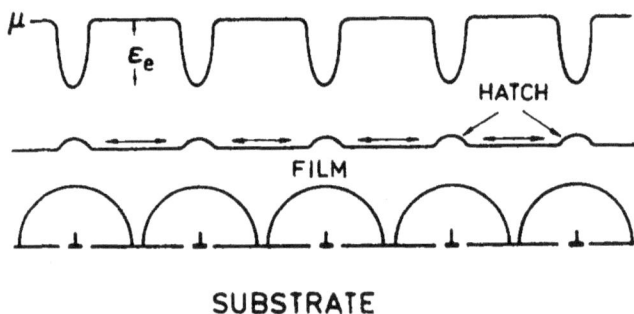

SUBSTRATE

Fig. 4.58. A schematic cross-sectional view of a cross hatch pattern. The upper curve illustrates a possible variation of the chemical potential of the crystal surface due to nonuniform distribution of misfit strain. The arrows show the direction of surface transport.

References

1. Adam, N. K., *The Physics and Chemistry of Surfaces* (Dover Pub. Inc., 1968).
2. Alerhand, O. L., Vanderbilt, D., Meade, R. D., and Joannopoulos, J. D., *Phys. Rev. Lett.* **61**, 1973 (1988).
3. Alerhand, O. L., Berker, A. N., Joannopoulos, J. D., Vanderbilt, D., Hamers, R. J., and Demuth, J. E., *Phys. Rev. Lett.* **64**, 2406 (1990).
4. Arfken, G., *Mathematical Methods for Physicists*, 2nd ed. (Academic, 1973).
5. Armstrong, R. D. and Harrison, J. A., *J. Electrochem. Soc.* **116**, 328 (1969).
6. Asai, M., Ueba, H., and Tatsuyama, C., *J. Appl. Phys.* **58**, 2577 (1985).
7. Ashu, P., Matthai, C. C., and Shen, T. H., *Surf. Sci.* **251/252**, 955 (1991).
8. Asonen, H., Barnes, C., Salokatve, A., and Vuoristo, A., *Appl. Surf. Sci.* **22/23**, 556 (1985).
9. Aspnes, D. E. and Ihm, J., *Phys. Rev. Lett.* **57**, 3054 (1986).
10. Aubry, S., *J. Phys.* **44**, 147 (1983).
11. Avrami, M., *J. Chem. Phys.* **7**, 1103 (1939).

12. Avrami, M., *J. Chem. Phys.* **8**, 212 (1940).
13. Avrami, M., *J. Chem. Phys.* **9**, 177 (1941).
14. Avron, J. E., Balfour, L. S., Kuper, C. G., Landau, L., Lipson, S. G., and Schulman, L. S., *Phys. Rev. Lett.* **45**, 814 (1980).
15. Babkin, A. V., Keshishev, K. O., Kopelovitch, O. B., and Parshin, A. Ya, *Sov. Phys. JETP Lett.* **39**, 633 (1984).
16. Bak, P., *Rep. Progr. Phys.* **45**, 587 (1982).
17. Bak, P. and Emery, V. J., *Phys. Rev. Lett.* **36**, 978 (1976).
18. Barbier, L. and Lapujoulade, J., *Surf. Sci.* **253**, 303 (1991).
19. Barbier, L., Khater, A., Salanon, B., and Lapujoulade, J., *Phys. Rev.* **B43**, 14730 (1991).
20. Barker, T. V., *J. Chem. Soc. Trans.* **89**, 1120 (1906).
21. Barker, T. V., *Mineral. Mag.* **14**, 235 (1907).
22. Barker, T. V., *Z. Kristallografiya* **45**, 1 (1908).
23. Barone, A., Esposito, F., Magee, C. J., and Scott, A. C., *Riv. Nuovo Cimento* **1**, 227 (1971).
24. Barrer, R. M., in *Surface Organometallic Chemistry: Molecular Approaches to Surface Catalysis*, eds. J. M. Basset, B. C. Gates, J. P. Candy, A. Choplin, M. Leconte, F. Quignard, and C. Santini (Kluwer, 1988), p. 221.
25. Bartha, J. W. and Henzler, M., *Surf. Sci.* **160**, 379 (1985).
26. Bauer, E., *Z. Kristallografiya* **110**, 372 (1958).
27. Bauer, E., Poppa, H., Todd, G., and Bonczek, F., *J. Appl. Phys.* **45**, 5164 (1974).
28. Bauer, E., Poppa, H., Todd, G., and Davis, P. R., *J. Appl. Phys.* **48**, 3773 (1977).
29. Bauer, E., *Appl. Surf. Sci.* **11/12**, 479 (1982).
30. Bean, J. C., *Science* **230**, 127 (1985).
31. Bean, J. C., Feldman, L. C., Fiory, A. T., Nakahara, S., and Robinson, I. K., *J. Vac. Sci. Technol.* **A2**, 436 (1984).
32. Becker, R. and Döring, W., *Ann. Phys.* **24**, 719 (1935).
33. Becker, A. F., Rosenfeld, G., Poelsema, B., and Comsa, G., *Phys. Rev. Lett.* **70**, 477 (1993).
34. Bennema, P., *J. Crystal Growth* **24/25**, 76 (1974).
35. Bennema, P. and Gilmer, G., in *Crystal Growth: An Introduction*, ed. P. Hartman (North Holland, 1973), p. 263.
36. Benson, G. C. and Shuttleworth, R., *J. Chem. Phys.* **19**, 130 (1951).
37. Benson, S., *The Foundations of Chemical Reactions* (McGraw-Hill, 1960).
38. Binnig, G., Rohrer, H., Gerber, Ch., and Weibel, E., *Appl. Phys. Lett.* **40**, 178 (1982a).
39. Binnig, G., Rohrer, H., Gerber, Ch., and Weibel, E., *Phys. Rev. Lett.* **49**, 57 (1982b).
40. Binnig, G. and Rohrer, H., *Surf. Sci.* **126**, 236 (1983).
41. Bolding, B. C. and Carter, E. A., *Surf. Sci.* **268**, 142 (1992).
42. Bollmann, W., *Phil. Mag.* **16**, 363, 383 (1967).
43. Bollmann, W., in *Grain Boundaries and Interfaces*, eds. P. Chaudhari and J. W. Matthews (North Holland, 1972), p. 1.

44. Bolmont, O., Chen, P., Cebenne, C. A., and Proix, F., *Phys. Rev.* **B24**, 4552 (1981).
45. Borovinski, L., and Tzindergozen, A., *Dokl. Akad. Nauk USSR* **183**, 1308 (1968).
46. Bostanov, V., *J. Crystal Growth* **42**, 194 (1977).
47. Bostanov, V., Russinova, R., and Budewski, E., *Comm. Dept. Chem.* (Bulg. Acad. Sci.) **2**, 885 (1969).
48. Bostanov, V., Staikov, G., and Roe, D. K., *J. Electrochem. Soc.* **122**, 1301 (1975).
49. Bostanov, V., Obretenov, W., Staikov, G., Roe, D. K., and Budewski, E., *J. Crystal Growth* **52**, 761 (1981).
50. Böttner, H., Schießl, U., and Tacke, M., *Superlatt. Superstruct.* **7**, 97 (1990).
51. Bragg, W. L. and Williams, E. J., *Proc. R. Soc.* **A145**, 699 (1934).
52. Brandes, H., *Z. Phys. Chem.* **126**, 198 (1927).
53. Brocks, G., Kelly, P. J., and Car, R., *Proc. 20th Int. Conf. on the Physics of Semiconductors*, eds. E. Anastassakis and J. Joannopoulos (World Scientific, 1991a), p. 127.
54. Brocks, G., Kelly, P. J., and Car, R., *Phys. Rev. Lett.* **66**, 1729 (1991b).
55. Brooks, H., *Metal Interfaces*, Am. Soc. Metals, Metals Park, Ohio (1952), p. 20.
56. Bruce, L. A. and Jaeger, H., *Phil. Mag.* **36**, 1331 (1977).
57. Bruce, L. A. and Jaeger, H., *Phil. Mag.* **A38**, 223 (1978a).
58. Bruce, L. A. and Jaeger, H., *Phil. Mag.* **A37**, 337 (1978b).
59. Buckle, E. R. and Ubbelohde, A. R., *Proc. R. Soc.* **A259**, 325 (1960).
60. Buckle, E. R. and Ubbelohde, A. R., *Proc. R. Soc.* **A261**, 196 (1961).
61. Budewski, E. and Bostanov, V., *Electrochim. Acta* **9**, 477 (1964).
62. Budewski, E., Bostanov, V., Vitanov, T., Stoynov, Z., Kotzeva, A., and Kaischew, R., *Electrochim. Acta* **11**, 1697 (1966).
63. Budewski, E., Vitanov, T., and Bostanov, V., *Phys. Status Solidi* **8**, 369 (1965).
64. Budewski, E., in *Comprehensive Treatise of Electrochemistry*, Vol. 7, eds. B. E. Conway, J. O'M. Bockris, E. Yeager, S. U. M. Khan, and R. E. White (Plenum, 1983), p. 399.
65. Burton, W. K. and Cabrera, N., *Disc. Faraday Soc.* **5**, 33 (1949).
66. Burton, W. K., Cabrera, N., and Frank, F. C., *Phil. Trans. R. Soc.* **243**, 299 (1951).
67. Cabrera, N. and Levine, M. M., *Philos. Mag.* **1**, 450 (1956).
68. Cabrera, N. and Vermilyea, D. A., in *Growth and Perfection of Crystals* (John Wiley, 1958), p. 393.
69. Cabrera, N. and Coleman, R. V., in *The Art and Science of Growing Crystals*, ed. J. J. Gilman (John Wiley, 1963), p. 3.
70. Cahn, J. W. and Hilliard, J. E., *J. Chem. Phys.* **28**, 258 (1958).
71. Cahn, J. W. and Hilliard, J. E., *J. Chem. Phys.* **31**, 688 (1959).
72. Carra, S., in *Epitaxial Electronic Materials*, eds. A. Baldereschi and C. Paorici (World Scientific, 1988), p. 71.

73. Carslaw, H. S. and Jaeger, J. C., *Conduction of Heat in Solids* (Clarendon, 1960).
74. Chadi, D. J., *Phys. Rev. Lett.* **43**, 43 (1979).
75. Chadi, D. J., *Phys. Rev. Lett.* **59**, 1691 (1987).
76. Chakraverty, B. K., *Surf. Sci.* **4**, 205 (1966).
77. Chang, L. L. and Ludeke, R., in *Epitaxial Growth*, Part A, ed. J. W. Matthews (Academic, 1975), p. 37.
78. Chang, K. H., Lee, C. P., Wu, J. S., Liu, D. G., Liou, D. C., Wang, M. H., Chen, L. J., and Marais, M. A., *J. Appl. Phys.* **70**, 4877 (1991).
79. Chen, J., Ming, N., and Rosenberger, F., *J. Chem. Phys.* **84**, 2365 (1986).
80. Chen, L. J., Cheng, H. C., and Lin, W. T., *Mat. Res. Soc. Symp. Proc.* **54**, 245 (1986).
81. Chen, L. J. and Tu, K. N., *Mater. Sci. Rep.* **6**, 53 (1991).
82. Chernov, A. A., *Usp. Fiz. Nauk* **73**, 277 (1961).
83. Chernov, A. A. and Lyubov, B., *Growth of Crystals* (USSR) **5**, 11 (1963).
84. Chernov, A. A., in *Annual Review of Materials Science*, Vol. 3, Palo Alto, California (Ann. Rev. Inc., 1973), p. 397.
85. Chernov, A. A., *Modern Crystallography III, Springer Series in Solid State Sciences*, Vol. 36 (Springer, 1984).
86. Chernov, A. A., *Contemp. Phys.* **30**, 251 (1989).
87. Chernov, A. A., Kuznetsov, Y. G., Smol'sky, I. L., and Rozhansky, V. N., *Kristallogra.* **31**, 1193 (1986) [*Sov. Phys. Crystallogr.* **31**, 705 (1986)].
88. Chernov, A. A. and Nishinaga, T., *Morphology of Crystals* (Terra Scientific, 1987), p. 207.
89. Chui, S. T. and Weeks, J. D., *Phys. Rev. Lett.* **40**, 733 (1978).
90. Christian, J. W., *The Theory of Transformations in Metals and Alloys, Part I: Equilibrium and General Kinetic Theory*, 2nd ed. (Pergamon, 1981).
91. Cohen, P. I., Petrich, G. S., Pukite, P. R., Whaley, G. J., and Arrott, A. S., *Surf. Sci.* **216**, 222 (1989).
92. Collins, F. C., *Z. Elektrochem.* **59**, 404 (1955).
93. Cormia, R. L., Price, F. P., and Turnbull, D., *J. Appl. Phys.* **37**, 1333 (1962).
94. Courtney, W. G., *J. Chem. Phys.* **36** 2009 (1962).
95. Curie, P., *Bull. Soc. Mineralog.* **8**, 145 (1885).
96. Davisson, C. and Germer, L. H., *Phys. Rev.* **30**, 707 (1927).
97. Defay, R., Prigogine, J., Bellemans, A., and Everett, D. H., *Surface Tension and Adsorption* (Longmans Green, 1966).
98. De Gennes, P. G., *J. Chem. Phys.* **48**, 2257 (1968).
99. Demianetz, L. N., Kuznetzov, V. A., and Lobachov, A. N., *Modern Crystallography III, Springer Series in Solid State Sciences*, Vol. 36 (Springer, 1984), p. 380.
100. De Miguel, J. J., Aumann, C. E., Kariotis, R., and Lagally, M. G., *Phys. Rev. Lett.* **67**, 2830 (1991).
101. Drechsler, M. and Nicholas, J. F., *Phys. Chem. Solids* **28**, 2609 (1967).
102. Dubnova, G. N. and Indenbom, V. L., *Sov. Phys. Crystallogr.* **11**, 642 (1966).
103. Dunning, W. J., in *Chemistry of the Solid State*, ed. W. E. Garner (Butterworths, 1955), p. 333.

104. Dunning, W. J., in *Nucleation*, ed. A. C. Zettlemoyer (Marcel Dekker, 1969), p. 1.
105. Dupré, A., *Théorie Mécanique de la Chaleur* (Gauthier-Villard, 1869), p. 369.
106. Emsley, J., *The Elements*, 2nd ed. (Oxford Univ. Press, 1991).
107. Fan, W. C., Ignatiev, A., and Wu, N. J., *Surf. Sci.* **235**, 169 (1990).
108. Farkas, L., *Z. Phys. Chem.* **125**, 239 (1927).
109. Farley, F. J., *Proc. R. Soc.* **212**, 530 (1952).
110. Farrow, R. F. C., *J. Electrochem. Soc.* **121**, 899 (1974).
111. Farrow, R. F. C., Parkin, S. S. P., Dobson, P. J., Neave, J. H., and Arrott, A. S., eds., *Thin Film Growth Techniques for Low-Dimensional Structures, NATO ASI Series, Series B: Physics*, Vol. 163 (Plenum, 1987).
112. Feder, J., Russel, K. C., Lothe, J., and Pound, G. M., *Adv. Phys.* **15**, 111 (1966).
113. Finch, G. I. and Quarrell, A. G., *Proc. R. Soc.* **A141**, 398 (1933).
114. Finch, G. I. and Quarrell, A. G., *Proc. Phys. Soc.* **46**, 148 (1934).
115. Fisher, D. S. and Weeks, J. D., *Phys. Rev. Lett.* **50**, 1077 (1983).
116. Fletcher, N. H., *J. Appl. Phys.* **35**, 234 (1964).
117. Fletcher, N. H., *Phil. Mag.* **16**, 159 (1967).
118. Fletcher, N. H. and Adamson, P. L., *Phil. Mag.* **14**, 99 (1966).
119. Fletcher, N. H. and Lodge, K. W., in *Epitaxial Growth*, Part B, ed. J. W. Matthews (Academic, 1975), p. 529.
120. Frank, F. C., *Disc. Faraday Soc.* **5**, 48 (1949a).
121. Frank, F. C., *Disc. Faraday Soc.* **5**, 67 (1949b).
122. Frank, F. C., in *Growth and Perfection of Crystals* (John Willey, 1958a), p. 3.
123. Frank, F. C., in *Growth and Perfection of Crystals* (John Willey, 1958b), p. 411.
124. Frank, F. C. and van der Merwe, J. H., *Proc. R. Soc. London* **A198**, 205 (1949a).
125. Frank, F. C. and van der Merwe, J. H., *Proc. R. Soc. London* **A198**, 216 (1949b).
126. Frank, F. C. and van der Merwe, J. H., *Proc. R. Soc. London* **A200**, 125 (1949c).
127. Frank, F. C., *J. Crystal Growth* **22**, 233 (1974).
128. Frankenheim, M. L., *Ann. Phys.* **37**, 516 (1836).
129. Franzosi, P., Salviati, G., Genova, F., Stano, A., and Taiariol, F., *Mat. Lett.* **3**, 425 (1985).
130. Franzosi, P., Salviati, G., Genova, F., Stano, A., and Taiariol, F., *J. Crystal Growth* **75**, 521 (1986).
131. Franzosi, P., Salviati, G., Scaffardi, M., Genova, F., Pellegrino, S., and Stano, A., *J. Crystal Growth* **88**, 135 (1988).
132. Frenkel, I. Ya, *Z. Phys.* (USSR) **1**, 498 (1932).
133. Frenkel, I. Ya, *J. Chem. Phys.* **1**, 200,538 (1939).
134. Frenkel, I. Ya and Kontorova, T., *J. Phys.* (USSR) **1**, 137 (1939).
135. Frenkel, I. Ya, *Kinetic Theory of Liquids* (Dover, 1955).
136. Frenken, J. W. M. and van der Veen, J. F., *Phys. Rev. Lett.* **54**, 134 (1985).

137. Fukuda, Y., Kohama, Y., and Ohmachi, Y., *Jap. J. Appl. Phys.* **29L**, 20 (1990).
138. Gallet, F., Nozieres, P., Balibar, S., and Rolley, E., *Europhys. Lett.* **2**, 701 (1986).
139. Gallet, F., Balibar, S., and Rolley, E., *J. Phys.* **48**, 369 (1987).
140. Gebhardt, M., in *Crystal Growth: An Introduction*, ed. P. Hartman (North Holland, 1973), p. 105.
141. Gibbs, J. W., *Am. J. Sci. Arts* **16**, 454 (1878).
142. Gibbs, J. W., *On the Equilibrium of Heterogeneous Substances, Collected Works* (Longmans, Green & Co., 1928).
143. Gilmer, G. H., Ghez, R., and Cabrera, N., *J. Crystal Growth* **8**, 79 (1971).
144. Gilmer, G., *J. Crystal Growth* **49**, 465 (1980a).
145. Gilmer, G., *Science* **208**, 355 (1980b).
146. Girifalco, L. A. and Weiser, V. G., *Phys. Rev.* **114**, 687 (1959).
147. Gowers, J. P., in *Thin Film Growth Techniques for Low-Dimensional Structures*, eds. R. F. C. Farrow, S. S. P. Parkin, P. J. Dobson, J. H. Neave, and A. S. Arrott, *NATO ASI Series, Series B: Physics*, Vol. 163 (Plenum, 1987), p. 471.
148. Grabow, M. H. and Gilmer, G. H., *Surf. Sci.* **194**, 333 (1988).
149. Gradmann, M., *Ann. Phys.* (Leipzig) **13**, 213 (1964).
150. Gradmann, U., *Ann. Phys.* (Leipzig) **17**, 91 (1966).
151. Gradmann, U., Kümmerle, W., and Tillmanns, P., *Thin Solid Films* **34**, 249 (1976).
152. Gradmann, U. and Tillmanns, P., *Phys. Status Solidi* **A44**, 539 (1977).
153. Gradmann, U. and Waller, G., *Surf. Sci.* **116**, 539 (1982).
154. Green, A. K., Prigge, S., and Bauer, E., *Thin Solid Films* **52**, 163 (1978).
155. Griffith, J. E. and Kochanski, G. P., *Critical Rev. Solid State Mater. Sci.* **16**, 255 (1990).
156. Grosse, W., *J. Inorg. Nucl. Chem.* **25**, 317 (1963).
157. Grünbaum, E., *Vacuum* **24**, 153 (1974).
158. Grünbaum, E., in *Epitaxial Growth*, Part B, ed. J. W. Matthews (Academic, 1975), p. 611.
159. Gushee, D. E., ed., *Nucleation Phenomena* (American Chemical Society, 1966).
160. Gutzow, I., *Fiz. Chim. Stekla* **1**, 431 (1975).
161. Gutzow, I. and Toschev, S., *Kristall. Technik* **3**, 485 (1968).
162. Haas, C., *Solid State Commun.* **26**, 709 (1978).
163. Haas, C., in *Current Topics of Materials Science*, Vol. 3, ed. E. Kaldis (North Holland, 1979), p. 1.
164. Halpern, V., *J. Appl. Phys.* **40**, 4627 (1969).
165. Hamers, R. J., Tromp, R. M., and Demuth, J. E., *Phys. Rev.* **B34**, 5343 (1986).
166. Hanbücken, M. and Neddermeyer, H., *Surf. Sci.* **114**, 563 (1982).
167. Hanbücken, M., Futamoto, M., and Venables, J. A., *Surf. Sci.* **147**, 433 (1984a).

168. Hanbücken, M., Futamoto, M., and Venables, J. A., *IOP Conference Series*, London, Bristol Institute of Physics, Chap. 9, p. 135.

169. Hanbücken, M. and LeLay, G., *Surf. Sci.* **168**, 122 (1986).

170. Harris, J. J., Joyce, B. A., and Dobson, P. J., *Surf. Sci.* **103**, L90 (1981a).

171. Harris, J. J., Joyce, B. A., and Dobson, P. J., *Surf. Sci.* **108**, L444 (1981b).

172. Hartman, P., in *Crystal Growth: An Introduction*, ed. P. Hartman (North Holland, 1973), p. 358.

173. Henzler, M. and Clabes, J., *Jap. J. Appl. Phys. Supl.* 2, Pt. 2, 389 (1974).

174. Henzler, M., Busch, H., and Friese, G., in *Kinetics of Ordering and Growth at Surfaces*, ed. M. G. Lagally, *NATO ASI Series, Series B: Physics*, Vol. 239 (Plenum, 1990), p. 101.

175. Herring, C., in *The Physics of Powder Metallurgy* (McGraw-Hill, 1951), p. 143.

176. Herring, C., in *Structure and Properties of Solid Surfaces* (Univ. of Chicago Press, 1953), p. 5.

177. Hertz, H., *Ann. Phys.* **17**, 177 (1882).

178. Heyer, H., Nietruch, F., and Stranski, I. N., *J. Crystal Growth* **11**, 283 (1971).

179. Heyraud, J. C. and J. J. Metois, *Acta Metall.* **28**, 1789 (1980).

180. Heyraud, J. C. and Metois, J. J., *Surf. Sci.* **128**, 334 (1983).

181. Hilliard, J. E., in *Nucleation Phenomena*, ed. D. E. Gushee (American Chemical Society, 1966), p. 79.

182. Hillig, W., *Acta Metall.* **14**, 1868 (1966).

183. Hirth, J. P. and Pound, G. M., *Condensation and Evaporation, Progress in Materials Science*, Vol. 11 (MacMillan, 1963).

184. Hirth, J. P. and Lothe, J., *Theory of Dislocations* (McGraw-Hill, 1968).

185. Hollomon, J. H. and Turnbull, D., in *Progress in Metal Physics*, Vol. 4, ed. B. Chalmers (Pergamon, 1953), p. 333.

186. Hollomon, J. H. and Turnbull, D., in *The Solidification of Metals and Alloys*, Am. Inst. Min. Metallurg. Eng. Symp., New York, 1951, p. 1.

187. Holt, D. B., *J. Phys. Chem. Solids* **27**, 1053 (1966).

188. Honjo, G. and Yagi, K., *J. Vac. Sci. Technol.* **6**, 576 (1969).

189. Honjo, G. and Yagi, K., in *Current Topics of Materials Science*, Vol. 6, ed. E. Kaldis (North Holland, 1980), p. 196.

190. Honnigmann, B., Gleichgewichts-und Wachstumsformen von Kristallen (Steinkopff, 1958).

191. Horng, C. T. and Vook, R., *J. Vac. Sci. Technol.* **11**, 140 (1974).

192. Hornstra, J., *J. Phys. Chem. Solids* **5**, 129 (1958).

193. Hoyt, J. J., *Acta Metall. Mater.* **38**, 1405 (1990).

194. Hu, S. M., *J. Appl. Phys.* **70**, R53 (1991).

195. Huntington, C. G., *Solid State Phys.* **7**, 213 (1958).

196. Jackson, K. A., in *Growth and Perfection of Crystals* (John Willey, 1958), p. 319.

197. Jackson, K. A., in *Nucleation Phenomena*, ed. D. E. Gushee (American Chemical Society, 1966), p. 35.

198. James, P. F., *Phys. Chem. Glasses* **15**, 95 (1974).

199. James, P. F., in *Nucleation and Crystallization in Glasses, Advances in Ceramics*, Vol. 4, eds. J. H. Simmons, D. R. Uhlmann, and G. H. Beall, Columbus, Ohio (1982), p. 1.
200. Janke, E., Emde, F., and Lösch, F., *Tafeln Höherer Funktionen, B. G. Teubner Verlagsgesellschaft* (Stuttgart, 1960).
201. Jayaprakash, C., Saam, W. F., and Teitel, S., *Phys. Rev. Lett.* 50, 2017 (1983).
202. Jentzsch, F., Froitzheim, H., and Theile, R., *J. Appl. Phys.* 66, 5901 (1989).
203. Jesser, W. A. and Kuhlmann-Wilsdorf, D., *Phys. Status Solidi* 19, 95 (1967).
204. Kaischew, R., *Z. Phys.* 102, 684 (1936).
205. Kaischew, R., *Ann. l'Univ. Sofia (Chem.)* 63, 53 (1946/7).
206. Kaischew, R., *Comm. Bulg. Acad. Sci. (Phys.)* 1, 100 (1950).
207. Kaischew, R., *Bull. Acad. Bulg. Sci. (Phys.)* 2, 191 (1951).
208. Kaischew, R., *Fortschr. Miner.* 38, 7 (1960).
209. Kaischew, R., Bliznakow, G., and Scheludko, A., *Comm. Bulg. Acad. Sci. (Phys. Ser.)* 1, 146 (1950).
210. Kaischew, R. and Mutaftschiev, B., *Electrochim. Acta* 10, 643 (1965).
211. Kaischew, R. and Budewski, E., *Contemp. Phys.* 8, 489 (1967).
212. Kaischew, R., *Selected Papers*, Bulg. Acad. Sci., Sofia, 1980.
213. Kaischew, R., *J. Crystal Growth* 51, 643 (1981).
214. Kamat, S. V. and Hirth, J. P., *J. Appl. Phys.* 67, 6844 (1990).
215. Kamke, E., *Differentialgleichungen: Lösungsmethoden u. Lösungen I. Gewönliche Differentialgleichungen*, Leipzig, 1959.
216. Kantrowitz, A., *J. Chem. Phys.* 19, 1097 (1951).
217. Kaplan, R., *Surf. Sci.* 93, 145 (1980).
218. Kaplan, I. G., *Theory of Molecular Interactions* (Elsevier, 1986).
219. Kariotis, R. and Lagally, M., *Surf. Sci.* 216, 557 (1989).
220. Kashchiev, D., *Surf. Sci.* 14, 209 (1969).
221. Kasper, E., Herzog, H. J., and Kibbel, H., *Appl. Phys.* 8, 199 (1975).
222. Kasper, E. and Herzog, H. J., *Thin Solid Films* 44, 357 (1977).
223. Kasper, E., *Appl. Phys.* A28, 129 (1982).
224. Kern, R., LeLay, G., and Metois, J. J., in *Current Topics in Materials Science*, Vol. 3, ed. E. Kaldis (North Holland, 1979), p. 128.
225. Keshishev, K. O., Parshin, A., and Babkin, A., *Zh. Eksp. Teor. Fiz.* 80, 716 (1981).
226. Kikuchi, R., in *Nucleation Phenomena* (Elzevier, 1977), p. 67.
227. Kirkwood, J. G. and Buff, F. P., *J. Chem. Phys.* 17, 338 (1949).
228. Knudsen, M., *Ann. Phys.* 29, 179 (1909).
229. Kohama, Y., Fukuda, Y., and Seki, M., *Appl. Phys. Lett.* 52, 380 (1988).
230. Kolb, E. D. and Laudise, R. A., *J. Am. Ceram. Soc.* 49, 302 (1966).
231. Kolb, E. D., Wood, D. L., Spencer, E. G., and Laudise, R. A., *J. Appl. Phys.* 38, 1027 (1967).
232. Kolmogorov, A. N., *Izv. Acad. Sci. USSR* (Otd. Phys. Math. Nauk) 3, 355 (1937).
233. Kossel, W., *Nachrichten der Gesellschaft der Wissenschaften Göttingen, Mathematisch-Physikalische Klasse*, Band 135, 1927.

234. Koutsky, J. A., Ph. D. thesis, Case Institute of Technology, Cleveland, Ohio, 1966.
235. Krasil'nikov, V., Yugova, T., Bublik, V., Drozdov, Y., Malkova, N., Shepenina, G., Hansen, K., and Rezvov, A., *Sov. Phys. Crystallogr.* **33**, 874 (1988).
236. Kratochvil, J. and Indenbom, V. L., *Czech. J. Phys.* **B13**, 814 (1963).
237. Kroemer, H., in *Heteroepitaxy of Silicon*, eds. J. C. C. Fan and J. M. Poate, *Mater. Res. Soc. Symp.* **67**, 3 (1986).
238. Kuznetsov, Y. G., Chernov, A. A., and Zakharov, N. D., *Kristallografiya* **31**, 1201 (1986); *Sov. Phys. Crystallogr.* **31** 709 (1986).
239. Lacmann, R., *Z. Kristallografiya* **116**, 13 (1961).
240. Land, D. V., People, R., Bean, J. C., and Sergent, A. M., *Appl. Phys. Lett.* **47**, 1333 (1985).
241. Landau, L. D., *Collected Works*, Vol. 2, Nauka, Moscow (1969), p. 119.
242. Laudise, R. A., *J. Am. Chem. Soc.* **81**, 562 (1959).
243. Laudise, R. A., *The Growth of Single Crystals* (Prentice-Hall, 1970).
244. Laudise, R. A. and Ballman, A. A., *J. Am. Chem. Soc.* **80**, 2655 (1958).
245. Laudise, R. A. and Ballman, A. A., *J. Phys. Chem.* **64**, 688 (1960).
246. Leamy, J. H., Gilmer, G. H., and Jackson, K. A., in *Surface Physics of Materials*, Vol. 1, ed. J. P. Blakeley (Academic, 1975), p. 121.
247. LeLay, G., *Surf. Sci.* **132**, 169 (1983).
248. LeLay, G., Chauret, A., Manneville, M., and Kern, R., *Appl. Surf. Sci.* **9**, 190 (1981).
249. Lennard-Jones, J. E., *Proc. R. Soc. London* **A106**, 463 (1924).
250. Lepage, J., Mezin, A., and Palmier, D., *J. Microsc. Spectrosc. Electron.* **9**, 365 (1984).
251. Levine, J. D., *Surf. Sci.* **34**, 90 (1973).
252. Lewis, B. and Campbell, D., *J. Vac. Sci. Technol.* **4**, 209 (1967).
253. Lewis, B. and Anderson, J. C., *Nucleation and Growth of Thin Films* (Academic, 1978).
254. Lighthill, M. J. and Whitham, G. B., *Proc. R. Soc.* **A229**, 281, 317 (1955).
255. Lin, W. T. and Chen, L. J., *J. Appl. Phys.* **59**, 3481 (1986).
256. Lin, D. S., Miller, T., and Chiang, T. C., *Phys. Rev. Lett.* **67**, 2187 (1991).
257. Lord, D. G. and Prutton, M., *Thin Solid Films* **21**, 341 (1974).
258. Lothe, J. and Pound, G. M., *J. Chem. Phys.* **36**, 2080 (1962).
259. Lothe, J. and Pound, G. M., in *Nucleation*, ed. A. C. Zettlemoyer (Marcel Dekker, 1969), p. 109.
260. Lu, Y., Zhang, Z., and Metiu, H., *Surf. Sci.* **257**, 199 (1991).
261. Lyubov, B. and Roitburd, A. L., in *Problemi Metallovedeniya i Fisiki Metallov*, Metallurgizdat, Moskow (1958), p. 91.
262. Machlin, E. S., *Trans. Am. Inst. Min. Metall. Pet. Eng.* **197**, 437 (1953).
263. Mackenzie, J. D., in *Modern Aspects of the Vitreous State* (Butterworths, 1960), p. 188.
264. Maiwa, K., Tsukamoto, K., and Sunagawa, I., *J. Crystal Growth* **102**, 43 (1990).
265. Marchand, M., Hood, K., and Caillé, A., *Phys. Rev.* **B37**, 1898 (1988).
266. Marchenko, V. I. and Parshin, A. Y., *Zh. Eksp. Teor. Fiz.* **79**, 257 (1980).

267. Marchenko, V. I., *Pis'ma Zh. Eksp. Teor. Fiz.* **33**, 397 (1981).
268. Marée, P. M., Barbour, J. C., and van der Veen, J. F., *J. Appl. Phys.* **62**, 4413 (1987).
269. Marée, P. M. J., Nakagawa, K., Mulders, F. M., van der Veen, J. F., and Kavanagh, K. L., *Surf. Sci.* **191**, 305 (1987).
270. Markov, I., *Thin Solid Films* **8**, 281 (1971).
271. Markov, I. and Kashchiev, D., *J. Cryst. Growth* **13,14**, 131 (1972a).
272. Markov, I. and Kashchiev, D., *J. Cryst. Growth* **16**, 170 (1972b).
273. Markov, I. and Kashchiev, D., *Thin Solid Films* **15**, 181 (1973).
274. Markov, I., Boynov, A., and Toschev, S., *Electrochim. Acta* **18**, 377 (1973).
275. Markov, I. and Toschev, S., *Electrodep. Surf. Treatm.* **3**, 385 (1975).
276. Markov, I. and Kaischew, R., *Thin Solid Films* **32**, 163 (1976a).
277. Markov, I. and Kaischew, R., *Kristall. Technik* **11**, 685 (1976b).
278. Markov, I., Stoycheva, E., and Dobrev, D., *Commun. Dept. Chem.* (Bulg. Acad. Sci.) **3**, 377 (1978).
279. Markov, I. and Karaivanov, V., *Thin Solid Films* **61**, 115 (1979).
280. Markov, I., *Mater. Chem. Phys.* **9**, 93 (1983).
281. Markov, I. and Milchev, A., *Surf. Sci.* **136**, 519 (1984a).
282. Markov, I. and Milchev, A., *Surf. Sci.* **145**, 313 (1984b).
283. Markov, I. and Milchev, A., *Thin Solid Films* **126**, 83 (1985).
284. Markov, I. and Stoyanov, S., *Contemp. Phys.* **28**, 267 (1987).
285. Markov, I., in *Crystal Growth and Characterization of Advanced Materials*, eds. A. N. Christensen, F. Leccabue, C. Paorici, and O. Vigil (World Scientific, 1988), p. 119.
286. Markov, I. and Trayanov, A., *J. Phys. C: Solid State Phys.* **21**, 2475 (1988).
287. Markov, I. and Trayanov, A., *J. Phys.: Condens. Matter* **2**, 6965 (1990).
288. Markov, I., *Surf. Sci.* **279**, L207. (1992).
289. Massies, J. and Linh, N. T., *J. Cryst. Growth* **56**, 25 (1982a).
290. Massies, J. and Linh, N. T., *Thin Solid Films* **90**, 113 (1982b).
291. Massies, J. and Linh, N. T., *Surf. Sci.* **114**, 147 (1982c).
292. Matthews, J. W., *Philos. Mag.* **6**, 1347 (1961).
293. Matthews, J. W., *Philos. Mag.* **8**, 711 (1963).
294. Matthews, J. W. and Crawford, J. L., *Thin Solid Films* **5**, 187 (1970).
295. Matthews, J. W., Mader, S., and Light, T. B., *J. Appl. Phys.* **41**, 3800 (1970).
296. Matthews, J. W. and Blakeslee, A. E., *J. Cryst. Growth* **27**, 118 (1974).
297. Matthews, J. W., ed., *Epitaxial Growth* (Academic, 1975).
298. Matthews, J. W., *J. Vac. Sci. Technol.* **12**, 126 (1975a).
299. Matthews, J. W., in *Epitaxial Growth*, Part B, ed. J. W. Matthews (Academic, 1975b), p. 560.
300. Matthews, J. W., Jackson, D. C., and Chambers, A., *Thin Solid Films* **26**, 129 (1975c).
301. Mayer, H., in *Advances in Epitaxy and Endotaxy*, eds. R. Niedermayer and H. Mayer (VEB Deutscher Verlag für Grund-stoffsindustrie, Leipzig, 1971), p. 63.
302. McMillan, W. L., *Phys. Rev.* **B14**, 1496 (1976).
303. Metois, J. J. and Heyraud, J. C., *J. Crystal Growth* **57**, 487 (1982).

304. Milchev, A., *Electrochim. Acta* **28**, 947 (1983).
305. Milchev, A., Stoyanov, S., and Kaischew, R., *Thin Solid Films* **22**, 22, 255 (1974).
306. Milchev, A. and Stoyanov, S., *J. Electroanal. Chem.* **72**, 33 (1976).
307. Milchev, A. and Malinowski, J., *Surf. Sci.* **156**, 36 (1985).
308. Milchev, A., *Phys. Rev.* **B33**, 2062 (1986).
309. Milchev, A. and Markov, I., *Surf. Sci.* **136**, 503 (1984).
310. Miller, D. C. and Caruso, R., *J. Cryst. Growth* **27**, 274 (1974).
311. Miyazaki, T., Hiramoto, H., and Okazaki, M., *Proc. 20th Int. Conf. Physics of Semiconductors*, eds. E. Anastassakis and J. Joannopoulos (World Scientific, 1991), p. 131.
312. Mo, Y. W., Kleiner, J., Webb, M. B., and Lagally, M. G., *Phys. Rev. Lett.* **66**, 1998 (1991).
313. Mönch, W., *Surf. Sci.* **86**, 672 (1979).
314. Moore, A. J. W., in *Metal Surfaces*, Am. Soc. for Metals, Metals Park, Ohio, 1963, p. 155.
315. Morse, P. M., *Phys. Rev.* **34**, 57 (1929).
316. Murthy, C. S. and Rice, B. M., *Phys. Rev.* **B41**, 3391 (1990).
317. Mutaftschiev, B., in *Adsorption et Croissance Cristalline*, CNRS, Paris, 1965, p. 231.
318. Myers-Beaghton, A. K. and Vvedensky, D. D., *Phys. Rev.* **B42**, 5544 (1990).
319. Narusawa, T., Gibson, W. M., and Hiraki, A., *Phys. Rev.* **B24**, 4835 (1981a).
320. Narusawa, T. and Gibson, W. M., *Phys. Rev. Lett.* **42**, 1459 (1981b).
321. Narusawa, T. and Gibson, W. M., *J. Vac. Sci. Technol.* **20**, 709 (1982).
322. Neave, J. H., Joyce, B. A., Dobson, P. J., and Norton, N., *Appl. Phys.* **31**, 1 (1983).
323. Neave, J. H., Joyce, B. A., and Dobson, P. J., *Appl. Phys.* **A34**, 179 (1984).
324. Neave, J. H., Dobson, P. J., Joyce, P. A., and Zhang, J., *Appl. Phys. Lett.* **47**, 100 (1985).
325. Nenow, D., *Prog. Crystal Growth Characterization* **9**, 185 (1984).
326. Nenow, D. and Dukova, E., *J. Crystal Growth* **3/4**, 166 (1968).
327. Nenow, D., Pavlovska, A., and Karl, N., *J. Crystal Growth* **67**, 587 (1984).
328. Nielsen, A., *Kinetics of Precipitation* (Pergamon, 1964).
329. Nielsen, A. E., in *Crystal Growth*, ed. H. S. Peiser (Pergamon, 1967), p. 419.
330. Nishioka, K. and Pound, G. M., in *Nucleation Phenomena* (Elzevier, 1977), p. 205.
331. Nordwall, H. J. de and Staveley, L. A. K., *J. Chem. Soc.* **224**, (1954).
332. Noziéres, P. and Gallet, F., *J. Phys. (Paris)* **48**, 353 (1987).
333. Ohno, T. R. and Williams, E. D., *Jap. J. Appl. Phys.* **28**, L2061 (1989a).
334. Ohno, T. R. and Williams, E. D., *Appl. Phys. Lett.* **55**, 2628 (1989b).
335. Oldham, W. G. and Milnes, A. G., *Solid State Electron.* **7**, 153 (1964).
336. Olsen, G. H., Abrahams, M. S., and Zamerowski, T. J., *J. Electrochem. Soc.* **121**, 1650 (1974).
337. Onsager, L., *Phys. Rev.* **65**, 117 (1944).
338. Oqui, N., *J. Mater. Sci.* **25**, 1623 (1990).
339. Ostwald, W., *Z. Phys. Chem.* **22**, 306 (1897).

340. Oxtoby, D. W., *J. Phys.: Condens. Matter* **4**, 7627 (1992).
341. Pandey, K. C., *Phys. Rev. Lett.* **47**, 1913 (1981).
342. Pashley, D. W., *Adv. Phys.* **5**, 173 (1956).
343. Pashley, D. W., *Adv. Phys.* **14**, 327 (1965).
344. Pashley, D. W., in *Recent Progress of Surface Science*, Vol. 3, eds. J. F. Danielli, A. C. Riddiford, and M. Rosenberg (Academic, 1970), p. 23.
345. Pashley, D. W., in *Epitaxial Growth*, Part A (Academic, 1975), p. 1.
346. Pauling, L. and Herman, Z. S., *Phys. Rev.* **B28**, 6154 (1983).
347. Paunov, M. and Harsdorff, M., *Z. Naturforsch.* **29**, 1311 (1974).
348. Pavlovska, A., *J. Crystal Growth* **46**, 551 (1979).
349. Pavlovska, A. and Nenow, D., *Surf. Sci.* **27**, 211 (1971a).
350. Pavlovska, A. and Nenow, D., *J. Crystal Growth* **8**, 209 (1971b).
351. Pavlovska, A. and Nenow, D., *J. Crystal Growth* **12**, 9 (1972).
352. Pavlovska, A. and Nenow, D., *J. Crystal Growth* **39**, 346 (1977).
353. Pavlovska, A., Faulian, K., and Bauer, E., *Surf. Sci.* **221**, 233 (1989).
354. Pehlke, E. and Tersoff, J., *Phys. Rev. Lett.* **67**, 465 (1991a).
355. Pehlke, E. and Tersoff, J., *Phys. Rev. Lett.* **67**, 1290 (1991b).
356. People, R. and Bean, J. C., *Appl. Phys. Lett.* **47**, 322 (1985).
357. People, R. and Bean, J. C., *Appl. Phys. Lett.* **49**, 229 (1986).
358. Ploog, K., in *Epitaxial Electronic Materials* (World Scientific, 1986), p. 261.
359. Poon, T. W., Yip, S., Ho, P. S., and Abraham, F. F., *Phys. Rev. Lett.* **65**, 2161 (1990).
360. Pötschke, G., Schröder, J., Günther, C., Hwang, R. Q., and Behm, R. J., *Surf. Sci.* **251/252**, 592 (1991).
361. Pound, G. M., Simnad, M. T., and Yang, L., *J. Chem. Phys.* **22**, 1215 (1954).
362. Powell, G. L. F. and Hogan, L. M., *Trans. Metallurg. Soc. AIME* **242**, 2133 (1968).
363. Probstein, R. F., *J. Chem. Phys.* **19**, 619 (1951).
364. Rawlings, K. J., Gibson, M. J., and Dobson, P. J., *J. Phys.* **D11**, 2059 (1978).
365. Reiss, H., *J. Chem. Phys.* **18**, 840 (1950).
366. Reiss, H., in *Nucleation Phenomena* (Elzevier, 1977), p. 1.
367. Robins, J. L. and Rhodin, T. N., *Surf. Sci.* **2**, 346 (1964).
368. Robins, J. L. and Donohoe, A. J., *Thin Solid Films* **12**, 255 (1972).
369. Robinson, V. N. E. and Robins, J. L., *Thin Solid Films* **5**, 313 (1970).
370. Robinson, V. N. E. and Robins, J. L., *Thin Solid Films* **20**, 155 (1974).
371. Roland, C. and Gilmer, G. H., *Phys. Rev. Lett.* **67**, 3188 (1991).
372. Roland, C. and Gilmer, G. H., *Phys. Rev.* **B46**, 13428 (1992a).
373. Roland, C. and Gilmer, G. H., *Phys. Rev.* **B46**, 13437 (1992b).
374. Rollmann, L. D., *Adv. Chem. Ser.* **173**, 387 (1979).
375. Rosenberger, F., *Fundamentals of Crystal Growth I: Macroscopic Equilibrium and Transport Concepts*, Springer Series in Solid State Physics, Vol. 5 (Springer, 1979).
376. Rosenberger, F., in *Interfacial Aspects of Phase Transformations*, ed. B. Mutaftschiev, *NATO Advanced Study Institutes Series, Series C: Mathematical and Physical Sciences*, Vol. 87 (Reidel, 1982), p. 315.
377. Royer, L., *Bull. Soc. Fr. Mineralog. Cristallogr.* **51**, 7 (1928).

378. Sakamoto, T., Kawai, N. J., Nakagawa, T., Ohta, K., and Kojima, T., *Appl. Phys. Lett.* **47**, 617 (1985a).

379. Sakamoto, T., Funabashi, H., Ohta, K., Nakagawa, T., Kawai, N. G., Kojima, T., and Bando, Y., *Superlatt. Microstruct.* **1**, 347 (1985b).

380. Sakamoto, T., Kawai, N. J., Nakagawa, T., Ohta, K., Kojima, T., and Hashiguchi, G., *Surf. Sci.* **174**, 651 (1986).

381. Sakamoto, T. and Hashiguchi, G., *Jap. J. Appl. Phys.* **25**, L78 (1986).

382. Sakamoto, T., Sakamoto, K., Nagao, S., Hashiguchi, G., Kuniyoshi, K., and Bando, Y., in *Thin Film Growth Techniques for Low-Dimensional Structures*, *NATO ASI Series, Series B: Physics*, Vol. 163 (Plenum, 1987), p. 225.

383. Sakamoto, T., Sakamoto, K., Miki, K., Okumura, H., Yoshida, S., and Tokumoto, H., in *Kinetics of Ordering and Growth at Surfaces*, ed. M. G. Lagally, *NATO ASI Series, Series B: Physics* (Plenum, 1990), p. 263.

384. Sakamoto, K., Miki, K., and Sakamoto, T., *J. Crystal Growth* **99**, 510 (1990).

385. Schlichting, H., *Boundary Layer Theory* (McGraw-Hill, 1968).

386. Scott, A. C., Chu, F. Y. F., and McLaughlin, D. W., *Proc. IEEE* **61**, 1443 (1973).

387. Sharma, B. L. and Purohit, R. K., *Semiconductor Heterojunctions* (Pergamon, 1974).

388. Sharples, A., *Polymer* **3**, 250 (1962).

389. Shaw, D. W., in *Heterostructures on Silicon: One Step Further with Silicon*, eds. Y. I. Nissim and E. Rosencher (Kluwer, 1989), p. 61.

390. Shi, G. and Seinfeld, J. H., *J. Chem. Phys.* **93**, 9033 (1990).

391. Shoji, K., Hyodo, M., Ueba, H., and Tatsuyama, C., *Jap. J. Appl. Phys.* **22**, 1482 (1983).

392. Sigsbee, R. A. and Pound, G. M., *Adv. Coll. Interface Sci.* **1**, 335 (1967).

393. Sigsbee, R. A., in *Nucleation*, ed. A. C. Zettlemoyer (Marcel Dekker, 1969), p. 151.

394. Skripov, V. P., Koverda, V. P., and Butorin, G. T., *Kristallografiya* **15**, 1219 (1970).

395. Skripov, V. P., Koverda, V. P., and Butorin, G. T., *Growth of Crystals*, Vol. 11 (Univ. of Erevan, 1975), p. 25.

396. Skripov, V. P., in *Current Topics in Materials Science*, Vol. 2, ed. E. Kaldis (North Holland, 1977), p. 327.

397. Smol'sky, I. L., Malkin, A. I., and Chernov, A. A., *Sov. Phys. Crystallogr.* **31**, 454 (1986).

398. Somorjai, G. A. and van Hove, M. A., *Progr. Surf. Sci.* **30**, 201 (1989).

399. Stern, O., *Z. Elektrochem.* **25**, 66 (1919).

400. Stillinger, F. and Weber, T. A., *Phys. Rev.* **B36**, 1208 (1987).

401. Stock, K. D. and Menzel, E., *J. Crystal Growth* **43**, 135 (1978).

402. Stock, K. D. and Menzel, E., *Surf. Sci.*, **91**, 655 (1980).

403. Stoop, L. C. A. and van der Merwe, J. H., *Thin Solid Films* **17**, 291 (1973).

404. Stowell, M. J., *Philos. Mag.* **21**, 125 (1970).

405. Stoyanov, S., *Thin Solid Films* **18**, 91 (1973).

406. Stoyanov, S., *J. Cryst. Growth* **24,25**, 293 (1974).

407. Stoyanov, S., in *Current Topics in Materials Science*, Vol. 3, ed. E. Kaldis (North Holland, 1979), p. 421.
408. Stoyanov, S. and Kashchiev, D., in *Current Topics in Materials Science*, Vol. 7, ed. E. Kaldis (North Holland, 1981), p. 69.
409. Stoyanov, S. and Markov, I., *Surf. Sci.* **116**, 313 (1982).
410. Stoyanov, S., *Surf. Sci.* **172**, 198 (1986).
411. Stoyanov, S., *Surf. Sci.* **199**, 226 (1988).
412. Stoyanov, S. and Michailov, M., *Surf. Sci.* **109**, 124 (1988).
413. Stoyanov, S., *J. Crystal Growth* **94**, 751 (1989).
414. Stoyanov, S., *Europhys. Lett.* **11**, 361 (1990).
415. Stranski, I. N., *Ann. Univ. Sofia* **24**, 297 (1927).
416. Stranski, I. N., *Z. Phys. Chem.* **36**, 259 (1928).
417. Stranski, I. N. and Kuleliev, K., *Z. Phys. Chem.* **A142**, 467 (1929).
418. Stranski, I. N. and Totomanow, D., *Z. Phys. Chem.* **A163**, 399 (1933).
419. Stranski, I. N. and Kaischew, R., *Z. Phys. Chem.* **B26**, 100 (1934a).
420. Stranski, I. N. and Kaischew, R., *Z. Phys. Chem.* **B26**, 114 (1934b).
421. Stranski, I. N. and Kaischew, R., *Z. Phys. Chem.* **B26**, 132 (1934c).
422. Stranski, I. N. and Kaischew, R., *Z. Phys. Chem.* **A170**, 295 (1934d).
423. Stranski, I. N. and Kaischew, R., *Ann. Phys.* **23**, 330 (1935).
424. Stranski, I. N., *Ann. Sofia Univ.* **30**, 367 (1936/7).
425. Stranski, I. N., Kaischew, R., and Krastanov, L., *Z. Phys. Chem.* **B23**, 158 (1933).
426. Stranski, I. N. and Krastanov, L., *Sitzungsber. Akad. Wissenschaft Wien* **146**, 797 (1938).
427. Strickland-Constable, R. F., *Kinetics and Mechanism of Crystallization* (Academic, 1968).
428. Y. Sugita, M. Tamura, and K. Sugawara, *J. Appl. Phys.* **40**, 3089 (1969).
429. Suliga, E. and Henzler, M., *J. Vac. Sci. Technol.* **A1**, 1507 (1983).
430. Swallin, R. A., *Thermodynamics of Solids* (Wiley, 1962).
431. Swartzentruber, B. S., Mo, Y. W., Kariotis, R., Lagally, M. G., and Webb, M. B., *Phys. Rev. Lett.* **65**, 1913 (1990).
432. Swendsen, R. H., *Phys. Rev.* **B17**, 3710 (1978).
433. Takayanagi, K., *Surf. Sci.* **104**, 527 (1981).
434. Tamman, G., *Der Glaszustand* (L. Voss, Leipzig, 1933).
435. Temkin, D., in *Mechanism and Kinetics of Crystallization*, Nauka i Technika, Minsk, 1964, p. 86.
436. Temkin, D., in *Growth of Crystals*, Vol. 5b, ed. N. N. Sheftal' (Consultant Bureau, 1968), p. 71.
437. Temkin, D. and Shevelev, V. V., *J. Crystal Growth* **52**, 104 (1981).
438. Temkin, D. and Shevelev, V. V., *J. Crystal Growth* **66**, 380 (1984).
439. Theodorou, G. and Rice, T. M., *Phys. Rev.* **B18**, 2840 (1978).
440. Thomas, D. G. and Staveley, L. A. K., *J. Chem. Soc.*, 4569 (1952).
441. Thomson, G. P. and Reid, A., *Nature* **119**, 80 (1927).
442. Timoshenko, S., *Theory of Elasticity* (McGraw-Hill, 1934).

443. Tkhorik, Yu A. and Khazan, L. S., *Plasticheskaya Deformaciya i Dislokacii Nesootvetstvia v Geteroepitaksial'nych Sistemakh*, Naukova Dumka, Kiev, 1983.
444. Toda, M., *J. Phys. Soc. Jpn.* **22**, 431 (1967).
445. Tolman, R. C., *J. Chem. Phys.* **17**, 333 (1949).
446. Toschev, S., Paunov, M., and Kaischew, R., *Commun. Dept. Chem.* (Bulg. Acad. Sci.) **1**, 119 (1968).
447. Toschev, S. and Markov, I., *J. Crystal Growth* **3/4**, 436 (1968).
448. Toschev, S. and Markov, I., *Ber. Bunsenges. Phys. Chem.* **73**, 184 (1969).
449. Toschev, S., Milchev, A., Popova, K., and Markov, I., *C. R. Acad. Bulg. Sci.* **22**, 1413 (1969).
450. Toschev, S. and Gutzow, I., *Kristall. Technik* **7**, 43 (1972).
451. Toschev, S., Stoyanov, S., and Milchev, A., *J. Crystal Growth* **13/14**, 123 (1972).
452. Toschev, S., in *Crystal Growth: An Introduction*, ed. P. Hartmann (North Holland, 1973), p. 1.
453. Tromp, R. M., Hamers, R. J., and Demuth, J. E., *Phys. Rev. Lett.* **55**, 1303 (1985).
454. Turnbull, D., *J. Appl. Phys.* **21**, 1022 (1950).
455. Turnbull, D. and Sech, R. E., *J. Appl. Phys.* **21**, 804 (1950).
456. Turnbull, D., in *Solid State Physics*, Vol. 3, eds. F. Seitz and D. Turnbull (Academic, 1956), p. 224.
457. Uhlmann, D. R. and Chalmers, B., in *Nucleation Phenomena*, ed. D. E. Gushee (American Chemical Society, 1966), p. 1.
458. van de Leur, R. H. M., Schellingerhout, A. J. G., Tuinstra, F., and Mooji, J. E., *J. Appl. Phys.* **64**, 3043 (1988).
459. van der Eerden, J. P., *J. Crystal Growth* **56**, 174 (1982).
460. van der Eerden, J. P., *Electrochim. Acta* **28**, 955 (1983).
461. van der Eerden, J. P. and Müller-Krumbhaar, H., *Phys. Rev. Lett.* **57**, 2431 (1986).
462. van der Merwe, J. H., *Proc. Phys. Soc. London* **A63**, 616 (1950).
463. van der Merwe, J. H., *J. Appl. Phys.* **34**, 117 (1963a).
464. van der Merwe, J. H., *J. Appl. Phys.* **34**, 123 (1963b).
465. van der Merwe, J. H., *J. Appl. Phys.* **41**, 4725 (1970).
466. van der Merwe, J. H., in *Treatise on Materials Science and Technology*, Vol. 2, ed. H. Herman (Academic, 1973), p. 1.
467. van der Merwe, J. H., *J. Microscopy* **102**, 261 (1974).
468. van der Merwe, J. H., in *Epitaxial Growth*, Part B, ed. J. W. Matthews (Academic, 1975), p. 494.
469. van der Merwe, J. H., in *CRC Critical Reviews in Solid State and Materials Science*, ed. R. Vancelow (CRC Press, Boca Raton, 1979), p. 209.
470. van der Merwe, J. H., Woltersdorf, J., and Jesser, W. A., *Mater. Sci. Eng.* **81**, 1 (1986).
471. Van Hove, J. M., Lent, C. S., Pukite, P. R., and Cohen, P. I., *J. Vac. Sci. Technol.* **B1**, 741 (1983).

472. Van Loenen, E. J., Elswijk, H. B., Hoeven, A. J., Dijkkamp, D., Lenssinck, J. M., and Dieleman, J., in *Kinetics of Ordering and Growth at Surfaces*, ed. M. Lagally (Plenum, 1990), p. 283.
473. Vegard, L., *Z. Phys.* 5, 17 (1921).
474. Vekilov, P. G., Kuznetsov, Yu G., and Chernov, A. A., *J. Crystal Growth* 121, 643 (1992).
475. Venables, J. A., in *Current Topics in Materials Science*, Vol. 2, ed. E. Kaldis (North Holland, 1979), p. 165.
476. Venables, J. A., Derrien, J., and Janssen, A. P., *Surf. Sci.* 95, 411 (1980).
477. Villain, J., in *Ordering in Strongly Fluctuating Condensed Matter Systems*, ed. T. Riste (Plenum, 1980), p. 222.
478. Vitanov, T., Sevastianov, E., Bostanov, V., and Budevski, E., *Elektrokhimiya (Sov. Electrochem., USSR)* 5, 451 (1969).
479. Vitanov, T., Popov, A., and Budevski, E., *J. Electrochem. Soc.* 121, 207 (1974).
480. Volmer, M. and Weber, A., *Z. Phys. Chem.* 119, 277 (1926).
481. Volmer, M., *Kinetik der Phasenbildung* (Theodor Steinkopf, 1939).
482. Volterra, V., *Ann. Ecole Norm. Super.* 24, 400 (1907).
483. Vook, R. W., *Int. Metals Rev.* 27, 209 (1982).
484. Vook, R. W., *Opt. Eng.* 23, 343 (1984).
485. Voronkov, V. V., *Sov. Phys. Crystallogr.* 15, 13 (1970).
486. Vvedensky, D. D., Clarke, S., Hugill, K. J., Wilby, M. R., and Kawamura, T., *J. Crystal Growth* 99, 54 (1990a).
487. Vvedensky, D. D., Clarke, S., Hugill, K. J., Myers-Beaghton, A. K., and Wilby, M. R., in *Kinetics of Ordering and Growth at Surfaces*, ed. M. G. Lagally (Plenum, 1990b), p. 297.
488. Wakeshima, H., *J. Chem. Phys.* 22, 1614 (1954).
489. Walton, A. G., *Formation and Properties of Precipitates* (Wiley, 1967).
490. Walton, A. G., in *Nucleation*, ed. A. C. Zettlemoyer (Marcel Dekker, 1969), p. 225.
491. Walton, D., *J. Chem. Phys.* 37, 2182 (1962).
492. Walton, D., in *Nucleation*, ed. A. C. Zettlemoyer (Marcel Dekker, 1969), p. 379.
493. Wang, M. H. and Chen, L. J., *J. Appl. Phys.* 71, 5918 (1992).
494. Weeks, J. D., in *Ordering in Strongly Fluctuating Condensed Matter Systems*, ed. T. Riste, *NATO Series B: Physics*, Vol. 50, 1992, p. 293.
495. Wilby, M. R., Clarke, S., Kawamura, T., and Vvedensky, D. D., *Phys. Rev.* B40, 10617 (1989).
496. Wilby, M. R., Ricketts, M. W., Clarke, S., and Vvedensky, D. D., *J. Crystal Growth* 111, 864 (1991).
497. Wilemski, G., *J. Chem. Phys.* 62, 3763 (1975a).
498. Wilemski, G., *J. Chem. Phys.* 62, 3772 (1975b).
499. Wilson, H. A., *Philos. Mag.* 50, 609 (1900).
500. Wolf, P. E., Gallet, F., Balibar, S., Rolley, E., and Nozirres, P., *J. Phys.* 46, 1987 (1985).
501. Woltersdorf, J., *Thin Solid Films* 85, 241 (1981).

502. Wood, C. E. C., *Surf. Sci. Lett.* **108**, L441 (1981).
503. Wulff, G., *Z. Kristallogr.* **34**, 449 (1901).
504. Yamaguchi, T. and Fujima, N., *J. Phys. Soc. Jpn.* **60**, 1028 (1991).
505. Yang, Y. and Williams, E. D., *Surf. Sci.* **215**, 102 (1989).
506. Young, T., *Trans. R. Soc. London* **95**, 65 (1805).
507. YaZeldovich, Ya B., *Acta Physicochim. URSS* **18**, 1 (1943).
508. Zeng, X. C. and Oxtoby, D. W., *J. Chem. Phys.* **95**, 5940 (1991).
509. Zettlemoyer, A. C., ed., *Nucleation* (Marcel Dekker, 1969).
510. Zhang, J. and Nancollas, G. H., *J. Crystal Growth* **106**, 181 (1990).
511. Zhang, Z., Lu, Y., and Metiu, H., *Surf. Sci. Lett.* **248**, L250 (1991a).
512. Zhang, Z., Lu, Y., and Metiu, H., *Surf. Sci. Lett.* **255**, L543 (1991b).
513. Zhang, Z., Lu, Y., and Metiu, H., *Surf. Sci. Lett.* **255**, L719 (1991c).
514. Zhdanov, G. S., *Kristallografiya* **21**, 706 (1976).

www.ingramcontent.com/pod-product-compliance
Lightning Source LLC
Chambersburg PA
BHW061231220326
599CB00028B/5393